Operator Theory: Advances and
Applications
Vol. 130

Editor:
I. Gohberg

Linear Operators and Matrices

The Peter Lancaster Anniversary Volume

Israel Gohberg
Heinz Langer
Editors

Springer Basel AG

Editors:

Prof. I. Gohberg
School of Mathematical Sciences
Raymond and Beverly Sackler
Faculty of Exact Sciences
Tel Aviv University
Ramat Aviv 69978
Israel

Prof. H. Langer
Mathematik
Technische Universität Wien
Wiedner Hauptstrasse 8–10/1411
1040 Wien
Austria

2000 Mathematics Subject Classification 47-06

A CIP catalogue record for this book is available from the
Library of Congress, Washington D.C., USA

Deutsche Bibliothek Cataloging-in-Publication Data

Linear operators and matrices : the Peter Lancaster anniversery volume /
Israel Gohberg ; Heinz Langer ed.. - Basel ; Boston ; Berlin : Birkhäuser, 2001
    (Operator theory ; Vol. 130)
    ISBN 978-3-0348-9467-8   ISBN 978-3-0348-8181-4 (eBook)
    DOI 10.1007/978-3-0348-8181-4

© 2002 Springer Basel AG
Originally published by Birkhäuser Verlag , Basel, Switzerland in 2002
Printed on acid-free paper produced from chlorine-free pulp. TCF ∞
Cover design: Heinz Hiltbrunner, Basel

ISBN 978-3-0348-9467-8

9 8 7 6 5 4 3 2 1   www.birkhauser-science.com

Contents

vi   Contents

Preface

In September 1998, during the 'International Workshop on Analysis and Vibrating Systems' held in Canmore, Alberta, Canada, it was decided by a group of participants to honour Peter Lancaster on the occasion of his 70th birthday with a volume in the series 'Operator Theory: Advances and Applications'. Friends and colleagues responded enthusiastically to this proposal and within a short time we put together the volume which is now presented to the reader. Regarding acceptance of papers we followed the usual rules of the journal 'Integral Equations and Operator Theory'. The papers are dedicated to different problems in matrix and operator theory, especially to the areas in which Peter contributed so richly. At our request, Peter agreed to write an autobiographical paper, which appears at the beginning of the volume. It continues with the list of Peter's publications. We believe that this volume will pay tribute to Peter on his outstanding achievements in different areas of mathematics.

I. Gohberg, H. Langer

Peter Lancaster *1929

Operator Theory:
Advances and Applications, Vol. 130, 1–7
© 2001 Birkhäuser Verlag Basel/Switzerland

My Life and Mathematics

Peter Lancaster

I was born in Appleby, a small county town in the north of England, on November 14th, 1929. I had two older brothers and was to have one younger sister. My family moved around the north of England as my father's work in an insurance company required. My secondary education was at Sir John Deane's Grammar School in Northwich, Cheshire, to age 16. Then it was completed at the Collegiate School, Liverpool, from 1946 to 1948. At school my first loves were art and mathematics. After an unsuccessful year in the School of Architecture of Liverpool University, I joined the honours mathematics program in the same university. Although I won scholarships, there was some financial insecurity and, at the time of this change of subject, my continued study was made possible by the personal intervention and support of the curate of my parish of West Derby, Liverpool, Rev. E. W. Pugh. I graduated from Liverpool University with an honours degree in Mathematics in 1952. While still an undergraduate I was married to Edna Hutchinson in 1951. We were to have three daughters.

At that time military conscription was normal after graduation, but employment in the aircraft industry was an acceptable alternative. So I worked as an aerodynamicist with the English Electric Corporation (later British Aerospace) at Warton Aerodrome, Lancashire, from 1952 to 1957. In this time I worked on aircraft vibration and stability problems. As well as mathematical and numerical analysis, this work included analogue computing and also (the infancy of) high speed digital computing. This experience provided me with a fund of problems for more careful analysis which strongly influenced my research career – to the present day. During this time I registered for an external M.Sc. program at the University of Liverpool under the supervision of Louis Rosenhead, FRS, Professor of Applied Mathematics. The degree was awarded in 1956 for a dissertation on the vibration phenomenon known as "flutter", including formulation of the model problem and methods of solution.

At the end of the required service in industry I applied for academic positions and was appointed Assistant Lecturer in Mathematics at the University of Malaya (then based in Singapore; formerly Raffles College and later to become the National University of Singapore). There were strong research traditions established in part by Professors A. Oppenheim in Pure Mathematcs (later to become Vice Chancellor and knighted for his services), and J. C. Cooke in Applied Mathematics. Both had recently left the department, but their tradition was maintained by the current Acting Head of Department, Richard K. Guy, who was to become a friend and colleague to the present day. Though the department was small, it was active and had the advantage of a good university library which had been built up from the founding of Raffles College in 1929. In this period I published several papers on vibrations, iterative methods for solution of matrix polynomial eigenvalue problems, and some more general questions of spectral structure and inverse problems for matrix polynomials (see papers 1–8). Achievements in these years were recognized with promotions in 1959 and 1961 (to Senior Lecturer). During this time I was strongly influenced by friendly and helpful correspondence with A. M. Ostrowski. Indeed, Ostrowski's style, publications, and assistance were a source of inspiration. In 1961 there seemed to be enough results for a Ph.D. dissertation, so I registered for a Ph.D. at the University of Singapore. The dissertation "On the Theory of Lambda-matrices with Applications to the Analysis of Vibrating Systems" was submitted and examined in 1962. One of the external examiners, Ian Sneddon FRS, of the University of Glasgow, recommended publication of an extended version of the dissertation as a monograph, and this appeared in 1966 (reference B1). After unexplained procedural delays, the Ph.D. was awarded in 1964. On the other hand, the monograph remains a frequently quoted work, particularly in mechanical engineering.

An important event in this "Singapore period" was the invitation to attend and present a paper at one of the early meetings in Gatlinburg, Tennessee, and known at that time as a "Matrix Symposium". This was in April, 1961, and was an excellent opportunity to meet Ostrowski personally as well as other leaders in the field. Some of these influenced my subsequent career and tastes in research. They included Lothar Collatz, Alston Householder, Gilbert Strang, Olga Taussky, Jack Todd, Dick Varga, and Jim Wilkinson. This was my first visit to North America, and therefore full of interest. The experience was coloured, however, by a tense situation at home as our two-year old daughter, Jane, went through a dangerous illness.

On deciding to move back to the west from Singapore in 1962, Edna and I were presented with an interesting choice: I had offers of appointment from my alma mater, Liverpool University, and from the fledgling Calgary campus of the University of Alberta (to become the University of Calgary in 1965). Our love of the outdoors, and impatience with the structured English society, swung the balance in favour of adventure and a new institution. So, with our two daughters, aged five and two, the move to Calgary was made in November of 1962.

In the following years my research activities continued in matrix theory, iterative methods, perturbation and spectral theory for matrix functions (papers 10–22). These were assisted by a one-year appointment (1965–1966) as a Visiting Associate Professor at the California Institute of Technology. This was under the aegis of Jack Todd and Olga Taussky.

I was aware of work on matrix and operator functions in Eastern Europe, and particularly the Soviet Union from the late fifties, and my first direct contact was probably with P. H. Müller of the TU Dresden. From this contact I learned of the work of his former student, Heinz Langer, and of Heinz' continuing work with M. G.

Krein. Fortunately, Israel Halperin made arrangements for Heinz to visit Canada for the 1966–1967 year. This included a visit to Calgary and a series of lectures on his work with Krein. At about this time the monograph B1, which includes some early results on spectral theory and factorization of matrix functions, came to Krein's attention and, through the good offices W. K. Hayman, an exchange of monographs was made in 1968. I was also impressed by the Krein-Gohberg book on non-selfadjoint operators when it appeared in 1969 (in English translation).

In 1968–1969 I was fortunate to spend a sabbatical leave with Ostrowski in Basel, Switzerland. This did not lead to active collaboration (he was, in general, a solitary worker), but he was a gracious host. At that time, one of my main interests was error analysis for Newton's method. I shared this with Jon Rokne, a graduate student who joined us in Basel (see 13, 16, 21, and 27). In November of 1968 I was able to attend an Oberwolfach meeting organized by Collatz and made the acquaintance of several members of his school. I was very impressed by Collatz' analytical approach to problems of engineering and numerical analysis. Among the connections made at that time were lifelong friends and occasional collaborators, Erich Bohl and Ludwig Elsner. Also, in the Spring of 1969 I visited Heinz Langer in Dresden and learned a lot about life in eastern Europe in the communist era. Lectures in Calgary during the sixties led to my writing a senior undergraduate text in matrix theory. This was published in 1969 (reference B2) and, in a second edition of 1985 with Miron Tismenetsky (B6), is still in print. The first edition was translated into Russian and 60,000 copies were printed. As a result, my name may be better known in the former Soviet Union than it is in the west.

During this period I took an active interest in approximation theory and, particularly, in the emerging development of spline functions and finite element methods (papers 24, 26, 31, 35, 41, 42, 49 and, later, B7). This was undertaken largely in collaboration with my student, David Watkins, and colleague, Kes Salkauskas. The work was advanced by a visit with Erich Bohl (in 1974) at the University of Münster which included a lecture series on topics of this kind.

A pivotal event in my career took place in 1975. Israel Gohberg had emigrated from the former Soviet Union in 1974 and, due to the initiative of Chandler Davis, visited Canada in early 1975. It was arranged that he should visit Calgary (in a snowy February) and this began a rich friendship and collaboration. I took a sabbatical leave in Europe in 1975–1976 and, while at the University of Dundee, we arranged to meet again. At this time I had just completed work which, to some degree, combined earlier ideas and showed how to organize the spectral data for a matrix polynomial in a concise and useful way (papers 36 and 37). Later in the year I visited Israel Gohberg and his Ph.D. student, Leiba Rodman, in Tel Aviv. From these beginnings a series of papers (38–40, 43, 44, 47, 50, 51, 56, 57, 61) on matrix functions evolved, and after many visits for collaboration in Canada and Israel, three monographs were written (B4, B5, and B8). The first was a thorough treatment of the theory of matrix polynomials based on our earlier papers, and the second and third (on indefinite scalar product spaces and invariant subspaces, respectively) developed topics of wider interest which grew naturally from the first to the second volume, and from the second to the third. These collaborations included Leiba Rodman's appointment as a Post Doctoral Fellow in Calgary from 1978 to 1980.

During the late seventies, partly influenced by discussions with Harald Wimmer of the University of Würzburg, my interest in systems theory and control developed. In particular, it was clear that our methods might apply to analysis of Riccati equations, and our first success in this direction, with Leiba Rodman, was the solution of a basic existence problem for algebraic Riccati equations in paper 45 (1980). This line of investigation continued with papers 64, 66, 69, 73. A systematic approach to systems theory using the techniques of linear algebra was one of the main themes of the monograph B8 (1986). I remember working on this during a long stay in Tel Aviv as part of a sabbatical leave in 1982–1983. This leave also included some months visiting the institute of Erich Bohl at the University of Konstanz where I lectured on systems theory, as well as some time with the lively group led by Rien Kaashoek at the Free University of Amsterdam. The book B10 (1995) on "Algebraic Riccati Equations", written with Leiba, was a high-point for me in this research direction. It contains a comprehensive account of necessary mathematical tools as well as eight chapters on "applications and connections" with problem areas in systems theory, optimal control, and statistics. This led to an interesting analysis of Newton's method when applied to matrix equations which was undertaken with graduate student Guo Chun Hua (115 and 119). Most recently, an invitation to present a lecture series in 1998 at Portugal's Centro Internacional de Matemática, Coimbra, led to new connections and experiences (see B12).

A productive line of investigation began around 1984 into structured matrices and their applications in tomography, seismic inversion, and signal processing, as well as the design of fast algorithms for these problem areas. This was stimulated by further collaboration with Israel Gohberg and with Post-Doctoral Fellow, Israel Koltracht (1984–1987). The papers 63, 70, 72, 74, 75, 79, 81, 82, 83, 86, 90 come under this heading and span the period 1986–1992.

With graduate student, Qiang Ye, we returned to some classical questions of variational characterization of spectra of matrix pencils and numerical linear algebra with papers 80, 85, 91, and 94. It turned out that paper 91 (1993), which appeared after some publication delays, contained an influential idea for further developments of factorization theory for matrix functions to be discussed below. Qiang represented my first significant connection with PR China and Guo Chun Hua, mentioned above, my most recent connection. The first edition of my matrix theory text is well-known in China (through a pirated edition) so, as in the USSR, my name was quite familiar in China. In the mid-eighties I met Sun Ji Guang in Bielefeld. He returned to Beijing and recommended to his M.Sc. student, Qiang, that he study with me in Calgary for the Ph.D. Later, Dai Hua of Nanjing University spent a year with me, and we worked intensively on matrix equations and numerical methods for multiple eigenvalues and inverse eigenvalue problems (104, 110, and 111).

In 1988 I had the privilege of organizing a conference in honour of Israel Gohberg's sixtieth birthday. This was held at the University of Calgary and, with the assistance of many helpers, it attracted participants from around the world. In particular, we were successful in bringing several participants from the USSR – and this was very unusual at the time; also somewhat unpredictable. On one occasion Israel Gohberg and I went to the airport to greet participants from the USSR. We were able to do that, but they were not all participants who had been invited! Soviet officialdom still took a hand in these affairs. The benefits of this occasion included a double anniversary volume (OT 40 and 41), and the opening of some new international connections for me.

Stimulated by conversations during and after this birthday celebration, I made a visit to the USSR in June of 1989 under an exchange program managed by the Steklov Institute and Queens University. This was a memorable experience including visits in Moscow (with A. G. Kostyuchenko and Andre Shkalikov), Leningrad (with V. N. Kublanovskaya), and Kishinev (with A. S. Markus, N. Krupnik, and others of the operator theory school of Kishinev and Odessa).

It was a revelation to be inside the Soviet Union for the first time. I was almost arrested for (allegedly) not paying enough bus fare in Moscow. I lost a camera in the famed Moscow subway but, to my amazement, we were able recover it next day at the "lost and found" office. I enjoyed warm hospitality in the homes of Kostyuchenko in Moscow and V. A. Yakubovich in Leningrad. But most remarkable was the reception in Kishinev. To my amazement I was met at the airport by a delegation of mathematicians, some with their spouses. This was the Gohberg-Krein school. I very quickly felt that they were all my cousins! I had a wonderful week of mathematics, socializing, local culture, and sight-seeing. It was on the agenda to meet with M. G. Krein but on the appointed day he was not sufficiently well. Sadly, he died later that year. I did get to Odessa (illegally) and enjoyed a swim in the Black Sea and the hospitality of the family of Ilya Spitkovski.

During this visit I had conversations with Alek Markus on topics involving matrix polynomials and these were to lead to another very fruitful collaboration that is still alive and well. There have been two main lines in this work, and both exploit the new notion of quasihyperbolic matrix or operator polynomials (which I first introduced in 1991). One line concerns the stability of gyroscopic vibrating systems (and was initiated with a graduate student, Larry Barkwell, in paper 88). This has led to a continuing interest in gyroscopic systems and collaborations with Wolfhard Kliem and Christian Pommer of the Danish Technical University among others (see 96, 117, 118, 122, 123, and 124). The second theme concerns factorization of operator functions with symmetries with respect to the imaginary line or the unit circle. I put papers 99, 102, 106–108 (with Markus and V. I. Matsaev), and 125 in this sequence.

In 1989–1990 I enjoyed a sabbatical leave in Australia and Isreal. I lectured at the University of Brisbane on fast algorithms for strucutred matrices and, at Monash University in Melbourne, began collaborative research with Eric Chu and

Alan Andrews (of La Trobe University) concerned with the smoothness and computation of parameter dependent eigenvalues (93, 101). The techniques of analytic perturbation theory (as developed by T. Kato and H. Baumgartel, for example, see also Chapter 19 of B8) were relevant here. They have always interested me and have played a role in several problem areas. Quite recently, work with Rostyk Hryniv (121) on functions with analytic dependence on two parameters has been exciting and has allowed us to make useful applications (see 123 and 127). By the way, Rostyk was a student of Andre Shkalikov who I first met in Moscow in 1989. Andre spent a couple of long visits in Calgary in the early nineties during which we studied the spectral theory underlying beam vibrations (97, see also 95). This history, and others in this biography, illustrate the great advantage of discussions with other scientists having different perspectives from one's own; they can lead to new research directions and a healthy mixing of different generations of investigators. In my experience, no connections work so well as face-to-face meetings.

The topic of gyroscopic systems which I took up in the early nineties, as well as recent work with Pedro Freitas of Lisbon (see 120) have brought me close to problems that I studied at the beginning of my research career in the 1950's. This is refreshing and, at the time of writing, I look forward to spending some months with host Peter Hagedorn in the Department of Applied Mechanics of the TU of Darmstadt where I hope to continue working in this direction. This is made possible by a Humboldt Research Award, and provides an opportunity to return to my roots in the theory of vibrating systems.

The evolution of my choice of research topics is reflected to some degree in the dissertations of the Ph.D. students I have supervised. They are:

M.V. Pattabhirahman, (1967)
  "A spectral decomposition for a polynomial operator".

J.G. Rokne, (1969)
  "Practical and theoretical studies in numerical error analysis".

D.S. Watkins, (1974)
  "Blending functions and finite elements".

J.R. Terray, (1975)
  "Numerical and theoretical studies of eigenvalue problems of mathematical physics".

G. W. Cross, (1975)
  "Square roots of linear transformations".

Farid O. Farid, (1989)
  "Spectral properties of diagonally dominant infinite matrices".

Qiang Ye, (1989)
  "Variational principles and numerical algorithms for symmetric matrix pencils".

P. Zizler, (1995)
>   "Linear operators in Krein Spaces".

Guo Chun Hua, (1998)
>   "Analysis and modification of Newton's method for algebraic Riccati equations".

Calgary
September, 2000

Operator Theory:
Advances and Applications, Vol. 130, 9–19
© 2001 Birkhäuser Verlag Basel/Switzerland

List of Publications of Peter Lancaster

Monographs and Textbooks

B1 Lancaster, P., *Lambda-matrices and Vibrating Systems*, Pergamon Press, 1966.

B2 Lancaster, P., *Theory of Matrices*, Academic Press, 1969, MR 39, 6885. MR 80a, 15001. MR 84b, 15004.

B3 Lancaster, P., *Mathematics: Models of the Real World*, Prentice-Hall, 1976.

B4 Gohberg, I., Lancaster, P. and Rodman, L., *Matrix Polynomials*, Academic Press, 1982. MR 84c, 15012.

B5 Gohberg, I., Lancaster, P. and Rodman, L., *Matrices and Indefinite Scalar Products*, Birkhäuser Verlag, 1983. MR 87j, 15001.

B6 Lancaster, P. and Tismenetsky, M., *Theory of Matrices*, 2nd edition, Academic Press, 1985. MR 87a, 15001.

B6 Lancaster, P. and Salkauskas, K., *Curve and Surface Fitting*, Academic Press, 1986.

B7 Gohberg, I., Lancaster, P. and Rodman, L., *Invariant Subspaces of Matrices with Applications*, John Wiley (Canadian Math. Soc. Monographs), 1986. MR 88a, 15001.

B8 Lancaster, P. and Rodman, L., *Solutions of the Continuous and Discrete Time Algebraic Riccati Equations: A Review*, Chapter 2 of *The Riccati Equation*, (Ed. Bittanti, Laub and Willems) Springer Verlag, 1991. MR 92d, 93008.

B9 Lancaster, P. and Rodman, L., *Algebraic Riccati Equations*, Oxford University Press, 1995.

B10 Lancaster, P. and Salkauskas, K., *Transform Methods in Applied Mathematics: An Introduction*, John Wiley, New York, 1996.

B11 Lancaster, P., *Lecture Notes on Linear Algebra, Control, and Stability*, Centro Internacional de Matemática, Coimbra, 1998. (59 pages) (2nd Edition, Dept. of Math. and Stat., University of Calgary, 1999. (72 pages)

Research Papers

1. Lancaster, P., *Free vibration and hysteretic damping*, J. Roy. Aero. Soc., **64** (1960), p. 229.

2. Lancaster, P., *Free vibrations of lightly damped systems by perturbation methods*, Quart. J. Mech. Appl. Math., **13** (1960), pp. 138–155. MR 22, 4153.

3. Lancaster, P., *Inversion of Lambda-Matrices and application to the theory of linear vibrations*, Arch. Rat. Mech. Anal., **6** (1960), pp. 105–114. MR 22, 8029.

4. Lancaster, P., *Expressions for damping matrices in linear vibration problems*, J. Aero. Sp. Sci., **28** (1961), p. 256.

5. Lancaster, P., *Direct solution of the flutter problem*, (British) Min. of Aviation R. and M., **3206** (1961).

6. Lancaster, P., *A generalised Rayleigh Quotient Iteration for Lambda matrices*, Arch. Rat. Mech. Anal., **8** (1961), pp. 209–322. MR 25, 2697.

7. Lancaster, P., *Some applications of the Newton-Raphson method to nonlinear matrix problems*, Proc. Roy. Soc. (London), Ser. A., **271** (1963). MR 27, 925.

8. Lancaster, P., *On regular pencils of matrices arising in the theory of vibrations*, Quart. J. Mech. App. Math., **16** (1963), pp. 253–257. MR 27, 162.

9. Lancaster, P., *Convergence of the Newton-Raphson method for arbitrary polynomials*, Mathematical Gazette, **48** (1964), pp. 291–295. MR 29, 5380.

10. Lancaster, P., *Bounds for latent roots in damped vibration problems*, SIAM Review, **6** (1964), pp. 121–126. MR 29, 6632.

11. Lancaster, P., *On eigenvalues of matrices dependent on a parameter*, Num. Math., **6** (1964), pp. 377–387. MR 30, 1606.

12. Lancaster, P., *Algorithms for Lambda-Matrices*, Num. Math., **6** (1964) pp. 388–394. MR 30, 1607.

13. Lancaster, P., *Error analysis for the Newton-Raphson method*, Num. Math., **9** (1966), pp. 55–68. MR 35, 1208.

14. Lancaster, P. and Webber, P.N., *Jordan Chains for Lambda-matrices*, Lin. Alg. & Appls., **1** (1968), pp. 563–569. MR 39, 228.

15. Pattabhiraman, M.V. and Lancaster, P., *Spectral properties of a polynomial operator*, Num. Math., **13** (1969), pp. 247–259. MR 40, 775.

16. Rokne, J. and Lancaster, P., *Automatic errorbounds for the approximate solution of equations*, Computing (1969), pp. 294–303. MR 41, 2966.

17. Lancaster, P. *Spektraleigenschaften von Operatorfunktionen*. Contribution to *Iterationsverfahren, Numerische Mathematik, Approximation Theorie*, Int. Schr. Num. Math., Vol. **15**, pp. 53–60, Birkhäuser, Basel (1970). MR 51, 6467.

18. Lancaster, P., *Explicit solutions of linear matrix equations*, SIAM Rev., **12** (1970), pp. 544–566.

19. Lancaster, P., *Jordan chains for Lambda-matrices II*, Aequationes Mathematicae, **5** (1970), pp. 290–293. MR 45, 1942.

20. Lancaster, P. *Some questions in the classical theory of vibrating systems*, Bull. Poly. Inst. Jassy, **17** (1971), pp. 125–132. MR 52, 2312.

21. Rokne, J. and Lancaster, P., *Complex interval arithmetic*, Comm. Assoc. Comp. Mach., **14** (1971), pp. 111–112.

22. Lancaster, P. and Farahat, H.K., *Norms on direct sums and tensor products*, Math. of Comp., **26** (1972), pp. 401–414. MR 46, 4229.

23. Lancaster, P., *A note on sub-multiplicative norms*, Num. Math., **19** (1972), pp. 206–208. MR 46, 4230.

24. Schmidt, E. and Lancaster, P., *L-splines for constant coefficient differential operators*, Proc. Manitoba Conf. on Num. Math., U. of Manitoba (1971). MR 49, 7656.

25. Cross, G.W. and Lancaster, P., *Square roots of complex matrices*, J. Lin. Mult. Alg., **1** (1974), pp. 288–293. MR 49, 5033.

26. Schmidt, E., Lancaster, P., and Watkins, D., *Bases of spline functions associated with constant coefficient differential operators*, SIAM J. Num. Anal., **12** (1975), pp. 630–645. MR 52, 8728.

27. Rokne, J. and Lancaster, P., *Complex interval arithmetic: algorithms*, Comp. J., **18** (1975), pp. 83–85.

28. Lancaster, P. and Wimmer, H.K., *Zur theorie der λ-matrizen*, Math. Nach., **68** (1975), pp. 325–330. MR 58, 28036.

29. Lancaster, P. and Terray, J., *Numerical solution of a boundary value problem arising in the study of heat transfer.* Contribution to *Numerische Losung von Differentialgleichungen*, Int. Schr. Num. Math., Vol. 27 pp. 303–308, Birkhäuser, Basel, 1975. MR 53, 4757.

30. Lancaster, P., *An efficient computation of the angle of latitude*, J. Can. Soc. Expl. Geophysicists, **11** (1975), pp. 72–73.

31. Lancaster, P., *Interpolation in a rectangle and finite elements of high degree*, J. Inst. Math. Appl., **18** (1976), pp. 65–77. MR 56, 13614.

32. Lancaster, P. and Watkins, D., *Interpolation in the plane and rectangular finite elements.* Contribution to *Numerische Behandlung von Differentialgleichungen, insbesondere mit der Methode der finiten Elemente*, Int. Schr. Num. Math., Vol. **31**, pp. 125–145, Birkhäuser, Basel, 1976. MR 58, 24849.

33. Terray, J. and Lancaster, P., *On the numerical calculation of eigenvalues and eigenvectors of operator polynomials*, J. Math. Anal. Appl., **60** (1977), pp. 370–378. MR 58, 24928.

34. Lancaster, P. and Rokne, J.G., *Solutions of nonlinear operator equations*, SIAM J. Math. Anal., **8** (1977), pp. 448–457. MR 55, 11078.

35. Watkins, D. and Lancaster, P., *Some families of finite elements*, J. Inst. Math. Appl., **19** (1977), pp. 385–397. MR 55, 11650.

36. Lancaster, P. *A fundamental theorem on Lambda-matrices with applications I: ordinary differential equations with constant coefficients*, Lin. Alg. & Appl., **18** (1977), pp. 189–211. MR 58, 5711.

37. Lancaster, P., *A fundamental theorem on Lambda-matrices with applications II: difference equations with constant coefficients*, Lin. Alg. & Appl., **18** (1977), pp. 213–222. MR 58, 5712.

38. Gohberg, I., Lancaster, P., and Rodman, L., *Spectral analysis of matrix polynomials I: canonical forms and divisors*, Lin. Alg. & Appl., **20** (1978), pp. 1–44. MR 57, 3155.

39. Gohberg, I., Lancaster, P., and Rodman, L., *Spectral analysis of matrix polynomials II: the resolvent form and spectral divisors*, Lin. Alg. & Appl., **21** (1978), pp. 65–88. MR 58, 16720.

40. Gohberg, I., Lancaster, P., and Rodman, L., *Representations and divisibility of operator polynomials*, Can. J. Math., **30** (1978), pp. 1045–1069. MR 80a, 47024.

41. Lancaster, P., *Composite methods for generating surfaces.* Contribution to *Polynomial and Spline Approximation*, ed. B.N. Sahney, Reidel Pub. Co. (1979), pp. 91–102. MR 81c, 41075.

42. Lancaster, P., *Moving weighted least squares methods.* Contribution to *Polynomial and Spline Approximation*, ed. B.N. Sahney, Reidel Pub. Co. (1979), pp. 103–120. MR 80c, 6503.

43. Gohberg, I., Lancaster, P. and Rodman, L., *Perturbation theory for divisors of operator polynomials*, SIAM J. Math. Anal., **10** (1979), pp. 1161–1183. MR 82b, 4701b.

44. Gohberg, I., Lancaster, P. and Rodman, L., *On selfadjoint matrix polynomials*, Integral Eq. and Op. Theory, **2** (1979), pp. 434–439. MR 80k, 15018.

45. Lancaster, P. and Rodman, L., *Existence and uniqueness theorems for the algebraic Riccati equation*, Int. J. Control, **32** (1980), pp. 285–309. MR 82c, 15017.

46. Barnett, S. and Lancaster, P., *Some properties of the Bezoutian for polynomial matrices*, J. Lin. Mult. Alg., **9** (1980), 99–110. MR 82c, 15014.

47. Gohberg, I., Lancaster, P. and Rodman, L., *Spectral analysis of selfadjoint matrix polynomials*, Annals of Math., **112** (1980), pp., 33–71. MR 82c, 15010.

48. Barnett, S. and Lancaster, P., *Matrices having striped inverse.* Progress in Cybernetics and Systems Research, Vol. 8, pp. 333–336. Edited by R. Trappl, G.J. Klir and F. Pichler, Hemisphere Publishing Corporation, Washington, D.C., 1981.

49. Lancaster, P. and Salkauskas, K., *Surfaces generated by moving least squares methods*, Math. Soc. Comp., **37** (1981), pp. 141–158. MR 83c, 65015.

50. Gohberg, I., Lancaster, P. and Rodman, L., *Factorization of self-adjoint matrix polynomials with constant signature*, J. Lin. Mult. Alg., **11** (1982), pp. 209–224. MR 83c, 15011.

51. Gohberg, I., Lancaster, P. and Rodman, L., *Perturbations of H self-adjoint matrices with applications to differential equations*, Integral Eq. and Op. Theory, **5** (1982), pp. 718–757.

52. Lancaster, P. and Pattabhiraman, M.V., *The local determination of Jordan bases for H-selfadjoint operators*, Lin. Alg. and Appl., **48** (1982), pp. 191–199. MR 84a, 15009.

53. Lancaster, P. and Rozsa, P., *On the matrix equation $AX + X^*A^* = C$*, SIAM J. Alg. and Discrete Methods, **4** (1983), pp. 432–436.

54. Lancaster, P. and Tismenetsky, M., *Inertia characteristics of self-adjoint matrix polynomials*, Lin. Alg. and Appl., **52** (1983), pp. 479–496.

55. Lancaster, P. and Tismenetsky, M., *Some extensions and modifications of classical stability tests for polynomials*, Inter. J. Control, **38** (1983), pp. 369–380.

56. Gohberg, I., Lancaster, P. and Rodman, L., *A sign-characteristic for self-adjoint meromorphic matrix functions*, Applicable Analysis, **16** (1983), pp. 165–185.

57. Gohberg, I., Lancaster, P. and Rodman, L., *A sign characteristic for rational matrix functions*, Mathematical Theory of Networks and Systems, pp. 363–369. Edited by P.A. Fuhrmann. Springer Verlag, Berlin, 1984.

58. Lancaster, P., Lerer, L. and Tismenetsky, M., *Factored forms for solutions of $AX - XB = C$ and $X - AXB = C$ in companion matrices*, Lin. Alg. and Appl., **62** (1984), pp. 19–49.

59. Lancaster, P. and Rozsa, P., *Eigenvectors of H-selfadjoint matrices*, Zeitschrift für Ang. Math. und Mech., **64** (1984), pp. 439–441.

60. Lancaster, P. and Maroulas, J., *The kernel of the bezoutian for operator polynomials*, J. Lin. Mult. Alg., **17** (1985), pp. 181–201.

61. Gohberg, I., Lancaster, P. and Rodman, L., *Perturbations of analytic hermitian matrix functions*, Applicable Analysis, **20** (1985), pp. 23–48. MR 87d, 15018.

62. Elsner, L. and Lancaster, P., *The spectral variation of pencils of matrices*, Jour. Computational Math., **3** (1985), pp. 262–274. MR 87j, 15029.

63. Koltracht, I. and Lancaster, P., *Condition numbers of Toeplitz and block-Toeplitz matrices*, Operator Theory: Advances and Applications, Vol. **18** (1986), pp. 271–300. MR 88i, 15015.

64. Gohberg, I., Lancaster, P. and Rodman, L., *On hermitian solutions of the symmetric algebraic Riccati equation*, SIAM J. of Control and Optim., Vol. **24** (1986), pp. 1323–1334. MR 88f, 93041.

65. Gohberg, I., Lancaster, P. and Rodman, L., *Quadratic matrix polynomials with a parameter*, Advances in Applied Math., **7** (1986), pp. 253–281. MR 88e, 47027.

66. Lancaster, P., Ran, A.C.M. and Rodman, L., *Hermitian solutions of the discrete algebraic Riccati equation*, Int. J. Control., **44** (1986), pp. 777–802. MR 87h, 93022.

67. Lancaster, P. *Common eigenvalues, divisors, and multiples of matrix polynomials: a review*, Lin. Alg. and Appl., **84** (1986), pp. 139–160. MR 87m, 15032.

68. Lancaster, P. and Maroulas, J., *Inverse eigenvalue problems for damped vibrating systems*, J. Math. Anal. Appl., **123** (1987), pp. 238–261. MR 88d, 34013.

69. Lancaster, P., Ran, A.C.M. and Rodman, L., *An existence and monotonicity theorem for the discrete algebraic matrix Riccati equation*, J. Lin. Mult. Alg., **20** (1987), pp. 353–361. MR 88j, 93033.

70. Gohberg, I., Kailath, T., Koltracht, I. and Lancaster, P., *Linear complexity parallel algorithms for linear systems of equations with recursive structure*, Lin. Alg. and Appl., **88** (1987), pp. 271–315. MR 88g, 65027.

71. Gohberg, I., Koltracht, I. and Lancaster, P., *Second order parallel algorithms for Fredholm integral equations with continuous displacement kernels*, Integral Eq. and Op. Theory, **10** (1987), pp. 577–594. MR 88m, 65205.

72. Koltracht, I. and Lancaster, P., *A definiteness test for Hankel matrices and their lower sub-matrices*, Computing, **39** (1987), pp. 19–26. MR 88h, 15046.

73. Lancaster, P. and Maroulas J., *Selective perturbation of spectral properties of vibrating systems using feedback*, Lin. Alg. and Appl., **98** (1988), pp. 309–330. MR 88k, 93060.

74. Koltracht, I. and Lancaster, P., *Generalized Schur parameters and the effects of perturbations*, Lin. Alg. and Appl., **105** (1988), 109–129.

75. Koltracht, I. and Lancaster, P., *Threshold algorithms for the prediction of reflection coefficients in a layered medium*, Geophysics, **53** (1988), 908–919.

76. Lancaster, P. and Ye, Q., *Inverse spectral problems for linear and quadratic matrix pencils*, Lin. Alg. and Appl., **107** (1988), 293–309.

77. Gohberg, I., Kaashoek, M.A. and Lancaster, P., *General Theory of regular matrix polynomials and band Toeplitz operators*, Integral Eq. and Op. Theory, **11** (1988), 776–882.

78. Farid, F.O. and Lancaster, P., *Spectral properties of diagonally dominant infinite matrices*, Part I, Proc. Roy. Soc. Edinburgh, **111A** (1989), 301–314.

79. Gohberg, I. Koltracht, I. and Lancaster, P., *On the numerical solution of integral equations with piecewise continuous displacement kernels*, Integral Eq. and Op. Theory, **12** (1989), 511–537.

80. Lancaster, P. and Ye, Q., *Variational properties and Rayleigh quotient algorithms for symmetric matrix pencils*, in the Gohberg Anniversary Collection, (Operator Theory and its Applications, Vol. **40**, Birkhäuser Verlag, Basel), (1989), 247–278. MR 91e, 65055.

81. Koltracht, I., Lancaster, P. and Smith, D., *The structure of some matrices arising in tomography*, Lin. Alg. and Appl., **130** (1990), 193–218. MR 91j, 65068.

82. Koltracht, I. and Lancaster, P., *Constraining strategies for linear iterative processes*, I.M.A. Jour. on Numerical Analysis, **10** (1990), 555–567. MR 91i, 65066.

83. Bruckstein, A., Kailath, T., Koltracht, I. and Lancaster, P., *On the reconstruction of layered media from reflection data*, SIAM J. Matrix Anal. and Appl., **12** (1991), 24–40. MR 92d, 86004.

84. Farid, F.O. and Lancaster, P., *Spectral properties of diagonally dominant infinite matrices, Part II*, Lin. Alg. and Appl. **143** (1991), 7–17. MR 91j, 47029.

85. Lancaster, P. and Ye, Q., *Variational and numerical methods for symmetric matrix pencils*, Bull. Australian Math. Soc., **43** (1991), 1–17. MR 91m, 65106.

86. Elsner, L., Koltracht, I. and Lancaster, P., *Convergence properties of ART and SOR algorithms*, Numer. Math., **59** (1991), pp. 91–106. MR 92i, 65065.

87. Jameson, A., Kreindler, E. and Lancaster, P., *Symmetric, positive semidefinite, and definite real solutions of $AX = XA^T$ and $AX = YB$*, Lin. Alg. and Appl., **160** (1992), 189–215. MR 92j, 15011.

88. Barkwell, L. and Lancaster, P., *Overdamped and gyroscopic vibrating systems*, Jour. of Appl. Mechanics, **59** (1992), 176–181. AMR **45 (5)**, #47.

89. Barkwell, L., Lancaster, P., and Markus, A.S., *Gyroscopically stabilized systems: a class of quadratic eigenvalue problems with real spectrum*, Canadian Jour. Math., **44** (1992), 42–53.

90. Gohberg, I., Koltracht, I. and Lancaster, P., *Second order parallel algorithms for piecewise smooth displacement kernels*, Integral Eq. and Op. Theory, **15** (1992), 16–29. MR 92h, 65199.

91. Lancaster, P., and Ye, Q., *Definitizable hermitian matrix pencils*, Aequationes Mathematicae, **46** (1993), 44–55.

92. Bohl, E. and Lancaster, P., *Perturbation of spectral inverses applied to a boundary layer phenomenon arising in chemical networks*, Lin. Alg. and Appl., **180** (1993), 35–59.

93. Andrew, A.L., Chu, K.W.E., and Lancaster, P., *Derivatives of eigenvalues and eigenvectors of matrix functions*, SIAM J. Matrix Anal. and Appl., **14** (1993), 903–926.

94. Lancaster, P., and Ye, Q., *Rayleigh-Ritz and Lanczos methods for symmetric matrix pencils*, Lin. Alg. and Appl., **185** (1993), 173–201.

95. Lancaster, P. Shkalikov, A., and Ye, Q., *Strongly definitizable linear pencils in Hilbert space*, Integral Eq. and Op. Theory, **17** (1993), 338–360.

96. Lancaster, P., Markus, A.S. and Ye, Q., *Low rank perturbations of strongly definitizable transformations and matrix polynomials*, Lin. Alg. and Appl., **197/198** (1994), 3–30.

97. Lancaster, P. and Shkalikov, A., *Damped vibrations of beams and related spectral problems*, Canadian Applied Math. Quarterly, **2** (1994), 45–90.

98. Lancaster, P. and Rodman, L., *Invariant neutral subspaces for symmetric and skew real matrix pairs*, Canadian Jour. of Math., **46** (1994), 602–618.

99. Lancaster, P., Markus, A.S., and Matsaev, V.I., *Definitizable operators and quasihyperbolic operator polynomials*, Jour. of Functional Analysis, **131** (1995), 1–28.

100. Lancaster, P. and Rodman, L., *Minimal symmetric factorizations of symmetric real and complex rational matrix functions*, Lin. Alg. and Appl. **220** (1995), 249–282.

101. Andrew, A.L., Chu, K.-w.E., and Lancaster, P., *On numerical solution of nonlinear eigenvalue problems*, Computing, **55** (1995), 91–111.

102. Krupnik, I., Lancaster, P. and Markus, A., *Factorization of selfadjoint quadratic matrix polynomials with real spectrum*, J. Lin. Mult. Alg., **39** (1995), 263–272.

103. Lancaster, P. and Rozsa, P., *The spectrum and stability of a vibrating rail supported by sleepers*, Computers and Mathematics with Applications, **31** (1996), 201–213.

104. Dai, H. and Lancaster, P., *Linear matrix equations from an inverse problem of vibration theory*, Lin. Alg. and Appl., **246** (1996), 31–47.

105. Krupnik, I., Lancaster, P. and Zizler, P., *Factorization of selfadjoint matrix polynomials with real spectrum*, J. Lin. Mult. Alg., **40** (1996), 327–336.

106. Lancaster, P., Markus, A.S. and Matsaev, V.I., *Perturbations of G-selfadjoint operators and operator polynomials with real spectrum*, in "Recent Developments in Operator Theory and its Applications", Proceedings of the Winnipeg Conference, OT **87**, Birkhäuser Verlag, Basel (1996), 207–221.

107. Lancaster, P., Markus, A.S. and Matsaev, V.I., *Definitizable G-unitary operators and their applications to operator polynomials*, in "Recent Developments

in Operator Theory and its Applications", Proceedings of the Winnipeg Conference, OT **87**, Birkhäuser Verlag, Basel (1996), 222–232.

108. Lancaster, P., Markus, A.S. and Matsaev, V.I., *Factorization of selfadjoint operator polynomials*, Jour. of Operator Theory, **35** (1996), 337–348.

109. Lancaster, P., Markus, A.S. and Zizler, P., *The order of neutrality for linear operators on indefinite inner product spaces*, Lin. Alg. and Appl., **259** (1997), 25–29.

110. Dai, H. and Lancaster, P., *Numerical methods for finding multiple eigenvalues of matrices depending on parameters*, Numerische Math., **76** (1997), 189–208.

111. Dai, H. and Lancaster P., *Newton's method for a generalized inverse eigenvalue problem*, Numerical Linear Algebra with Applications, **4** (1997), 1–21.

112. Krupnik, I. and Lancaster, P., *H-selfadjoint and H-unitary matrix pencils*, SIAM J. Matrix Anal. and Appl., **19** (1998), 307–324.

113. Krupnik, I. and Lancaster, P., *Minimal pencil realizations of rational matrix functions with symmetries*, Canadian Math. Bull. **41** (1998), 178–186.

114. Krupnik, I. and Lancaster, P., *Linearizations, realization, and scalar products for regular matrix polynomials*, Linear Algebra and its Applications, **272** (1998), 45–57.

115. Guo, Chun-Hua and Lancaster, P., *Analysis and modification of Newton's method for algebraic Riccati equations*, Math. Comp., **67** (1998), 1089–1105.

116. Lancaster, P., Maroulas, J. and Zizler, P., *The numerical range of self-adjoint matrix polynomials*, Operator Theory: Advances and Applications, **106** (1998), 291–304.

117. Lancaster, P. and Kliem, W., *Comments on stability properties of conservative gyroscopic systems*, Jour. of Appl. Mech., **66** (1999), 272–273.

118. Lancaster, P. and Zizler, P., *On the stability of gyroscopic systems*, Jour. of Appl. Mech. **65** (1998), 519–522.

119. Guo, Chun-Hua and Lancaster, P., *Iterative solution of two matrix equations*, Math. Comp., **68** (1999), 1589–1603.

120. Freitas, P. and Lancaster, P., *On the optimal spectral abscissa for a system of linear oscillators*, SIAM J. Matrix Anal. and Appl. **21** (1999), 195–208.

121. Hryniv, R.O. and Lancaster, P., *On the perturbation of analytic matrix functions*, Integral Eq. and Op. Theory, **34** (1999), 325–338.

122. Lancaster, P., *Strongly stable gyroscopic systems*, Elec. Jour. of Linear Algebra, **5** (1999), pp. 53–66.

123. Hryniv, R.O., Lancaster, P. and Renshaw, A.A., *A stability criterion for parameter dependent gyroscopic systems*, Jour. of Appl. Mech., **66** (1999), 660–664.

124. Hryniv, R.O., Lancaster, P., Kliem, W. and Pommer, C., *A precise bound for gyroscopic stabilization*, Zeit. für Ang. Math. u. Mech., **80** (2000), 507–516.

125. Lancaster, P. and Markus, A. S., *A note on factorization of analytic matrix functions*, Operator Theory: Advances and Applications, **122** (2001), 323–330.

126. Hryniv, R. and Lancaster, P., *Stabilization of gyroscopic systems*, Zeit. für Ang. Math. u. Mech., (to appear).

127. Lancaster, P. and Psarrakos, P., *The numerical range of selfadjoint quadratic matrix poynomials*, SIAM J. Matrix Anal. and Appl., (to appear).

128. Lancaster, P. and Psarrakos, P., *Normal and seminormal eigenvalues of matrix functions*, Integral Eq. and Op. Theory, (to appear).

Expositions, Surveys, etc.

129. Lancaster, P., *Some problems in aero-elasticity*, Bull. Malayan Math. Soc., **5**, (1958), pp. 27–33.

130. Lancaster, P., *Some reflections on the "Easter Conference on Mathematics"*, Bull. Malayan Math. Soc., **6** (1959), pp. 69–72.

131. Lancaster, P., *The damped harmonic oscillator*, Bull. Malayan Math. Soc., **6** (1959), pp. 96–99.

132. Lancaster, P., *Approaching the speed of sound*, NABLA: Bull. Malayan Math. Soc., **7** (1960), pp. 52–60.

133. Lancaster, P., *A garden swing*, NABLA: Bull. Malayan Math. Soc., **7** (1960), pp. 65–67.

134. Lancaster, P., *Ill-conditioned equations*, NABLA: Bull. Malayan Math. Soc., **7** (1960), pp. 165–167.

135. Lancaster, P., *Symmetric transformations of the companion matrix*, NABLA: Bull. Malayan Math. Soc., **8** (1961), pp. 146–148.

136. Lancaster, P., *On Cooke's extension of Simpson's rule*, NABLA: Bull. Malayan Math. Soc., **8** (1961), pp. 182–184.

137. Lancaster, P., *Some numerical methods for solving equations: 1 and 2*, NABLA: Bull. Malayan Math. Soc., **9** (1962), pp. 45–53 and 85–95.

138. Lancaster, P., *A Centennial Expedition*, Can. Alp. Jour., **51** (1968), pp. 209–211.

139. Lancaster, P., *Applied mathematics in the undergraduate curriculum*, Notes of the Can. Math. Congress, Vol. **5**, No. 2 (1972), pp. 6–8.

140. Lancaster, P., *A review of numerical methods for eigenvalue problems nonlinear in the parameter*, Contribution to Numerik und Anwendungen von

Eigenwertaufgaben und Verzweigungsproblemen, Int. Schr. Num. Math., **38** (1977), pp. 43–67. MR 58, 19116.

141. Lancaster, P., *Solving quadratic equations*, Menemui Matematik, **1** (1979), 23–26. MR 81g 65001.

142. Lancaster, P., *The transition from high school to university mathematics*, Notes of the Can. Math. Soc., **12** (1980), pp. 8–21.

143. Lancaster, P., *A review of recent results on factorization of matrix and operator-valued functions*, Nonlinear Anal. and Applic., pp. 117–139. Edited by S.P. Singh and J.H. Burry, Marcel Dekker, New York, 1982. MR 84c, 15015.

144. Lancaster, P., *Generalized Hermitian matrices: a new frontier for numerical analysis?*, Proc. 9th Dundee Biennial Conf. on Num. Anal., pp. 179–189, Springer Verlag, Lecture Notes in Math., Vol. **912**, 1982.

145. Lancaster, P., *Mathematics and Society*, The Canadian Encyclopedia, pp. 1099–1100, Hurtig Publishers, Edmonton, 1986. Second Edition, 1988.

146. Lancaster, P., *Obituary: A.M. Ostrowski*, Aequationes Mathematicae, **33** (1987), pp. 121–122. MR 88i, 01113.

147. Lancaster, P., *What are universities for?*, Position paper for the conference of the same title, University of Calgary, 1988.

148. Lancaster, P., *Quadratic eigenvalue problems*, Linear Alg. and Appl., **150** (1991), 499–506.

149. Andrew, A.L., Chu, K.W.E., and Lancaster, P., *Sensitivities of eigenvalues and eigenvectors of problems nonlinear in the eigenparameter*, Applied Mathematics Letters, **5(3)**, (1992), 69–72.

150. Lancaster, P., *Spectra and stability of quadratic eigenvalue problems*, Proceedings of the International Workshop on the Recent Advances in Applied Mathematics, Kuwait University, 1996.

151. Lancaster, P., *The role of the Hamiltonian in the solution of algebraic Riccati equations*, In "Dynamical Systems, Control, Coding, Computer Vision", Eds. Picci, G., and Gillian, D. S., Birkhäuser, Basel, 1999, 157–172.

Operator Theory:
Advances and Applications, Vol. 130, 21–22
© 2001 Birkhäuser Verlag Basel/Switzerland

Forty-four Years with Peter Lancaster

Richard K. Guy

Abstract. Our mathematical paths have rarely crossed, but for his inspiring example of what a mathematician is, and for his close friendship, I shall always be indebted to Peter.

We had tried to recruit Peter Lancaster to the University of Malaya in Singapore two years earlier, but the Appointments Board of the Association of Universities of the British Commonwealth assembled a committee, including Harold Davenport and Hans Heilbronn, who evidently did not believe that mathematics other than number theory was worth pursuing.

However, two years later, Peter's second attempt was successful. Alexander Oppenheim had become Vice-Chancellor, Jack Cooke had departed to Farnborough, and it was left to me, as acting Head of Department, to welcome Peter. Louise and I were in England and called on the Lancaster family in Preston in 1957 when their eldest daughter Jane was three weeks old. Peter was working as an aerodynamicist for English Electric, having studied for a Master's degree with Louis Rosenhead in Liverpool, His main concern was with vibrating systems, particularly wings of aircraft, and the underlying matrices were to be at the back of much of his later work.

We were a small department, which meant that mathematics, as it should be, was regarded as a single subject. Everyone taught everything, and Peter sat in on an honours course that I gave in Linear Algebra, using Mirsky's *Algebra* and Halmos's *Finite-dimensional Vector Spaces*, though it didn't need much hindsight to see that our roles should really have been interchanged. At least the lecturer learned something.

It was usual, under the British Commonwealth system, for staff members to pursue higher degrees in their own time, getting what assistance they could from colleagues. There was noone to help Peter gain his PhD, in fact only an administration which turned out to be quite hostile. Fortunately Peter's brilliance could not be dimmed, his work caught the attention of Alexander Ostrowski, he presented a paper at a conference in Oak Ridge organized by Alston Householder, and when the atmosphere in Singapore in 1962 was no longer conducive to his development, he moved to the University of Alberta in Calgary, which four years later became The University of Calgary.

Meanwhile Eric Milner had moved to Reading and I had gone to the newly-forming Indian Institute of Technology in Delhi. The Lancaster family, now including daughter Jill, called on us in India on their way to Calgary. A few years later, when the monsoon was overdue in Delhi, Edna Lancaster sent us a picture of the family, now including Joy, picnicking in the snow by Spray Lake. Louise urged me to take the job that was offered at Calgary, and we rejoined them in 1965. Eric Milner was also persuaded to come to Calgary in 1967, so the three of us had another thirty years together until Eric unfortunately lost his battle with cancer.

Peter held visiting positions at the California Institute of Technology, 1965–66 and at the University of Basel, 1968–69 and a Senior Visiting Fellowship of the British Science Research Council at the University of Dundee, 1975–76 as well as shorter stays at the Universities of Munster, Tel Aviv (where he was Toeplitz Lecturer) and Konstanz, and a Dozor Visiting Fellowship at Ben Gurion University.

I leave it to those more familiar with his field to speak of his professional work. He has supervised many graduate students, and, as the present volume testifies, has collaborated with people round the world, including several distinguished mathematicians. He was President of the Canadian Mathematical Society at a time when his energy and vision were usefully brought to bear in the movement which has since continued and made the Society much more well known to the international community, and much more conscious of the wider role that mathematics plays. He was elected to a Fellowship of the Royal Society of Canada in 1984. He has recently been an active proponent of the Pacific Institute for Mathematical Sciences and a member of the comparatively small group that has propelled it in a short time to become an internationally recognized institution. Last year he received the prestigious Research Award from the Alexander von Humboldt Foundation.

But there is much more to Peter than a mere mathematician. He originally began training as an architect, and his skills served him well when he chaired the committee which oversaw the building of the first library block at the University of Calgary. In his youth he was a keen basketball player, and he successfully coached the University of Malaya team when he was in Singapore. It was there that we stayed awake for 48 hours, driving to the Trengganu coast and back, and working through the night to help save the giant leathey turtles.

He is also a pianist and a photographer: some fine prints of the Rocky Mountains adorn his home. The mountains have also been the scene for many an enjoyable climb or ski. What a privilege and stimulus it has been to have such a friend and colleague and companion down the years.

Operator Theory:
Advances and Applications, Vol. 130, 23–27
© 2001 Birkhäuser Verlag Basel/Switzerland

Peter Lancaster, my Friend and Co-author

Israel Gohberg

The first time I heard about the mathematician Peter Lancaster was from Heinz Langer. In 1967–68 Heinz spent a year in Toronto University at the invitation of Professor Israel Halperin. During his visit to Canada he was also invited by Peter Lancaster to spend a short period in Calgary. After his return Heinz told us about the work of Peter, about his book on vibrations of systems and about his personality. He also brought the book to Odessa and Peter's results were often quoted in the seminars and discussions, and very soon Peter became popular in Odessa.

In July 1974 I immigrated to Israel. At the end of that year I visited the USA at the invitation of my colleagues from Stony Brook: R.G. Douglas and J. Pincus. In the beginning of 1975 I was a guest of Seymour Goldberg in the University of Maryland, College Park. During my latter visit I was invited by Chandler Davis to visit the University of Toronto and by Peter Lancaster to visit Calgary.

From the start I was impressed by Peter. We discussed mathematics and started to look for an area for joint work. After my return to Israel in March I obtained my first Ph.D. student in Tel Aviv University. He was Leiba Rodman. I started to work with him on matrix polynomials. The problem which we were interested in was the problem of reconstruction of a matrix polynomial from its spectral data. The spectral data consists of the eigenvalues, eigenvectors and chains of generalized eigenvectors. We were looking for something which would generalize for matrix polynomials the Jordan form for matrices. I thought that this problem would fit Leiba because he had a much better education in algebra than in analysis. In 1975–76 we obtained our first results. The results were difficult to formulate and to present. On my second visit to Calgary Peter and I continued to look for an area of common interest and Peter showed me a number of his recent papers. Reading those papers I understood that some of Peter's results could form the piece of information which we were missing in the work with Leiba. We invited Peter to join our team. Peter accepted and soon visited Tel Aviv University and we started to work actively together.

In 1978 our first two papers were published in Linear Algebra and its Applications. These papers contained the solution of the above mentioned problem, and much more. They contained important formulas for the polynomials via the spectral data results for factorization matrix polynomials including spectral factorization and applications to differential and integral equations. We continued to work on matrix polynomials, but we already understood that our papers could serve as a basis for a book on matrix polynomials. This book was written in a relatively short period of time when Leiba spent his post doctoral period at the University of Calgary as a guest of Peter. During this time our other friends also joined us in our studies. I have in mind H. Bart, M.A. Kaashoek and L. Lerer. Some of their results were also included in the book.

During the writing of this book the authors became interested in matrix polynomials which have important applications. I have in mind the theory of matrix polynomials with selfadjoint coefficients. Our first results in this area were also included in the book.

In particular, whilst developing the theory of matrix polynomials with self-adjoint coefficients we discovered new invariants for such polynomials. We studied these invariants and in the process of this work we got the feeling that we had a good basis for a second book "Matrices and indefinite scalar products".

An important part of this second book was dedicated to problems of perturbations in the presence of an indefinite scalar product. It also contained important applications to differential equations. Included are deep results of M.G.Krein, I.M.Gelfand and V.B.Lidsky, which were obtained within our framework.

The work on our books was pleasant and all authors enjoyed it. Leiba spent long periods in Calgary and I visited Calgary for one to two months each year. Peter also visited us in Tel Aviv many times. Each of us worked with full dedication and responsibility. Many times during our work we had differences of opinion. I do not remember a single case where we were not able to reach a concensus after a friendly discussion. Except once, during a hike when we could not reach a concensus as to whether we should have lunch immediately or wait till the end of the hike. Peter's taste, his knowledge, talent and command of English combined with the working capability, breakthrough power and talent of Leiba, were very important for our team. Peter's wife Edna, and their three daughters (Jane, Jill, and Joy) were an integral part of our team. This warm family displayed outstanding hospitality and understanding of my and Leiba's Russian-Jewish mentality.

I would like to tell separately about the wonderful weekends that we spent in the Lancasters' cottage in the Rocky Mountains, between Calgary and Banff. It is a nice small cottage built of wood in the middle of the wilderness. All around is wild forest and a nearby river visited by beavers and bears. How many people enjoyed the wonderful hospitality of the Lancasters in the modern comfort of this cottage. How many picnics, barbeques and dinners. The Lancasters have a book containing remarks of their guests. In this book there is a very long list in many languages of mathematicians and other guests. Often we stayed overnight at the cottage. The nights spent there were beautiful. Around the brightly burning fireplace the company gathered, continuing discussions in mathematics, projects and plans; rarely did we allow ourselves a discussion on general topics or politics. Very often during those visits we made long and difficult hikes. I am proud that I introduced the Lancasters to mushroom hunting. We often enjoyed the combination of hiking and mushroom hunting, followed by mushroom dinners on our return. During our walks and hikes we also discussed many issues, but mathematics of our books and

Peter and Edna in the Rocky Mountains
Photograph taken by the author

our research was one of the main topics. My wife, Bella, and I will never forget the wedding of Jane (Peter's eldest daughter) to Ken held in a small hotel in the middle of the Rocky Mountain wilderness. It was one of the nicest events that we have ever attended. The atmosphere was extremely friendly, the background was beautiful and the food was excellent. The folk dancing was very enjoyable and fitting for the event.

Peter and Edna highly appreciated their English heritage. This did not exclude criticism of some aspects of it. The family consider themselves Canadians but every year Peter and Edna, sometimes together with their children, would visit their parents and other relatives in England. I had the privilege of meeting Peter's entire family: his parents, brothers and sister. Once after a conference in Great Britain organized by N. Young, Peter took me to Appleby to meet the family. Such a warm family; I very much enjoyed meeting them, as well as an English brunch with the family.

During our work on the second book the plan of the third book was born. The plan was to present matrix theory, geometrical, algebraical, analytical and topological aspects from one point of view, namely from the point of view of invariant subspaces. This book is "Invariant subspaces of matrices with applications". In linear algebra the set of tools is so large that often the invariant subspaces are lost. Our plans were to put invariant subspaces as the basis of the entire theory. I would like to note that all our books contain applications to system theory.

In the late '80s my former doctoral student, Israel Koltracht from the Weizmann Institute of Science, spent a successful two year post doctoral period in Calgary with Peter. This led to the cooperation of the three of us in a new area in numerical analysis for structured matrices and integral operators. We published a number of papers in this area, and again the expertize and experience of Peter was important and the work enjoyable.

In 1988 Peter, Rien Kaashoek and myself published a large paper. It is in fact a small monograph (more than one hundred pages), which contains the complete spectral theory of regular matrix polynomials and applications to band block Toeplitz matrices. In fact it contains also the most complete theory for inversion of Toeplitz band block matrices.

Peter Lancaster organized many conferences in Calgary and its environs. These conferences were always well prepared and interesting and were very important in view of the far distance Calgary is from the main mathematical centers. Peter understood this and made every effort to overcome the problem. I would like to mention especially the conference held in Calgary in 1988. This was a comparatively large conference with a very good representation from all over the world including a large group of Soviet and Eastern European participants. I am proud and grateful that this conference was dedicated to my sixtieth birthday. It was certainly a milestone in the development of operator theory and its applications.

Though now emeritus and retired from his academic duties, Peter still continues as usual his research, organization of conferences and other work that he likes. Our joint work continues even today and there are new plans and new projects.

Peter and the author of these lines
in a small hotel in the Rocky Mountains

Recently Peter went through a tragically difficult period; he lost Edna, his wife and best friend. He and the entire family took it very hard. Peter is very strong and on a recent visit to him I saw how he was struggling with those difficulties; there were the first signs that he is recovering and starting a new life.

Operator Theory:
Advances and Applications, Vol. 130, 29–41
© 2001 Birkhäuser Verlag Basel/Switzerland

The Joint Numerical Range of Bordered and Tridiagonal Matrices

Maria Adam and John Maroulas

Dedicated to Professor Peter Lancaster, with special appreciation.

Abstract. Let A_m $(m = 1, \ldots, k)$ be $n \times n$ matrices, the joint numerical range is defined by
$$\text{JNR}[A_1, \ldots, A_k] = \{ (x^* A_1 x, \ldots, x^* A_k x) : x \in \mathbf{C}^n , \|x\| = 1 \}.$$
In this paper, some geometric properties of JNR are presented, when the hermitian A_m are bordered or $(2\mu - 1)-$ diagonal matrices and the convexity of JNR is investigated.

1. Introduction

Let $\mathcal{M}_n$ be the algebra of $n \times n$ complex matrices. For the matrices $A_1, \ldots, A_k \in \mathcal{M}_n$, the *joint numerical range*, is the set

$$\text{JNR}[A_1, \ldots, A_k] = \{ (x^* A_1 x, \ldots, x^* A_k x) : x \in \mathbf{C}^n , \|x\| = 1 \}. \qquad (1.1)$$

It is also called *k-dimensional field of k matrices* [HJ, p.85] and it will be denoted by

$$\text{JNR}[A_m]_{m=1}^k .$$

Clearly, for $k = 1$ the joint numerical range is identified with the numerical range of the matrix A_1. Also, for any matrix $A \in \mathcal{M}_n$

$$\text{JNR} \left[\frac{A + A^*}{2} , \frac{A - A^*}{2i} \right] = \text{NR}[A].$$

The joint numerical range is not necessarily convex, but the convexity of joint numerical range is known for hermitian matrices when

$$n = k = 2 \quad \textbf{and} \quad n \geq 3, \, k \leq 3.$$

The convex hull of joint numerical range will be denoted by $\text{Co}\{\text{JNR}[A_m]_{m=1}^k\}$.

The size of $\text{JNR}[A_m]_{m=1}^k$ is measured by the smallest sphere centered at the origin that contains it. Clearly, the radius of the sphere, known as the *numerical radius*, is defined by:

$$r_k = sup\{ |z| : z \in \text{Co}\{\text{JNR}[A_m]_{m=1}^k\}\}.$$

Recently a connection of $\text{Co}\{\text{JNR}[A_m]_{m=1}^k\}$ with the numerical range of matrix polynomial

$$P(\lambda) = A_k \lambda^{k-1} + \ldots + A_2 \lambda + A_1$$

has been presented in [PT]. Precisely, it is noticed that

$$\begin{aligned} \text{NR}[P(\lambda)] &= \{\lambda \in \mathbf{C} : x^* P(\lambda) x = 0, \; x \in \mathbf{C}^n \setminus \{0\}\} \\ &= \{\lambda \in \mathbf{C} : c_k \lambda^{k-1} + \ldots + c_2 \lambda + c_1 = 0, \\ &\qquad (c_1, c_2, \ldots, c_k) \in \text{Co}\{\text{JNR}[A_m]_{m=1}^k\}\}. \end{aligned}$$

In this paper, we investigate geometric properties of joint numerical range of special matrices which occur in graph theory. In the next section the first statement is that the numerical range of $n \times n$ bordered matrix is an elliptic disk. Afterwards, this idea is undertaken for the joint numerical range of a family of hermitian bordered matrices to be hyperellipsoid with nonempty interior.

The last section is devoted to the same problem for 3×3 tridiagonal hermitian matrices and the according results are presented.

2. The NR of bordered matrices

Using the term *"bordered matrix"* we consider the matrices of the form

$$S = \begin{bmatrix} a_{11} & a_{12} & a_{13} & \cdots & a_{1n} \\ a_{21} & 0 & 0 & \cdots & 0 \\ a_{31} & 0 & & & \\ \vdots & \vdots & & \mathbf{O} & \\ a_{n1} & 0 & & & \end{bmatrix} \tag{2.1}$$

which correspond to "star" graphs [BR]. Clearly, for $n = 3$ interchanging first and second rows and columns, the resulting matrix PSP^T is the tridiagonal matrix,

$$\begin{bmatrix} 0 & a_{21} & 0 \\ a_{12} & a_{11} & a_{13} \\ 0 & a_{31} & 0 \end{bmatrix}$$

where P is permutation matrix. Generalizing the result in [C] on the numerical range of tridiagonal matrices with zero diagonals, we say:

Lemma 2.1. *The numerical range of matrix*

$$A = \begin{bmatrix} a & a_{12} & 0 \\ a_{21} & a_{22} & a_{23} \\ 0 & a_{32} & a \end{bmatrix} \tag{2.2}$$

is an elliptical disk, centered at the point $\frac{1}{2}(a + a_{22})$ of the complex plane.

Proof. It is enough to consider the matrix

$$A_0 = \begin{bmatrix} 0 & a_{12} & 0 \\ a_{21} & \hat{a}_{22} & a_{23} \\ 0 & a_{32} & 0 \end{bmatrix} \quad ; \quad \hat{a}_{22} = a_{22} - a$$

since $\text{NR}[A] = \text{NR}[A_0] + a$. For the matrix A_0, after algebraic manipulations, we verify that

$$\sigma(A_0) = \{\, 0, \; \frac{\hat{a}_{22} + \Delta^{1/2}}{2}, \; \frac{\hat{a}_{22} - \Delta^{1/2}}{2} \,\} \text{ with } \Delta = \hat{a}_{22}^2 + 4(a_{21}a_{12} + a_{23}a_{32}) \quad \text{and}$$

I. $q = tr(A_0^* A_0) - \sum_{j=1}^{3} |\lambda_j|^2 \geq (|a_{21}| - |a_{12}|)^2 + (|a_{23}| - |a_{32}|)^2 > 0$

II. $q\, tr A_0 + \sum_{j=1}^{3} |\lambda_j|^3 \lambda_j - tr(A_0^* A_0^2) = 0$

III. $(|\lambda_2| + |\lambda_3|)^2 - |\lambda_2 - \lambda_3|^2 \leq q$

Therefore, by [KRS], the $\text{NR}[A_0]$ is an ellipse with center $\frac{1}{2}\hat{a}_{22}$.   □

Following using Lemma 2.1 and a result in [KRS] we say:

Proposition 2.2. *The numerical range of S in (2.1) is an elliptic disk centered at the point* $\frac{a_{11}}{2}$ *with axes of length* $(\alpha \pm |\beta|)^{1/2}$, *where*

$$\alpha = \frac{|a_{11}|^2}{2} + \sum_{\substack{i,j=1 \\ i \neq j}}^{n} |a_{ij}|^2 \quad , \quad \beta = \frac{a_{11}^2}{2} + 2\sum_{j=2}^{n} a_{1j}a_{j1}. \tag{2.3}$$

Proof. Denoting by $\hat{\alpha} = [\, a_{21}\, a_{31} \,\ldots\, a_{n1}\,]^T$ and $Q = I - 2\dfrac{yy^*}{y^*y}$ the Householder matrix, for $y = \hat{\alpha} - k_1 e_1$, then

$$Q\hat{\alpha} = k_1 e_1 \quad ; \quad k_1 = \frac{\|\hat{\alpha}\|}{|a_{21}|} a_{21}$$

and

$$RSR^* = \begin{bmatrix} a_{11} & b_{12} & \cdots & b_{1n} \\ k_1 & 0 & \cdots & 0 \\ 0 & & & \\ \vdots & & \mathbf{O} & \\ 0 & & & \end{bmatrix}$$

where $R = diag(1, Q)$. If $R_1 = diag(1, 1, Q_1)$, with Q_1 the Householder matrix such that

$$Q_1 \begin{bmatrix} \bar{b}_{13} \\ \vdots \\ \bar{b}_{1n} \end{bmatrix} = k_2 e_1 \quad ; \quad k_2 = \frac{\|\,[\,b_{13}, \,\ldots, \,b_{1n}\,]\,\|}{|b_{13}|} \bar{b}_{13}$$

then

$$(R_1 R)S(R_1 R)^* = \hat{S} \oplus O_{n-3} \; ; \quad \hat{S} = \begin{bmatrix} a_{11} & b_{12} & \bar{k}_2 \\ k_1 & 0 & 0 \\ 0 & 0 & 0 \end{bmatrix}$$

Clearly $0 \in \text{NR}[\hat{S}]$, $\text{NR}[S] = \text{NR}[\hat{S}]$ and by Lemma 2.1, the $\text{NR}[\hat{S}]$ is an ellipse centered at the point $a_{11}/2$.

To calculate the axes of ellipse, by the elements of S in (2.1), consider its focii

$$\lambda_{1,2} = \frac{a_{11} \pm \sqrt{\Delta}}{2} \quad \text{with} \quad \Delta = a_{11}^2 + 4k_1 b_{12} \quad \text{and the quantity}$$

$$\delta = tr(\hat{S}^* \hat{S}) - \sum |\lambda_{\hat{S}}|^2 = |k_1|^2 + |k_2|^2 + |b_{12}|^2 + \frac{|a_{11}|^2}{2} - \frac{|\Delta|}{2}.$$

Clearly,

$$\delta = \frac{|a_{11}|^2}{2} + \sum_{j=2}^{n} |a_{j1}|^2 + \sum_{j=2}^{n} |b_{1j}|^2 - \frac{1}{2} \left| a_{11}^2 + 4[b_{12} \ldots b_{1n}] k_1 e_1 \right|$$

$$= \frac{|a_{11}|^2}{2} + \sum_{j=2}^{n} |a_{j1}|^2 + \| [a_{12} \ldots a_{1n}] Q^* \|^2 - \frac{1}{2} \left| a_{11}^2 + 4[a_{12} \ldots a_{1n}] Q^* Q \hat{\alpha} \right|$$

$$= \frac{|a_{11}|^2}{2} + \sum_{j=2}^{n} \left(|a_{j1}|^2 + |a_{1j}|^2 \right) - \frac{1}{2} \left| a_{11}^2 + 4[a_{12} \ldots a_{1n}] \hat{\alpha} \right|$$

$$= \alpha - |\beta|$$

and $|\lambda_2 - \lambda_1|^2 = |\Delta| = 2|\beta|$. Therefore, the axes of the ellipse are

$$\delta^{1/2} \quad \text{and} \quad \left(\delta^2 + |\lambda_2 - \lambda_1|^2 \right)^{1/2} = (\alpha + |\beta|)^{1/2}. \qquad \square$$

Since for any $z \in \text{NR}[S]$, $\left| z - \frac{a_{11}}{2} \right| \leq \frac{(\alpha + |\beta|)^{1/2}}{2}$, we obtain:

Corollary 2.3. *The numerical radius of the bordered matrix S in (2.1) is equal to*

$$\frac{|a_{11}| + (\alpha + |\beta|)^{1/2}}{2}.$$

Evidently

$$\hat{S} = \begin{bmatrix} & & & a_{n1} & & \\ & O & & \vdots & & O \\ & & & a_{n-s+1,1} & & \\ a_{1n} & \cdots & a_{1,n-s+1} & a_{11} & a_{12} & \cdots & a_{1,n-s} \\ & & & a_{21} & & \\ & O & & \vdots & & O \\ & & & a_{n-s,1} & & \end{bmatrix}$$

is similar to S by a permutation matrix and is the sub-direct sum of two bordered matrices S_1, S_2. The ellipses $E_i = \text{NR}[S_i]$ have the point $\dfrac{a_{11}}{2}$ as their common center and their axes are equal to $(\alpha_i \pm |\beta_i|)^{1/2}$, where

$$\alpha_1 = \frac{|a_{11}|^2}{2} + \sum_{\substack{i,j=1 \\ i \neq j}}^{n-s} |a_{ij}|^2 \quad , \quad \beta_1 = \frac{a_{11}^2}{2} + 2\sum_{j=2}^{n-s} a_{1j}a_{j1}$$

$$\alpha_2 = \frac{|a_{11}|^2}{2} + \alpha - \alpha_1 \qquad , \qquad \beta_2 = \frac{a_{11}^2}{2} + \beta - \beta_1.$$

Since

$$\hat{S} = \begin{bmatrix} & \vdots & O \\ \cdots & a_{11}/2 & \\ O & & O \end{bmatrix} + \begin{bmatrix} O & & O \\ & a_{11}/2 & \cdots \\ O & & \vdots \end{bmatrix} = \hat{S}_1 + \hat{S}_2$$

we remark that:

$$\text{Co}\{\text{NR}[S_1] \cup \text{NR}[S_2]\} \subseteq \text{NR}[S] \subseteq \text{NR}[\hat{S}_1] + \text{NR}[\hat{S}_2].$$

3. The JNR of bordered matrices

Given the $n \times n$ hermitian matrices $A_1, A_2, \ldots, A_k$, we consider the matrix

$$M = h_1 A_1 + h_2 A_2 + \ldots + h_k A_k \tag{3.1}$$

for any unit vector $\mathbf{h} = (h_1, h_2, \ldots, h_k) \in \mathbf{R}^k$.

The approximation of $\text{JNR}[A_m]_{m=1}^k$ by convex polyhedra is owed by the next lemma.

Lemma 3.1. *The support plane*

$$\varepsilon : h_1 x_1 + h_2 x_2 + \ldots + h_k x_k = \lambda_{max}(M)$$

of the surface $\partial\, \text{Co}\{JNR[A_m]_{m=1}^k\}$ *is tangent at the point*

$$\rho_0 = (w_0^* A_1 w_0, \ w_0^* A_2 w_0, \ldots, \ w_0^* A_k w_0)$$

where w_0 *is the corresponding unit eigenvector of* $\lambda_{max}(M)$.

This statement is a generalization of the cases $k = 2$ or $k = 3$ in [AT] and it is confirmed in a similar way. If the $\text{JNR}[A_m]_{m=1}^k$ is not convex then there exists a support plane ε having more than one common points with $\text{Co}\{JNR[A_m]_{m=1}^k\}$ and evidently all curves of $\partial\, \text{Co}\{JNR[A_m]_{m=1}^k\}$ are not differentiable at any point.

The investigation of convexity of JNR has been treated by Li-Poon and there the problem to characterize and determine the maximal linearly independent convex family, for the JNR of hermitian matrices, has been put forward. It has been proved in [LP, Th.2.3] that, for a linearly independent k-tuple of $n \times n$ hermitian matrices A_m, if the $\text{JNR}[A_m]_{m=1}^k$ is convex then

$$dim(span\{I, A_1, \ldots, A_k\}) \leq 2n - 1.$$

Following we consider a family of *hermitian bordered* matrices

$$
S_m = \begin{bmatrix} a_{m1} & a_{m2} & \cdots & a_{mn} \\ \bar{a}_{m2} & 0 & \cdots & 0 \\ \vdots & \vdots & & \mathbf{O} \\ \bar{a}_{mn} & 0 & & \end{bmatrix} \quad ; \quad m = 1, \ldots, k \tag{3.2}
$$

for $n \geq 3$ and $3 \leq k \leq 2n-1$. The restriction of k comes from the observations that $S_m \in W = span\{E_{11}, E_{1j} + E_{j1}, i(E_{1j} - E_{j1}) : 2 \leq j \leq n\}$, where $\{E_{ij}\}$ is the standard basis of $\mathcal{M}_n$ and [LP]

$$
JNR[S_m]_{m=1}^k \text{ is convex} \Leftrightarrow
$$

$$
\Leftrightarrow JNR[E_{11}, E_{1j} + E_{j1}, i(E_{1j} - E_{j1}); 2 \leq j \leq n] \text{ is convex}.
$$

Proposition 3.2. *Let $\{S_m\}_{m=1}^k$ be a linearly independent family of matrices in (3.2). The boundary of $Co\{JNR[S_m]_{m=1}^k\}$ is an hyperellipsoid in $\mathbf{R}^k$ with center $\frac{1}{2}(a_{11}, a_{21}, \ldots, a_{k1})$.*

Proof. For any unit vector $(h_1, \ldots, h_k) \in \mathbf{R}^k$ we consider as in (3.1), the hermitian matrix

$$
M = h_1 S_1 + h_2 S_2 + \ldots + h_k S_k.
$$

Since

$$
|\lambda I - M| = \lambda^n - \lambda^{n-1} \sum_{m=1}^k h_m a_{m1} - \lambda^{n-2} \sum_{j=2}^n \left| \sum_{m=1}^k h_m a_{mj} \right|^2
$$

we have

$$
\lambda_{max}(M) = \lambda_0 = \frac{1}{2} \left(\sum_{m=1}^k h_m a_{m1} + \Delta^{1/2} \right)
$$

where $\Delta = \left(\sum_{m=1}^k h_m a_{m1} \right)^2 + 4 \sum_{j=2}^n \left| \sum_{m=1}^k h_m a_{mj} \right|^2 > 0$. The corresponding unit eigenvector of M is

$$
w_0 = \lambda_0^{-1/2} \Delta^{-1/4} \left[\lambda_0 \quad \sum_{m=1}^k h_m \bar{a}_{m2} \quad \cdots \quad \sum_{m=1}^k h_m \bar{a}_{mn} \right]^T. \tag{3.3}
$$

Hence, by the lemma 3.1, the point $(w_0^* S_1 w_0, \ldots, w_0^* S_k w_0)$ belongs to

$$
\partial Co\{JNR[S_m]_{m=1}^k\}
$$

and the r-coordinate of this point is equal to

$$
w_0^* S_r w_0 = \frac{a_{r1}}{2} + \frac{1}{\sqrt{\Delta}} \left[\frac{a_{r1}}{2} \sum_{m=1}^k h_m a_{m1} + \right. \tag{3.4}
$$

$$
\left. + \sum_{j=2}^n \left[(\sum_{m=1}^k h_m a_{mj}) \bar{a}_{rj} + (\sum_{m=1}^k h_m \bar{a}_{mj}) a_{rj} \right] \right]. \tag{3.5}
$$

Setting in (3.5)

$$\pi_r = 2 \left[\frac{a_{r1}^2}{4} + \sum_{j=2}^{n} |a_{rj}|^2 \right] \Delta^{-1/2}$$

$$\sigma_{r,l} = \sigma_{l,r} = 2 \left[Re(\sum_{j=2}^{n} a_{rj}\bar{a}_{lj}) + \frac{a_{r1}a_{l1}}{4} \right] \Delta^{-1/2}$$

$$
\begin{aligned}
h_1 &= cos\theta_1 sin\theta_2 \dots sin\theta_{k-1} \\
h_2 &= sin\theta_1 sin\theta_2 \dots sin\theta_{k-1} \\
h_3 &= cos\theta_2 sin\theta_3 \dots sin\theta_{k-1} \quad ; \quad 0 \le \theta_i \le 2\pi \ , \ i = 1,2,\dots, k-1 \\
h_4 &= cos\theta_3 sin\theta_4 \dots sin\theta_{k-1} \\
&\vdots \\
h_k &= cos\theta_{k-1},
\end{aligned}
$$

by the k equations

$$
\begin{aligned}
h_1\pi_1 + h_2\sigma_{12} + \dots + h_k\sigma_{1k} &= w_0^* S_1 w_0 - \frac{a_{11}}{2} = x_1 \\
h_1\sigma_{21} + h_2\pi_2 + \dots + h_k\sigma_{2k} &= w_0^* S_2 w_0 - \frac{a_{21}}{2} = x_2 \\
&\vdots \qquad\qquad\qquad \vdots \\
h_1\sigma_{k1} + h_2\sigma_{k2} + \dots + h_k\pi_k &= w_0^* S_k w_0 - \frac{a_{k1}}{2} = x_k,
\end{aligned}
\qquad (3.6)
$$

due to $\|\mathbf{h}\| = 1$, the parameters $\theta_1,\dots,\theta_{k-1}$ are eliminated and we obtain the quadratic form

$$g_{11}x_1^2 + g_{22}x_2^2 + \dots + g_{kk}x_k^2 + 2\sum_{i=1}^{k-1}\sum_{j=1}^{k-i} g_{i(i+j)}x_i x_{i+j} = g_0^2. \qquad (3.7)$$

In (3.7) $\quad g_{mm} = v_m \circ v_m$, $\quad g_{i(i+j)} = v_i \circ v_{i+j}$; $\quad v_m = \begin{bmatrix} \tau_{m1} & \tau_{m2} & \dots & \tau_{mk} \end{bmatrix}$ where τ_{ms} is the cofactor of ms-element of matrix

$$G = \begin{bmatrix} \pi_1 & \sigma_{12} & \dots & \sigma_{1k} \\ \sigma_{21} & \pi_2 & \dots & \sigma_{2k} \\ \vdots & & \ddots & \vdots \\ \sigma_{k1} & \sigma_{k2} & \dots & \pi_k \end{bmatrix}$$

and $g_0 = det\, G$. The matrix

$$Q = \begin{bmatrix} g_{11} & g_{12} & \dots & g_{1k} \\ g_{21} & g_{22} & \dots & g_{2k} \\ \vdots & & \ddots & \vdots \\ g_{k1} & g_{k2} & \dots & g_{kk} \end{bmatrix}$$

of quadratic form (3.7) is a real nonsingular Gram matrix. Since for the system equations in (9) $rank G = k$, we have $rank(adj G) = rank[v_1 \ldots v_k] = k$ and hence $Q > 0$. Therefore the form in (3.7) is an hyperellipsoid in $\mathbf{R}^k$ centered at the point $\frac{1}{2}(a_{11}, a_{21}, \ldots, a_{k1})$. $\qquad\square$

Furthermore, by Proposition 3.2, since every point on the boundary of

$$Co\{JNR[S_m]_{m=1}^k\}$$

is an extreme point, it is worth noting:

Corollary 3.3. *The boundary points of $Co\{JNR[S_m]_{m=1}^k\}$ are identified with the shell (outer boundary) of $JNR[S_m]_{m=1}^k$.*

Note that, if $S_1, \ldots, S_k$ are not linearly independent matrices and let $S_k = \sum_{i=1}^{k-1} l_i S_i$, then after some algebraic manipulation we have that the last column and row of G is equal to zero and consequently the quadratic form (3.7) degenerates to zero.

Proposition 3.4. *If the origin is not an outer boundary point of $JNR[S_m]_{m=1}^k$, then the interior of $JNR[S_m]_{m=1}^k$ is nonempty.*

Proof. By the bordered matrices $S_1, S_2, \ldots, S_k$ we consider the matrix polynomial

$$P(\lambda) = S_k \lambda^{k-1} + \ldots + S_2 \lambda + S_1.$$

Since any vector $x \in span\{e_2, \ldots, e_n\}$ is common isotropic of S_m, i.e. $x^* S_m x = 0$, then $NR[P(\lambda)] = \mathbf{C}$ and especially $(0, \ldots, 0)$ is inside $JNR[S_m]_{m=1}^k$. Denoting by

$$\mathcal{L}(t) =$$

$$\{(c_1, c_2, \ldots, c_k) : c_k t^{k-1} + \ldots + c_2 t + c_1 = 0, \ (c_1, \ldots, c_k) \in JNR[S_m]_{m=1}^k\}$$

for any $t \in NR[P(\lambda)]$, $\mathcal{L}(t)$ is a connected set in [PT] and $(0, \ldots, 0) \in \mathcal{L}(t)$. Thus, for an outer boundary point $w = (w_1, \ldots, w_k)$ of $JNR[S_m]_{m=1}^k$ we define the polynomial

$$q_w(\lambda) = w_k \lambda^{k-1} + \ldots + w_2 \lambda + w_1$$

which is annihilating at t_0 and due to connectivity of $\mathcal{L}(t_0)$ there exists a curve $\mathcal{K}$ joining the origin and w. All points of the curve $\mathcal{K}$ belong also to $JNR[S_m]_{m=1}^k$, since by the continuous function

$$\Phi : \{x \in \mathbf{C}^n : \|x\| = 1\} \longrightarrow JNR[S_m]_{m=1}^k$$

defined by

$$\Phi(x) = (x^* S_1 x, \ldots, x^* S_k x)$$

we have

$$\Phi[\{x \in \mathbf{C}^n : \|x\| = 1, \ x^* P(t_0) x = 0\}] = \mathcal{L}(t_0) \qquad\square$$

If we don't use the hypothesis for the origin, then the $JNR[S_m]_{m=1}^k$ might be a shell without interior, as we see in the next example.

Example 3.5. Let the bordered matrices

$$S_1 = \begin{bmatrix} 1 & 0 \\ 0 & 0 \end{bmatrix}, \quad S_2 = \begin{bmatrix} 0 & 1 \\ 1 & 0 \end{bmatrix}, \quad S_3 = \begin{bmatrix} 0 & i \\ -i & 0 \end{bmatrix}$$

and $x = [\, t, \; e^{i\theta}\sqrt{1-t^2}\,]^\top, \; t \in [0,1]$. Then

$$
\begin{aligned}
\mathrm{JNR}[S_1, S_2, S_3] &= \{\, (t^2, \; 2t\sqrt{1-t^2}\cos\theta, \; -2t\sqrt{1-t^2}\sin\theta) \; : \\
&\qquad 0 \le t \le 1, \; 0 \le \theta \le 2\pi \,\} \\
&= \{\, (x, y, z) \; : \; 4(x-1/2)^2 + y^2 + z^2 = 1 \,\}.
\end{aligned}
$$

Clearly, $(0,0,0) \in \partial\,\mathrm{JNR}[S_1, S_2, S_3]$ and the $\mathrm{JNR}[S_1, S_2, S_3]$ is the surface of ellipsoid.

Following the Propositions 3.2 and 3.4 we conclude that:

Proposition 3.6. *If $S_1, \ldots, S_k$ be an k-tuple of linearly independent hermitian bordered matrices with $a_{m1} = 0$ $(m = 1, \ldots, k)$, then the $JNR[S_m]_{m=1}^k$ is a convex set.*

Proof. Clearly, $S_m \in V = span\{E_{1j} + E_{j1}, i(E_{1j} - E_{j1}) \; : \; 2 \le j \le n\}$ and $dim(V \cup \{I\}) = 2n-1$. It is enough to see that the points $t\,(w_0^* S_1 w_0, \ldots, w_0^* S_k w_0)$, $t \in [0,1]$ belong to $\mathrm{JNR}[S_m]_{m=1}^k$. Setting $\hat{S}_m = tS_m$ and $\hat{M} = h_1\hat{S}_1 + \ldots + h_k\hat{S}_k$, we have

$$\lambda_{max}(\hat{M}) = t\,\lambda_{max}(M) = t\lambda_0, \qquad \hat{\Delta} = t^2\Delta$$

and w_0 in (3.3) is the corresponding unit eigenvector of $\lambda_{max}(\hat{M})$. Hence $(w_0^* t S_1 w_0, \ldots, w_0^* t S_k w_0)$ is interior point of $\mathrm{JNR}[S_m]_{m=1}^k$. $\qquad\square$

Finally, since $dim(V \cup \{I\}) = 2n - 1$, we confirm that the matrices $\{S_m\}$ in (3.2) with $a_{m1} = 0$ form a maximal linearly independent family with maximum number of elements, such that the $\mathrm{JNR}[S_m]_{m=1}^k$ is convex.

4. The NR of $(2\mu - 1)$-diagonal matrices

Compared with serial systems, parallel systems permit more freedom of expression in problem analysis and programming. A foundation in the skills of thinking in parallel is basic to the understanding of such systems. Computations within each level are performed in parallel and this simple idea is the basis of recursion, where the total computation is repeatedly divided into separate computations of equal complexity that can be executed in parallel. The natural means of carrying out these operations is to use a μ-tree inter connection of processors. These trees correspond to the $(2\mu - 1)$-diagonal matrix $T \in \mathcal{M}_n$, [M]:

$$
T = \begin{bmatrix}
a & a_{12} & \cdots & a_{1,\mu} & 0 & & \\
a_{21} & a_{22} & \cdots & a_{2,\mu} & a_{2,\mu+1} & & \mathbf{O} \\
\vdots & \vdots & & & & \ddots & \\
a_{\mu 1} & a_{\mu 2} & \cdots & a_{\mu\mu} & & & a_{n-\mu+1,n} \\
0 & a_{\mu+1,2} & & & & & \vdots \\
& \ddots & & & a_{n-1,n-1} & & a_{n-1,n} \\
\mathbf{O} & & a_{n,n-\mu+1} & \cdots & a_{n,n-1} & & a
\end{bmatrix}
$$

Clearly, for any unit vector $x = x_1 e_1 + (x_2 e_2 + \ldots + x_{n-1} e_{n-1}) + x_n e_n = x_1 e_1 + \xi + x_n e_n = x_1 e_1 + \|\xi\| \zeta + x_n e_n = [e_1 \ \ \zeta \ \ e_n][x_1 \ \ \|\xi\| \ \ x_n]^T$, where $\zeta = \dfrac{\xi}{\|\xi\|}$, we obtain

$$
x^* T x = \delta^* \begin{bmatrix}
e_1^* T e_1 & e_1^* T \zeta & e_1^* T e_n \\
\zeta^* T e_1 & \zeta^* T \zeta & \zeta^* T e_n \\
e_n^* T e_1 & e_n^* T \zeta & e_n^* T e_n
\end{bmatrix} \delta \ ; \ \delta = [x_1 \ \ \|\xi\| \ \ x_n]^T , \ \|\delta\| = 1 \quad (4.1)
$$

and consequently

$$
\mathrm{NR}[T] = \bigcup_{e_1, \zeta, e_n} \mathrm{NR}\left(\begin{bmatrix}
e_1^* T e_1 & e_1^* T \zeta & e_1^* T e_n \\
\zeta^* T e_1 & \zeta^* T \zeta & \zeta^* T e_n \\
e_n^* T e_1 & e_n^* T \zeta & e_n^* T e_n
\end{bmatrix} \right).
$$

In (4.1), note that,

$$
e_1^* T e_1 = e_n^* T e_n = a \ , \ \ e_1^* T e_n = e_n^* T e_1 = 0
$$

i.e. the matrix is tridiagonal with diagonal entries a, $\zeta^* T \zeta$, a. Hence, by Lemma 2.1 we have:

Corollary 4.1. *The numerical range of $(2\mu - 1)$-diagonal matrices is union of ellipses, which arise from 3×3 compression matrix in (4.1).*

The above tridiagonal matrix is more appropriate than the 2×2 matrix for the approximation of $\mathrm{NR}[T]$ and this is illustrated in the following example.

Example 4.2. Let the matrix

$$
T = \begin{bmatrix}
3 & 1 & 2 & 0 & 0 \\
2 & 0 & 3 & 5 & 0 \\
1 & 1 & 1 & 0 & 0 \\
0 & 0 & 1 & -3 & 2 \\
0 & 0 & -3 & 3 & 3
\end{bmatrix}
$$

In the next figure, $\mathrm{NR}[T]$ on the left is approximated by 6 ellipses (NR of 3×3 tridiagonal matrices) and on the right the number of ellipses is double (NR of 2×2 matrices). The MATLAB procedure is presented before the references.

Moreover, due to the transformation of a 3×3 bordered matrix to a tridiagonal matrix and by Propositions 3.2 and 3.6, we have:

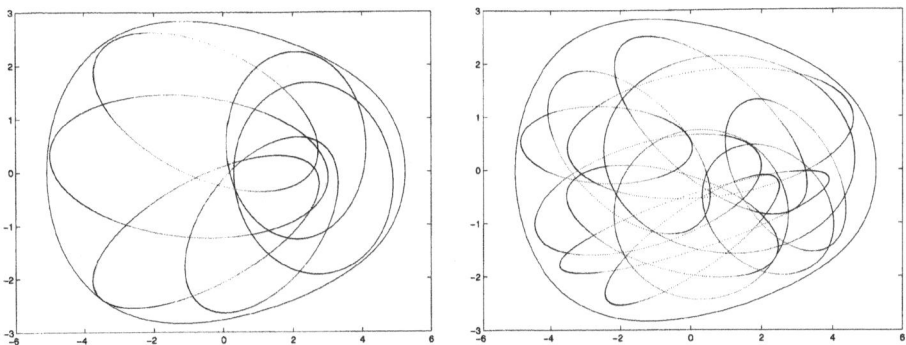

FIGURE 1. *Approximation of* NR[T].

Proposition 4.3. *For* $3 \leq k \leq 5$, *the* $\partial\, JNR[T_m]_{m=1}^k$ *of the linearly independent hermitian and tridiagonal matrices*

$$
T_m = \begin{bmatrix} f_m & a_m & 0 \\ \bar{a}_m & b_m & c_m \\ 0 & \bar{c}_m & f_m \end{bmatrix}, \tag{4.2}
$$

is an hyperellipsoid in $\mathbf{R}^5$. *If* $f_m = b_m = 0$, *then the* $JNR[T_m]_{m=1}^k$ *is convex set.*

The presentation of JNR as a union of joint numerical ranges of matrices of dimension less than n is a convenient way to approach the problem, as noted for the numerical range of matrices in [MA]. In fact:

Lemma 4.4. *For the matrices* $A_1, A_2, \ldots, A_k \in \mathcal{M}_n$

$$
JNR[A_m]_{m=1}^k = \bigcup_{\substack{\xi_1,\ldots,\xi_\tau \\ \tau < n}} JNR\left(\begin{bmatrix} \xi_1^* A_m \xi_1 & \cdots & \xi_1^* A_m \xi_\tau \\ \vdots & & \vdots \\ \xi_\tau^* A_m \xi_1 & \cdots & \xi_\tau^* A_m \xi_\tau \end{bmatrix} \right)_{m=1}^k \tag{4.3}
$$

where $\xi_1, \ldots, \xi_\tau$ *run over all sets by* τ *orthonormal vectors of* $\mathbf{C}^n$.

Proof. For any unit vector x, the vector $y = P^* x$, where $P = [\xi_1\ \xi_2\ \cdots\ \xi_\tau]$, $P^* P = I_\tau$, is also unit and

$$
x^* A_m x = y^* \begin{bmatrix} \xi_1^* A_m \xi_1 & \cdots & \xi_1^* A_m \xi_\tau \\ \vdots & & \vdots \\ \xi_\tau^* A_m \xi_1 & \cdots & \xi_\tau^* A_m \xi_\tau \end{bmatrix} y.
$$

Therefore, the (4.3) is obvious.   □

One easily deduces the following corollary by the Lemma 4.4 and Proposition 4.3:

Corollary 4.5. *For a k-tuple linearly independent $n \times n$ hermitian $(2\mu - 1)$-diagonal matrices the $JNR[T_m]_{m=1}^k$ is union of hyperellipsoids in $\mathbf{R}^5$, where $3 \le k \le n^2 - (n - \mu)(n - \mu + 1)$.*

If the matrices T_m in (4.2) are real and not symmetric, denoting by

$$Re-\text{JNR}[T_m]_{m=1}^k$$

the joint numerical range of the family $T_1, \ldots, T_k$, for real unit vectors $x \in \mathbf{R}^3$, then [HJ, p.85]

$$Re-\text{JNR}[T_m]_{m=1}^k = Re-\text{JNR}\left[\frac{T_m + T_m^T}{2}\right]_{m=1}^k$$

Therefore, by Proposition 4.3 we have:

Corollary 4.6. *The $Re-JNR[T_m]_{m=1}^k$ for 3×3 linearly independent tridiagonal matrices $T_m = [t_{ij}]$ with $t_{11} = t_{33}$, is an hyperellipsoid in $\mathbf{R}^5$.*

MATLAB procedure:

Step 1. Introduce the $n \times n$ matrix T.

Step 2. Introduce the arbitrary linearly independent vectors x_1, $x_2 \in \mathbf{C}^n$.

Step 3. Orthonormalize this set to ξ_1, ξ_2.

Step 4. Calculate the matrix $A = [\xi_i^* T \xi_j]$, $i, j = 1, 2$.

Step 5. Illustrate the numerical range of matrix A.

Step 6. Repeat this procedure for some other set of vectors.

Acknowledgements. The authors wish to thank Prof. Chi-Kwong Li for his remarks. Part of the paper has been presented in the 5th Workshop on "Numerical Ranges and Radii" June 2000 – Nafplio, Greece. Research supported by ICCS under grant "Archimedes".

References

[AT]   Y.H.Au-Yeung and N.K.Tsing, An extension of the Hausdorff-Toeplitz theorem on the numerical range, *Proc. Amer. Math. Soc.* **89** (1983), 215–218.

[BR]   R.A. Brualdi and H.J. Ryser, *Combinatorial Matrix Theory*, Cambridge University Press 1991.

[C]   M.T.Chien, On the Numerical Range of Tridiagonal Operators, *Linear Algebra and Applics.* **246** (1996), 203–214.

[HJ]   R.Horn and C.R.Johnson, *Topics in matrix analysis*, Cambridge University Press 1991.

[KRS] D.S.Keeler, L.Rodman and I.M.Spitkovsky, The numerical range of 3×3 matrices, *Linear Algebra and Applics.* **252** (1997), 115–139.

[LP]   C-K.Li and Y-T.Poon, Convexity of the Joint Numerical Range, *SIAM J. on Matrix Analysis and Applics.* **21** (2000), 668–678.

[M] J.J. Modi, *Parallel Algorithms and Matrix Computation*, Clarendon Press, Oxford 1990.

[MA] J.Maroulas and M.Adam, Compressions and Dilations of Numerical Ranges, *SIAM J. on Matrix Analysis and Applics.* **21** (1999), 230–244.

[PT] P.Psarrakos and M.Tsatsomeros, On the relation between the numerical range and the joint numerical range of matrix polynomial, *Electronic Linear Algebra* **6** (2000), 20–30.

Department of Mathematics
National Technical University
Zografou Campus Athens (15780)

1991 Mathematics Subject Classification. Primary 15A60; Secondary 47A12

Received July 27, 2000

Operator Theory:
Advances and Applications, Vol. 130, 43–54

Iterative Computation of Higher Derivatives of Repeated Eigenvalues and the Corresponding Eigenvectors

Alan L. Andrew and Roger C.E. Tan

To Peter Lancaster on his 70th birthday

Abstract. This paper is concerned with iterative methods for computing partial derivatives of eigenvalues and eigenvectors of matrix-valued functions of several real variables. First, an analysis is given of a previously announced method which computes mixed partial derivatives of simple eigenvalues and the corresponding eigenvectors and also second order mixed partial derivatives of repeated eigenvalues. Next a new method for computing third order partial derivatives of repeated eigenvalues and second order derivatives of the corresponding eigenvectors is presented and its key properties are established. Efficiency and numerical stability are considered as well as theoretical convergence.

1. Introduction

We consider here the eigenvalue problem

$$A(\mathbf{t})\mathbf{x}_i(\mathbf{t}) = \lambda_i(\mathbf{t})\mathbf{x}_i(\mathbf{t}), \quad i = 1, \ldots, n, \tag{1.1}$$

where the (real or complex) $n \times n$ matrix $A(\mathbf{t})$ (and consequently the eigenvalues, $\lambda_i(\mathbf{t})$, and eigenvectors, $\mathbf{x}_i(\mathbf{t})$), are functions of a vector parameter, $\mathbf{t} = (t_1, \ldots, t_\rho) \in \mathbb{R}^\rho$. The variation of $\lambda_i(\mathbf{t})$ and $\mathbf{x}_i(\mathbf{t})$ as $\mathbf{t}$ is varied continuously has received much attention from both mathematicians [8, 16, 19, 22] and engineers [1, 9, 12, 17, 23, 24] who, motivated by applications to the optimal design of structures[14] and to the solution of inverse eigenvalue problems (including model updating [18]), have suggested many numerical algorithms for computing the values of partial derivatives of both $\lambda_i(\mathbf{t})$ and $\mathbf{x}_i(\mathbf{t})$ with respect to the components of $\mathbf{t}$. Such numerical algorithms and their properties are the subject of this paper. Our main concern is problems in which eigenvalues are repeated or very tightly clustered. Such examples occur as a structure approaches an optimum [14] or as a result of some structural symmetries, and it is known [5, 7] that methods designed for repeated eigenvalues are more stable for such clustered eigenvalues than methods designed for well separated eigenvalues.

Most work on the numerical computation of derivatives of eigenvalues and eigenvectors has considered only first derivatives, but recently several authors have considered second derivatives of eigenvectors corresponding to *simple* eigenvalues, and both modal expansion methods [9] and direct methods [3, 12] have been proposed. However, the only paper of which we are aware that gives a satisfactory method of computing second derivatives of eigenvectors corresponding to *repeated* eigenvalues is [5], which considers the Hermitian case for $\rho = 1$. That paper examines a rather general class of methods, all of which include as one step the computation of a particular solution of a system of equations having infinitely many solutions. However, in [5] methods for obtaining this particular solution are discussed only briefly, and iterative methods are not considered at all.

For small matrices, direct methods are generally to be preferred, but finite element analysis often produces large sparse matrices for which iterative methods are well suited when, as is usually the case, derivatives of only a few of the eigenvalues and the corresponding eigenvectors are required. This has motivated several papers devoted to iterative computation of first derivatives of simple eigenvalues and the corresponding eigenvectors ([1, 20, 23] and the references there). First derivatives of repeated eigenvalues and the corresponding eigenvectors are considered in [7] which also briefly reviews the related literature.

Sometimes both first and higher derivatives are used to obtain a first approximation to eigenvalues and eigenvectors of $A(\mathbf{t} + \delta\mathbf{t})$, for some known $\delta\mathbf{t}$, when the eigenvalues and eigenvectors of $A(\mathbf{t})$ are known [9]. Usually quite crude approximations of the derivatives are adequate for this purpose. Iterative methods can produce these estimates efficiently with only a few iterations.

When an eigenvalue is multiple for all $\mathbf{t}$ throughout some neighbourhood, the derivatives of the corresponding eigenvectors are not uniquely defined in that neighbourhood. We are concerned instead with the case, common in optimization problems [14], in which an eigenvalue is multiple at an isolated point $\mathbf{t} = \mathbf{t}_0$, but, for all $\mathbf{t} \neq \mathbf{t}_0$ in some neighbourhood of $\mathbf{t}_0$, all eigenvalues are simple. In this case, continuity requires a unique choice of basis for the eigenspace corresponding to the multiple eigenvalue. For all i, following [7], we define $\mathbf{x}_i$ to be the continuous eigenvector corresponding to λ_i, with the normalizing condition

$$\mathbf{x}_i^*(\mathbf{t})\mathbf{x}_i(\mathbf{t}) = 1, \tag{1.2}$$

which is usually imposed in the engineering literature. As in [3, 5], our results are readily modified to deal with alternative normalizing conditions. We assume that $\lambda_1(\mathbf{t}_0) = \cdots = \lambda_r(\mathbf{t}_0)$ is an eigenvalue of multiplicity r and that the $\mathbf{x}_i(\mathbf{t}_0)$ are linearly independent, so that $A(\mathbf{t}_0)$ must be nondefective.

The Hermitian case, which is very common in engineering applications, is probably the most important application of our results, although derivatives of repeated eigenvalues and the corresponding eigenvectors of nondefective non-Hermitian matrices have been the subject of several recent papers in the engineering literature [24]. Many (but not all) engineering applications involving non-Hermitian matrices arise from the linearization of some problem in which the eigenvalue parameter appears quadratically. When computing derivatives of eigenvalues and eigenvectors of such problems, it is generally best to work directly with the original quadratic formulation. A method of computing second derivatives of *simple* eigenvalues and the corresponding eigenvectors of problems with rather general nonlinear dependence on the eigenparameter is described in [3]. Indeed, many results relevant to the present work have been generalized to problems with nonlinear dependence on the eigenparameter by P. LANCASTER and his co-workers. In addition to [3], the list includes work on simultaneous computation of simple eigenvalues and the corresponding eigenvectors and their first derivatives [4], computation of repeated eigenvalues of matrices dependent on a parameter [11], and sufficient conditions for analyticity of multiple eigenvalues and the corresponding eigenvectors [15].

We assume throughout that all required partial derivatives exist. For some commonly satisfied sufficient conditions for this see [3, 5, 8, 13, 15, 16, 19]. Like [7], Section 3 involves no mixed partial derivatives and is effectively dealing with functions of a single real variable. Then if, throughout some neighbourhood of $\mathbf{t}_0$, $A(\mathbf{t})$ is Hermitian and is an analytic function of t_j when all other components of $\mathbf{t}$ are held constant, it follows from a result of RELLICH [8] that its eigenvalues will all be analytic functions of t_j and its eigenvectors may be chosen so that they are also analytic functions of t_j.

Like the algorithms of [5], the algorithms studied here can be extended to problems with several distinct repeated eigenvalues but, since such multiple sets of repeated eigenvalues are relatively rare in applications, we do not treat them here. We assume that, for all $i > r$, $\lambda_i(\mathbf{t}_0)$ is a simple eigenvalue. Eigenvalues are not labelled in order of magnitude: we do *not* assume the repeated eigenvalue to be a dominant eigenvalue. However, in applications the repeated eigenvalues are most likely to be near the boundary of the spectrum and our methods are best suited to this case. Our algorithms allow for the computation of the derivatives of an arbitrary number, $s \geq r$, of eigenvalues and the corresponding eigenvectors, with the repeated and simple eigenvalues being treated simultaneously.

Notation. Throughout this paper, as in [6, 7], the subscript ",j" denotes the partial derivative with respect to the jth argument, t_j, the subscript ",jl" denotes the second order partial derivative with respect to t_j and t_l, and so on. Henceforth, all quantities are assumed to be evaluated at $\mathbf{t}_0$ unless specifically stated otherwise, and the argument $\mathbf{t}_0$ is omitted to reduce clutter. Matrices are denoted by capital letters, their elements by (subscripted) lower case letters and their columns by (subscripted) boldface lower case letters. For any letter w, unless explicitly stated otherwise, w_{ip} is the ipth element, and $\mathbf{w}_i$ the ith column, of a matrix W. In our algorithms, X is the $n \times s$ matrix whose ith column is $\mathbf{x}_i$ ($i = 1, \ldots, s$), $\Lambda = \text{diag}(\lambda_1, \ldots, \lambda_s)$ and $\hat{X}$ is any $n \times s$ matrix whose ith column is $\mathbf{x}_i$ if $i > r$, and whose first r columns span the same space as $\mathbf{x}_1, \ldots, \mathbf{x}_r$.

In Section 2 we analyse the properties of a recently proposed iterative method [6] which computes the mixed partial derivatives, $\lambda_{i,jl}$, $\lambda_{i,jlp}$ and $\mathbf{x}_{i,jl}$, of simple eigenvalues and the corresponding eigenvectors. We show that it also gives the mixed second order partial derivatives, $\lambda_{i,jl}$, of repeated eigenvalues (but not the corresponding eigenvectors). In Section 3, as promised in [7], we present and analyse a new method for computing third order derivatives, $\lambda_{i,jjj}$, of repeated eigenvalues and second order derivatives, $\mathbf{x}_{i,jj}$, of the corresponding eigenvectors. That method uses the results of Section 2, but it does not allow the computation of mixed partial derivatives. We also require the condition

$$\lambda_{i,j} \neq \lambda_{p,j} \quad \text{whenever} \quad r \geq i \neq p \leq r. \tag{1.3}$$

The analysis of Sections 2 and 3 considers only the results of exact computation. The performance of our algorithms in the presence of round-off and truncation errors is considered in Section 4, which also considers the efficiency of our algorithms. In principle, our methods could be adapted to compute derivatives of arbitrarily high order using ideas of [5], but since these higher derivatives are required less often in applications and since the competitiveness of iterative methods decreases with the order of the derivatives required, we do not pursue this.

2. Mixed partial derivatives

Our algorithms do not require X as input but compute X in Step 1 using $\hat{X}$ and Y, an $n \times r$ matrix whose columns span the left eigenspace corresponding to the multiple eigenvalue. In the Hermitian case, we may take the columns of Y to be the first r columns of $\hat{X}$. Algorithm 2.1 also uses A, $A_{,j}$, $A_{,l}$, $A_{,jl}$ and Λ. It computes the mixed partial derivatives $\lambda_{i,jl}$, $i = 1, \ldots, s$, and, for *simple* eigenvalues, the corresponding $\mathbf{x}_{i,jl}$. Since (1.2) determines $\mathbf{x}_i$ only to within a scalar multiple of unit modulus and this may vary continuously in the complex case, a further restriction is required to determine $\mathbf{x}_{i,j}$ and $\mathbf{x}_{i,jl}$ uniquely [2, 3]. Algorithm 2.1 gives the value which minimizes the Euclidean norms of $\mathbf{x}_{i,j}$ and $\mathbf{x}_{i,jl}$, which, as shown in [3], is the one satisfying

$$\mathbf{x}_i^* \mathbf{x}_{i,j} = 0 \tag{2.1}$$

and

$$\mathbf{x}_i^* \mathbf{x}_{i,jl} = -\text{Re}(\mathbf{x}_{i,j}^* \mathbf{x}_{i,l}). \tag{2.2}$$

For simple eigenvalues, $\lambda_{i,jl}$ may also be computed using the well-known formula $\lambda_{i,jl} = \mathbf{y}_i^*(A_{,jl}\mathbf{x}_i + A_{,j}\mathbf{x}_{i,l} + A_{,l}\mathbf{x}_{i,j} - \lambda_{i,j}\mathbf{x}_{i,l} - \lambda_{i,l}\mathbf{x}_{i,j})/(\mathbf{y}_i^*\mathbf{x}_i)$, where $\mathbf{y}_i$ is a left eigenvector corresponding to λ_i. Algorithm 2.1 computes $\lambda_{i,jl}$ for repeated eigenvalues also, while for simple eigenvalues it also gives mixed second order derivatives of the corresponding eigenvectors. These may be used, if desired, to compute $\lambda_{i,jlp} = \mathbf{y}_i^*(A_{,jlp}\mathbf{x}_i + A_{,jl}\mathbf{x}_{i,p} + A_{,lp}\mathbf{x}_{i,j} + A_{,pj}\mathbf{x}_{i,l} + A_{,j}\mathbf{x}_{i,lp} + A_{,l}\mathbf{x}_{i,pj} + A_{,p}\mathbf{x}_{i,jl} - \lambda_{i,jl}\mathbf{x}_{i,p} - \lambda_{i,lp}\mathbf{x}_{i,j} - \lambda_{i,pj}\mathbf{x}_{i,l} - \lambda_{i,j}\mathbf{x}_{i,lp} - \lambda_{i,l}\mathbf{x}_{i,pj} - \lambda_{i,p}\mathbf{x}_{i,jl})/(\mathbf{y}_i^*\mathbf{x}_i)$.

In principle, the quantities required in Step 1 of Algorithm 2.1 may be computed by any method, for example methods described in [5]. However, since iterative methods are likely to be competitive for computing second derivatives only for problems for which they are also competitive for computing first derivatives, it is recommended that the iterative method of [7] be used. This ensures that (2.1) is satisfied [7] and it also assists the choice in Step 2 of the shift parameter σ. It is easily shown that the best choice of σ for the iteration of Step 3 is exactly the same as in the algorithms of [7]. Even a crude approximation of the optimum shift can speed convergence, but if a choice is made, perhaps interactively, using the method of [7] in Step 1, the same σ should be used for the rest of the algorithm. Theorem 2.2 shows that the ultimate convergence properties of Algorithm 2.1 do not depend on the choice of $F(0)$ in Step 2. Nevertheless, a good choice can improve the accuracy obtained with a fixed number of iterations. Our method enables advantage to be taken of any prior information, such as a truncated modal expansion [9] or an estimate for $X_{,jl}$ for a neighbouring value of $\mathbf{t}$.

Algorithm 2.1. Step 1: *Using the iterative method of [7], or otherwise, compute* X, $X_{,j}$ *and* $\Lambda_{,j}$. *If* $l \neq j$, *compute* $X_{,l}$ *and* $\Lambda_{,l}$ *similarly.*

Step 2: *Compute* $\Gamma = A_{,jl}X + A_{,j}X_{,l} + A_{,l}X_{,j} - X_{,l}\Lambda_{,j} - X_{,j}\Lambda_{,l}$. *Select small real numbers* ϵ_1, ϵ_2, *an integer* k_0 *and an origin shift,* σ. *Select an initial approximation,* $F(0)$, *to the* $n \times s$ *matrix* $X_{,jl}$.

Step 3: *Until either* $\|F(k+1) - F(k)\| < \epsilon_1$ *and* $\|N(k+1) - N(k)\| < \epsilon_2$, *for some convenient norms, or else* $k > k_0$, *compute successively, for* $k = 0, 1, 2, \ldots$,

$$V(k) = \Gamma + (A - \sigma I)F(k) \tag{2.3}$$
$$N(k) = (X^*X)^{-1}X^*[V(k) - F(k)(\Lambda - \sigma I)] \tag{2.4}$$
$$F(k+1) = [V(k) - XN(k)](\Lambda - \sigma I)^{-1}. \tag{2.5}$$

Step 4: *For* $i = 1, \ldots, s$, *the final value of* $n_{ii}(k)$ *computed in Step 3 is the accepted value of* $\lambda_{i,jl}$. *For this final value of* k, *and for each* $i = 1, \ldots, s$ *for which* λ_i *is simple, compute* $\mathbf{c}_i(k) = (c_{1i}(k), \ldots, c_{si}(k))^T$ *where*

$$c_{pi}(k) = n_{pi}(k)/(\lambda_i - \lambda_p) \quad if \quad p \neq i, \tag{2.6}$$
$$c_{pi}(k) = 0 \quad if \quad p = i. \tag{2.7}$$

Compute $\mathbf{w}_i(k) = \mathbf{f}_i(k) + X\mathbf{c}_i(k)$. *Then the accepted value of* $\mathbf{x}_{i,jl}$ *is* $\mathbf{w}_i(k) - [\mathbf{x}_i^*\mathbf{w}_i(k) + \mathrm{Re}(\mathbf{x}_{i,j}^*\mathbf{x}_{i,l})]\mathbf{x}_i$.

Theorem 2.2. *Let* A *be nondefective and let* $(X^*X)^{-1}, (\Lambda - \sigma I)^{-1}, A_{,j}, A_{,l}, A_{,jl}$, $\Lambda_{,j}, \Lambda_{,l}, \Lambda_{,jl}, X_{,j}, X_{,l}$, *and* $X_{,jl}$ *all exist, let* (1.3) *be satisfied, let all quantities be computed without roundoff and let all quantities computed in Step 1 be computed exactly, i.e. with neither roundoff nor truncation error. Then, for all* $F(0)$, *there* *exist scalars b_{pi} and vectors $\mathbf{b}_{pi}$ such that, for $i = 1, \ldots, s$,*

$$\mathbf{f}_i(k) = \mathbf{x}_{i,jl} - \sum_{p=1}^{s} b_{pi}\mathbf{x}_p + \sum_{p=s+1}^{n} ((\lambda_p - \sigma)/(\lambda_i - \sigma))^k \mathbf{b}_{pi}. \qquad (2.8)$$

If in addition

$$|(\lambda_p - \sigma)/(\lambda_i - \sigma)| < 1 \quad \text{whenever} \quad p > s \geq i, \qquad (2.9)$$

then $V(k)$, $F(k)$ and $N(k)$ approach limits $V(\infty)$, $F(\infty)$ and $N(\infty)$ as $k \to \infty$. If these limits are used in Step 4 of Algorithm 2.1 in place of specific $V(k)$, $F(k)$ and $N(k)$, then the values of $\lambda_{i,jl}$ computed in Step 4 are the correct values for all eigenvalues (including the repeated eigenvalues, if any) and, for all simple eigenvalues, the values of $\mathbf{x}_{i,jl}$ computed in Step 4 are also the correct values.

Proof. It follows from (2.3), (2.4) and (2.5) that

$$\begin{aligned} F(k+1) = \ & [A_{,jl}X + A_{,j}X_{,l} + A_{,l}X_{,j} - X_{,l}\Lambda_{,j} - X_{,j}\Lambda_{,l} + (A - \sigma I)F(k) \\ & -X(X^*X)^{-1}X^*(A_{,jl}X + A_{,j}X_{,l} + A_{,l}X_{,j} - X_{,l}\Lambda_{,j} \\ & -X_{,j}\Lambda_{,l} + AF(k) - F(k)\Lambda)](\Lambda - \sigma I)^{-1}. \end{aligned} \qquad (2.10)$$

Taking partial derivatives of (1.1) with respect to t_j and then t_l shows that

$$A_{,jl}X + A_{,j}X_{,l} + A_{,l}X_{,j} + AX_{,jl} = X\Lambda_{,jl} + X_{,j}\Lambda_{,l} + X_{,l}\Lambda_{,j} + X_{,jl}\Lambda. \qquad (2.11)$$

It follows that the constant sequence $F(k) = X_{,jl}$ satisfies (2.10), and hence the general solution of the recursive relation (2.10) is $F(k) = X_{,jl} + Z(k)$, where

$$Z(k+1) = \{(A - \sigma I)Z(k) - X(X^*X)^{-1}X^*[AZ(k) - Z(k)\Lambda]\}(\Lambda - \sigma I)^{-1}. \qquad (2.12)$$

Since the $\mathbf{x}_i$ are linearly independent, there exist scalars $d_{pi}(k)$ such that the ith column $\mathbf{z}_i(k)$ of $Z(k)$ satisfies $\mathbf{z}_i(k) = \sum_{p=1}^{n} d_{pi}(k)\mathbf{x}_p$. Now, by (2.12),

$$X^*Z(k+1) = X^*Z(k), \qquad (2.13)$$

and, since premultiplication by $X(X^*X)^{-1}X^*$ represents a projection onto the space spanned by the columns of X,

$$d_{pi}(k+1) = d_{pi}(k)(\lambda_p - \sigma)/(\lambda_i - \sigma), \qquad p = s+1, \ldots, n. \qquad (2.14)$$

A routine computation, described more fully in the proof of Theorem 2.1 of [20], shows that (2.8) follows from (2.13) and (2.14). Also, by (2.3), (2.11) and (2.4),

$$V(k) = X\Lambda_{,jl} + X_{,jl}(\Lambda - \sigma I) + (A - \sigma I)Z(k),$$

$$N(k) = \Lambda_{,jl} + (X^*X)^{-1}X^*[AZ(k) - Z(k)\Lambda]$$

and hence, since, by (2.8),

$$\mathbf{z}_i(k) = -\sum_{p=1}^{s} b_{pi}\mathbf{x}_p + \sum_{p=s+1}^{n} ((\lambda_p - \sigma)/(\lambda_i - \sigma))^k \mathbf{b}_{pi},$$

it follows that (2.9) implies convergence with $\mathbf{f}_i(\infty) = \mathbf{x}_{i,jl} - X\mathbf{b}_i$ and

$$n_{pi}(\infty) = \lambda_{i,jl}\delta_{ip} + \sum_{t=1}^{s}(\lambda_i - \lambda_t)b_{ti}\delta_{pt} = \lambda_{i,jl}\delta_{ip} + b_{pi}(\lambda_i - \lambda_p), \qquad (2.15)$$

where $n_{pi}(\infty)$ is the element in the (p,i)th position of $N(\infty)$ and δ_{ip} is the Kronecker δ. It follows that, for $i = 1,\ldots,s$, $n_{ii}(\infty) = \lambda_{i,jl}$ and, by (2.6), $b_{pi} = c_{pi}(\infty)$ when $\lambda_p \neq \lambda_i$. Hence the result follows from the construction in Step 4, which also ensures that the computed $\mathbf{x}_{i,jl}$ satisfies (2.2). $\qquad\qquad\Box$

3. Eigenvectors corresponding to repeated eigenvalues

Mixed partial derivatives of repeated eigenvalues and the corresponding eigenvectors often do not exist and, even when they do, the eigenvector derivatives are more difficult to compute. The case $l = j$ is easier however. Existence is then easier to establish and second derivatives, $\mathbf{x}_{i,jj}$, of eigenvectors corresponding to repeated eigenvalues are easily computed by the following algorithm provided the repeated eigenvalues satisfy (1.3).

Recall that we have assumed that there is a single multiple eigenvalue at t_0, its multiplicity is r and the eigenvalues are labelled so that $\lambda_1(t_0) = \cdots = \lambda_r(t_0)$. Algorithm 3.1 requires that (2.9) be satisfied for convergence in Step 1, and it requires (1.3) for the unique determination of the columns of X in Step 1 and the division in Step 4. It uses as input $A, A_{,j}, A_{,jj}, A_{,jjj}, \hat{X}, Y$, and Λ, and it produces as output $\mathbf{x}_{i,jj}$ and $\lambda_{i,jjj}$, $i = 1,\ldots,r$, in addition to the quantities computed in Step 1.

Algorithm 3.1. Step 1: *Using Algorithm 2.1 with $l = j$, compute X, $X_{,j}$, $\Lambda_{,j}$, $\Lambda_{,jj}$ and the limits, $F(\infty)$ and $N(\infty)$, of the sequences $\{F(k)\}$ and $\{N(k)\}$. (The main part of Step 4 of Algorithm 2.1 need not be carried out if the additional derivatives computed there are not required.)*

Step 2: *Compute the submatrix C_2, defined as the last $s - r$ rows of the $s \times r$ matrix*

$$C = (c_{ip}) = \begin{pmatrix} C_1 \\ C_2 \end{pmatrix}$$

where

$$c_{ip} = n_{ip}(\infty)/(\lambda_p - \lambda_i), \quad i = r+1,\ldots,s, \ p = 1,\ldots,r.$$

Step 3: *Compute the $r \times r$ matrix M, with ipth element m_{ip}, where*

$$\begin{aligned} M &= (Y^* X_1)^{-1} Y^* [A_{,jjj} X_1 + 3A_{,jj} X_{1,j} - 3X_{1,j}\Lambda_{1,jj} - 3F_1(\infty)\Lambda_{1,j} \\ &\quad + 3A_{,j}(F_1(\infty) + X_2 C_2)], \end{aligned}$$

$X_1(= (\mathbf{x}_1,\ldots,\mathbf{x}_r))$ and $F_1(\infty)$ are the matrices formed by the first r columns of X and $F(\infty)$ respectively, $X_2 = (\mathbf{x}_{r+1},\ldots,\mathbf{x}_s)$ and Λ_1, $\Lambda_{1,j}(= \operatorname{diag}(\lambda_{1,j},\ldots,\lambda_{r,j}))$ and $\Lambda_{1,jj}$ are formed from the top left $r \times r$ blocks of Λ, $\Lambda_{,j}$ and $\Lambda_{,jj}$ respectively. Then, for $i = 1,\ldots,r$, m_{ii} is the accepted value of $\lambda_{i,jjj}$.

Step 4: *Compute the remaining elements of C as*

$$c_{ip} = m_{ip}/[3(\lambda_{p,j} - \lambda_{i,j})], \quad i, p = 1, \ldots, r, \; i \neq p,$$

and

$$c_{ii} = -\mathbf{x}_{i,j}^* \mathbf{x}_{i,j} - \mathbf{x}_i^* \mathbf{f}_i(\infty) - \sum_{\substack{p=1 \\ p \neq i}}^{s} c_{pi} \mathbf{x}_i^* \mathbf{x}_p, \quad i = 1, \ldots r$$

where $\mathbf{f}_i(\infty)$ is the ith column of $F(\infty)$. For $i = 1, \ldots, r$, compute $\mathbf{f}_i(\infty) + X\mathbf{c}_i$. This is the accepted value of $\mathbf{x}_{i,jj}$.

Theorem 3.2. *Let A be nondefective and let $(X^*X)^{-1}$, $(\Lambda - \sigma I)^{-1}$, $A_{,j}$, $A_{,jj}$, $A_{,jjj}$, $\Lambda_{,j}$, $\Lambda_{,jj}$, $\Lambda_{,jjj}$, $X_{,j}$, $X_{,jj}$ and $X_{,jjj}$ exist, let $\lambda_1(\mathbf{t}_0) = \cdots = \lambda_r(\mathbf{t}_0)$, where $r \leq s$, and let all other eigenvalues be simple at $\mathbf{t}_0$. Also let (2.9) and (1.3) be satisfied, let all quantities be computed without roundoff and let the exact values of Y, $A_{,j}$, $A_{,jj}$, X, $X_{1,j}$, $\Lambda_{1,j}$ and $\Lambda_{1,jj}$ and the exact limits $N(\infty)$ and $F(\infty)$ of the sequences $\{N(k)\}$ and $\{F(k)\}$ be used in Steps 2 to 4 of Algorithm 3.1 instead of the approximations produced by Step 1. Then, for $i = 1, \ldots, r$ the approximations to $\lambda_{i,jjj}$ and $\mathbf{x}_{i,jj}$ computed by Algorithm 3.1 will also be the exact values.*

Proof. It follows from (2.9) and Theorem 2.2 that there exists an $r \times r$ matrix B_1 and an $(s - r) \times r$ matrix B_2 such that $X_{1,jj} = F_1(\infty) + X_1 B_1 + X_2 B_2$, and hence $Y^* X_{1,jj} = Y^*(F_1(\infty) + X_1 B_1)$, since $Y^* X_2 = 0$. Hence it follows from differentiating $AX_1 = X_1 \Lambda_1$ three times with respect to t_j and premultiplying by Y^* using $Y^* A = \lambda_1 Y^*$ and rearranging, that

$$\begin{aligned} \Lambda_{1,jjj} &= (Y^* X_1)^{-1} Y^* [A_{,jjj} X_1 + 3A_{,jj} X_{1,j} + 3A_{,j} F_1(\infty) + 3A_{,j}(X_1 B_1 \\ &\quad + X_2 B_2) - 3X_{1,j} \Lambda_{1,jj} - 3F_1(\infty) \Lambda_{1,j} - 3X_1 B_1 \Lambda_{1,j}]. \end{aligned}$$

As in the proof of Theorem 2.2, it follows from (2.15) and the definition of C_2 in Step 2 of Algorithm 3.1 that $C_2 = B_2$. Hence, since $Y^* A_{,j} X_1 = Y^* X_1 \Lambda_{1,j}$, a simple calculation, using the definition of M in Step 3 and the hypothesis that exact quantities are used, shows that

$$M = \Lambda_{1,jjj} + 3(B_1 \Lambda_{1,j} - \Lambda_{1,j} B_1).$$

Since $\Lambda_{1,jjj}$ is diagonal and the diagonal elements of $B_1 \Lambda_{1,j} - \Lambda_{1,j} B_1$ are zero, it follows that, for $i = 1 \ldots, r$, the computed values of $\lambda_{i,jjj}$, which are the diagonal elements of M, are the exact values, and the off-diagonal elements of B_1 are the same as the corresponding elements of C_1. Since it is also readily checked that, for $i = 1, \ldots, r$, the computed value of c_{ii} ensures that (2.2) is satisfied, we may take $B_1 = C_1$. The result follows. □

Instead of computing c_{ii} in Step 4, we could alternatively proceed as at the end of Algorithm 2.1, initially setting $c_{ii} = 0$ and then applying a correction.

4. Implementation

Most iterative methods that have been proposed for the computation of derivatives of eigenvalues and eigenvectors require the solution of a system of linear equations within the main iterative loop [1, 23]. Although the triangular decomposition need only be done once, this is still a much more expensive operation than any required within the main iterative loops of our algorithms. The iterative loops of our algorithms are Steps 1 and 3 of Algorithm 2.1 and Step 1 of Algorithm 3.1. The most expensive operation they contain is the premultiplication of an $n \times s$ matrix by the $n \times n$ matrix $A - \sigma I$, which is normally sparse. Not only is it easy to take advantage of sparsity but, even if A were not sparse, the method is well suited to parallel computation. We have not attempted parallel implementation of our algorithms, but, as described below, we have implemented another technique which greatly increases efficiency – extrapolation.

Theorem 2.2 proves more than convergence. The form of the last terms in (2.8) shows that convergence in Algorithm 2.1 may be accelerated by applying one of the ε-algorithms or the H algorithm to the sequence $\{F(k)\}$ as described in detail for the iterative computation of first derivatives in [7]. In the absence of roundoff, this would produce, in less than $2n$ iterations, the exact limit $F(\infty)$ of this sequence required for Algorithm 2.1 to produce the exact value of $\mathbf{x}_{i,jl}$. When also $N(\infty)$ in Algorithm 3.1 is obtained by substituting this $F(\infty)$ for $F(k)$, first in (2.3) and then, with the resulting $V(\infty)$, in (2.4), Algorithm 3.1 would also produce exact results. These exact results would be obtained even when (2.9) is not satisfied. In the presence of roundoff however, (2.9) remains important as it avoids the stability problems encountered by these extrapolation methods when applied to divergent sequences. (See the discussion of [6, 7, 20].) Conditions under which it is advantageous to add extra columns to X to increase the stability of the extrapolation are the same as those described for the corresponding computation of first derivatives in [7], since the sequences involved in the two processes have identical convergence properties.

Most of the rest of the discussion in [7], on stability of the algorithms presented there for computing first derivatives, also applies equally well to the algorithms presented here. Again the Hermitian case is simplest. This case is discussed more fully in [5], which however does not consider problems specific to iterative methods. An additional reason why the stability analysis of iterative methods is simpler in the case of Hermitian matrices is that, by (2.13) and (2.14), orthogonality of the eigenvectors ensures that $\|F(k) - X_{,jl}\|_2 \to 0$ *monotonically* as $k \to \infty$, whenever (2.9) is satisfied. For eigenvalues treated as simple, the only extra condition required for stability in the Hermitian case is that they be well separated from other eigenvalues. Clusters of close eigenvalues treated as equal should also be well separated from eigenvalues outside the cluster. The suggestions in [7], about when close eigenvalues should be regarded as equal, apply also to the algorithms presented here. When A is Hermitian (or, more generally, normal) and all eigenvalues not treated as equal are well separated, the eigenspace corresponding to the clustered eigenvalues will be well-conditioned, as will Λ and the eigenvectors corresponding to the simple eigenvalues. Thus all quantities input to Algorithm 2.1 can be computed very accurately by standard methods. (In applications, $A(\mathbf{t})$ is normally known in closed form [7]. Computation of partial derivatives of A may then be done analytically and is not a significant source of error in our algorithms.) For clustered or repeated eigenvalues, stability of Algorithm 2.1 requires an additional condition, namely that whenever two eigenvalues, λ_p and λ_q, say, are treated as equal, the partial derivatives $\lambda_{p,j}$ and $\lambda_{q,j}$ should be well separated. As shown in [7], this ensures that the quantities computed in Step 1 of Algorithm 2.1 may be stably computed by the methods of [7]. It also ensures that the calculations in Step 4 of Algorithm 3.1 do not encounter small denominators, which could have had large relative errors. Similarly, the requirement that other eigenvalues be well separated avoids small denominators in Step 4 of Algorithm 2.1 and Step 2 of Algorithm 3.1. In principle, our algorithms can be modified, using a technique similar to that described in [5], to cope with the case in which close eigenvalues also have close derivatives, but, as iterative methods are likely to be less competitive in such cases, we omit the details. Algorithm 2.1 remains useful in the case $r = 1$, when all eigenvalues are simple and the condition (1.3) is vacuous.

Stability is more difficult to establish in the non-Hermitian case. Our requirement that the eigenvectors be linearly independent, though adequate with exact computation, is not sufficient to prevent them sometimes being so *nearly* linearly dependent that roundoff causes problems of the type discussed in [2, pp. 215–216] for a simpler but related problem. Because of this, and because the results proved in Sections 2 and 3 refer only to the results of exact computation, we tested both algorithms on the numerical examples described in [7]. These have the clustered eigenvalues well separated from the other eigenvalues and the clustered eigenvalues have well separated derivatives, but otherwise we are not aware of any reason why they should be unusually easy. In particular, none of them are close to Hermitian matrices and in most cases the repeated eigenvalues are non-dominant.

For some of our examples, the eigenvalues and eigenvectors and their derivatives are known in closed form and hence the errors in our computed solutions could be calculated exactly. Since, except in the trivial case of constant eigenvectors, such examples are difficult to construct, these examples are small. Nevertheless, our results in these cases, all of which were close to machine precision accuracy, do give an independent check of the correctness of our algorithms. Our other examples involved larger randomly generated matrices and, although the error could no longer be computed explicitly, they did provide a check on the acceleration methods described above. Since the form of the error terms in (2.8) is the same as in the corresponding result in [7], the theory of the extrapolation methods [7, 10] predicts that, for each example, the rate of convergence will be the same as for the algorithms tested in [7]. Our numerical results confirmed that this result was not altered significantly by roundoff. Typically, less than $n/2$ iterations were required to produce results agreeing with the final limit to about 7 decimal places. More details of some of our numerical tests are given in the Technical Report [21]. Results of some additional numerical tests of Algorithm 2.1 with $l \neq j$, for problems with *simple* eigenvalues, are given in [6].

Acknowledgements. Part of this work was done while the second author was visiting La Trobe University, and was supported by a grant from the Australian Research Council and a grant from the Academic Research Fund of the National University of Singapore.

References

[1] Alvin, K. F., Efficient computation of eigenvector sensitivities for structural dynamics, *AIAA J.* **35** (1997), 1760–1766.

[2] Andrew, A. L., Iterative computation of derivatives of eigenvalues and eigenvectors, *J. Inst. Math. Appl.* **24** (1979), 209–218.

[3] Andrew, A. L., Chu, K.-W. E., Lancaster, P., Derivatives of eigenvalues and eigenvectors of matrix functions, *SIAM J. Matrix Anal. Appl.* **14** (1993), 903–926.

[4] Andrew, A. L., Chu, K.-W. E., Lancaster, P., On the numerical solution of nonlinear eigenvalue problems, *Computing* **55** (1995), 91–111.

[5] Andrew, A. L., Tan, R. C.-E., Computation of derivatives of repeated eigenvalues and the corresponding eigenvectors of symmetric matrix pencils, *SIAM J. Matrix Anal. Appl.* **20** (1998), 78–100.

[6] Andrew, A. L., Tan, R. C.-E., Computation of mixed partial derivatives of eigenvalues and eigenvectors by simultaneous iteration, *Comm. Numer. Methods Engrg.* **15** (1999), 641–649.

[7] Andrew, A. L., Tan, R. C.-E., Iterative computation of derivatives of repeated eigenvalues and the corresponding eigenvectors, *Numer. Linear Algebra Appl.* **7** (2000), 151–167.

[8] Baumgärtel, H., *Analytic perturbation theory for matrices and operators*, Birkhäuser, Basel 1985.

[9] Brandon, J. A., Second order design sensitivities to asses the applicability of sensitivity analysis, *AIAA J* **29** (1991), 135–139.

[10] Brezinski, C., Redivo Zaglia, M., *Extrapolation methods: theory and practice*, North-Holland, Amsterdam 1991.

[11] Dai, H., Lancaster, P., Numerical methods for finding multiple eigenvalues of matrices depending on parameters, *Numer. Math.* **76** (1997), 189–208.

[12] Friswell, M. I., Calculation of second and higher order derivatives, *J. Guidance Control Dyn.* **18** (1995), 919-921.

[13] Gohberg, I. C., Lancaster, P., Rodman, L., *Matrices and indefinite scalar products*, Birkhäuser, Basel 1983.

[14] Haug, E. J., Choi, K. K., Komkov, V., *Design sensitivity analysis of structural systems*, Academic Press, New York 1986.

[15] Hryniv, R., Lancaster, P., On the perturbation of analytic matrix functions, *Integral Equations Oper. Theory* **34** (1999), 325–338.

[16] Lancaster, P., Tismenetsky, M., *The theory of matrices, 2nd ed.*, Academic Press, New York 1985.

[17] Mills-Curran, W. C., Calculation of eigenvector derivatives for structures with repeated eigenvalues, *AIAA J.* **26** (1988), 867–871.

[18] Mottershead, J. E., Friswell, M. I., Model updating in structural dynamics: a survey, *J. Sound Vibration* **167** (1993), 347–375.

[19] Sun, J.-G., Multiple eigenvalue sensitivity analysis, *Linear Algebra Appl.* **137/138** (1990), 183–211.

[20] Tan, R. C.-E., Andrew, A. L., Computing derivatives of eigenvalues and eigenvectors by simultaneous iteration, *IMA J. Numer. Anal.* **9** (1989), 111–122.

[21] Tan, R. C.-E., Andrew, A. L., *Some numerical tests of algorithms for computing derivatives of eigenvalues and eigenvectors*, Math. Res. Paper 98-23, La Trobe University, Melbourne 1999.

[22] Vishik, M. I., Lyusternik, The solution of some perturbation problems for matrices and selfadjoint or non-selfadjoint differential equations I, *Russian Math. Surveys* **15** (1960), 1–73.

[23] Zhang, O., Zerva, A., Accelerated iterative procedure for calculating eigenvector derivatives, *AIAA J.* **35** (1997), 340–348.

[24] Zhang, Y.-Q., Wang, W.-L., Eigenvector derivatives of generalized nondefective eigenproblems with repeated eigenvalues, *Trans. ASME J. Engrg. Gas Turbines Power* **117** (1995), 207–212.

Mathematics Department
La Trobe University
Bundoora, Victoria 3083
Australia

Mathematics Department
National University of Singapore
Singapore 119260

1991 Mathematics Subject Classification. Primary 65F15; Secondary 65F10

Received May 23, 2000

Operator Theory:
Advances and Applications, Vol. 130, 55–82
© 2001 Birkhäuser Verlag Basel/Switzerland

Colligations in Pontryagin Spaces with a Symmetric Characteristic Function

D. Alpay, T.Ya. Azizov, A. Dijksma and J. Rovnyak

To Peter Lancaster with best wishes on the occasion of his 70-th birthday

Abstract. A symmetry in the characteristic function of a colligation is investigated for its effect on the main operator of the colligation.

1. Introduction

In this paper we study operator-valued generalized Schur functions which satisfy a symmetry condition. Such a condition imposes restrictions on the main operators in associated canonical colligations. We characterize these restrictions and give examples where they are satisfied.

We follow the notation and formulation of results on generalized Schur functions in [2]. For the convenience of the reader we recall a few basic notions. Assume that $S \in \mathbf{S}_\kappa(\mathfrak{F}, \mathfrak{G})$, where $\mathfrak{F}$ and $\mathfrak{G}$ are Pontryagin spaces having the same negative index. This means that (i) S is a holomorphic operator valued function with values in $\mathfrak{L}(\mathfrak{F}, \mathfrak{G})$, the space of bounded linear operators from $\mathfrak{F}$ to $\mathfrak{G}$, (ii) the domain of holomorphy $\Omega(S)$ of S is a region in the open unit disk $\mathbb{D}$ of the complex plane containing the origin, and (iii) if

$$K_S(w,z) = \frac{1_{\mathfrak{G}} - S(z)S(w)^*}{1 - z\bar{w}}, \qquad z, w \in \Omega(S),$$

then all Hermitian matrices of the form

$$\left(\langle K_S(w_j, w_i)g_j, g_i \rangle \right)_{i,j=1}^n, \qquad n \in \mathbb{N}, \ g_1, \ldots, g_n \in \mathfrak{G}, \ w_1, \ldots, w_n \in \Omega(S),$$

have at most κ negative eigenvalues and at least one such matrix has precisely κ negative eigenvalues (counted according to multiplicity). The function S admits a representation

$$S(z) = D + zC(1 - zA)^{-1}B, \qquad z \in \Omega(S), \tag{1.1}$$

where the right side is the characteristic function of a unitary colligation of bounded operators

$$U = \begin{pmatrix} A & B \\ C & D \end{pmatrix} : \begin{pmatrix} \mathfrak{H} \\ \mathfrak{F} \end{pmatrix} \to \begin{pmatrix} \mathfrak{H} \\ \mathfrak{G} \end{pmatrix} \tag{1.2}$$

in which $\mathfrak{H}$ is a Pontryagin space with negative index $\mathrm{ind}_- \mathfrak{H} \geq \kappa$. We call such a representation a **unitary realization** for S. The colligation (1.2) is called a **Julia colligation** if, in addition to being unitary, it satisfies the equivalent conditions $\ker B = \{0\}$ and $\ker C^* = \{0\}$. If the unitary colligation is **closely connected**, that is,

$$\overline{\mathrm{span}}\left\{\mathrm{ran}\,(1_\mathfrak{H} - zA)^{-1}B,\ \mathrm{ran}\,(1_\mathfrak{H} - zA^*)^{-1}C^* : z \in \Omega(S)\right\} = \mathfrak{H},$$

then $\mathrm{ind}_- \mathfrak{H} = \kappa$ and the colligation is determined up to isomorphism (see [2, Theorems 2.1.2 and 2.1.3]). Among the isomorphic unitary closely connected realizations of $S(z)$ is the special **canonical unitary realization** which is described in [2, Theorem 2.3.1]. The state space $\mathfrak{H}$ in this realization is the reproducing kernel Pontryagin space $\mathfrak{D}(S)$ associated with the kernel

$$D_S(w, z) = \begin{pmatrix} K_S(w, z) & \dfrac{S(z) - S(\bar{w})}{z - \bar{w}} \\[2ex] \dfrac{\tilde{S}(z) - \tilde{S}(\bar{w})}{z - \bar{w}} & K_{\tilde{S}}(w, z) \end{pmatrix}.$$

In this formula, the function $\tilde{S}(z)$ is defined by

$$\tilde{S}(z) = S(\bar{z})^*, \qquad z \in \Omega(\tilde{S}) = \overline{\Omega(S)},$$

and $\tilde{S}(z)$ belongs to $\mathbf{S}_\kappa(\mathfrak{G}, \mathfrak{F})$.

We consider the symmetry condition

$$\left.\begin{aligned} \tilde{S}(z) &= X S(z) X, \qquad z \in \Omega(S) \cap \Omega(\tilde{S}), \\ &\text{where } X \in \mathfrak{L}(\mathfrak{G}, \mathfrak{F}) \text{ is a unitary operator,} \end{aligned}\right\} \tag{1.3}$$

and its effect on the main operator A of a closely connected unitary realization of $S(z)$. The condition (1.3) is characterized in Theorem 3.1. It implies that JA is selfadjoint in $\mathfrak{H}$ for some bounded operator J on $\mathfrak{H}$ which is unitary as well as selfadjoint. Theorem 3.4 states that this condition of J–selfadjointness of A by itself is sufficient for (1.3) if the associated colligation U is also a Julia colligation. In Theorem 3.6 we consider the canonical unitary realization of $S(z)$ and describe the connection between the operator X in (1.3) and the operator J on its state space $\mathfrak{D}(S)$. We also consider coisometric closely outer connected realizations of $S(z)$ (see Theorem 3.2 and Corollary 3.7). They play a role when they are unitary, and hence isomorphic to the unitary closely connected colligations. The state space of the canonical coisometric realization is the reproducing kernel Pontryagin space $\mathfrak{H}(S)$ with kernel $K_S(w, z)$; it is simpler than the space $\mathfrak{D}(S)$.

Our motivation for the present note is the paper [11] by A. Lubin, who considered the same problem in the setting of Hilbert spaces thereby extending work of P. A. Fuhrmann [8]. In [11] S is a Schur function on the open unit disk whose values are contractions from one Hilbert space to another and A is the main operator in the Sz.-Nagy and Foiaş model for S. The derivation of the relation between the signature operator and the operator X in (1.3) in [11] is based on a lifting theorem; here we use de Branges' complementation theory which yields a convenient characterization of the state space $\mathfrak{D}(S)$ in the canonical unitary colligation.

Our approach leads to results similar to those of Lubin but valid in the class of generalized Schur functions having κ negative squares whose values are bounded operators between two Pontryagin spaces with the same negative index. In contrast, in [2, pp. 119–121] the effect of a symmetry of the form $S(z) = V^*S(-z)U$ on A is examined, where U, V are unitary operators on $\mathfrak{F}$, $\mathfrak{G}$, respectively. A similar problem related to minimal realizations of rational functions is considered in [1]. Finally, we remark that although much is known from [2] about the realization properties of functions in the classes $\mathbf{S}_\kappa(\mathfrak{F}, \mathfrak{G})$, the detailed structure of such functions is not well understood exce in special cases. The subclass of functions with symmetries discussed in this paper provides new examples of functions in $\mathbf{S}_\kappa(\mathfrak{F}, \mathfrak{G})$, and such examples may eventually contribute to a better understanding of the full class.

In Section 2 we provide the necessary preliminaries, in Section 3 we prove the main theorems, and in Section 4 we give examples.

2. Preliminaries

We assume throughout that $\mathfrak{F}$ and $\mathfrak{G}$ are Pontryagin spaces having the same negative index. If $S \in \mathbf{S}_\kappa(\mathfrak{F}, \mathfrak{G})$, we denote by $\mathfrak{H}(S)$ and $\mathfrak{H}(\tilde{S})$ the reproducing kernel Pontryagin spaces with reproducing kernels $K_S(w, z)$ and $K_{\tilde{S}}(w, z)$ defined above. The first of these spaces consists of $\mathfrak{G}$-valued functions, the second of $\mathfrak{F}$-valued functions, and both spaces have negative index equal to κ. We first study the consequences of the symmetry (1.3) on the spaces $\mathfrak{H}(S)$ and $\mathfrak{H}(\tilde{S})$.

Lemma 2.1. *Assume that S satisfies the symmetry condition (1.3). Then the mapping $\mathcal{X}$ of multiplication by X, which associates to any $h \in \mathfrak{H}(S)$ the function $\mathcal{X}h$ defined by $(\mathcal{X}h)(z) = X(h(z))$, is a unitary operator from $\mathfrak{H}(S)$ onto $\mathfrak{H}(\tilde{S})$. Its adjoint is multiplication by X^*.*

The lemma is a special case of [2, Theorem 1.5.7] and follows from the identity

$$K_S(w, z) = X^* K_{\tilde{S}}(w, z) X.$$

An important operator in realization theory is the mapping Λ from $\mathfrak{H}(S)$ into $\mathfrak{H}(\tilde{S})$ which is defined by

$$\left(\Lambda\left(\sum_j K_S(w_j, \cdot)g_j\right)\right)(z) = \sum_j \frac{\tilde{S}(z) - \tilde{S}(\bar{w}_j)}{z - \bar{w}_j} g_j \tag{2.1}$$

for any $w_1, \ldots, w_n \in \Omega(S) \cap \Omega(\tilde{S})$ and $g_1, \ldots, g_n \in \mathfrak{G}$. By [2, Theorem 3.4.1] it is a continuous bicontraction with adjoint given by

$$\left(\Lambda^*\left(\sum_j K_{\tilde{S}}(w_j, \cdot)f_j\right)\right)(z) = \sum_j \frac{S(z) - S(\bar{w}_j)}{z - \bar{w}_j} f_j$$

for any $w_1, \ldots, w_n \in \Omega(S) \cap \Omega(\tilde{S})$ and $f_1, \ldots, f_n \in \mathfrak{F}$.

Lemma 2.2. *If S satisfies the symmetry condition* (1.3), *then* $\Lambda^* \mathcal{X} = \mathcal{X}^* \Lambda$ *and* $\mathcal{X} \Lambda^* = \Lambda \mathcal{X}^*$.

Proof. If $h(\cdot) = K_S(w, \cdot)g$ for some $w \in \Omega(S) \cap \Omega(\tilde{S})$ and $g \in \mathfrak{G}$, then $(\mathcal{X}h)(z) = XK_S(w,z)g = K_{\tilde{S}}(w,z)Xg$, and so

$$(\Lambda^* \mathcal{X} h)(z) = \frac{S(z) - S(\bar{w})}{z - \bar{w}} Xg = X^* \frac{\tilde{S}(z) - \tilde{S}(\bar{w})}{z - \bar{w}} g = (\mathcal{X}^* \Lambda h)(z).$$

By linearity and continuity, $\Lambda^* \mathcal{X} = \mathcal{X}^* \Lambda$. The second relation follows from the first by the unitarity of $\mathcal{X}$. $\qquad\square$

The state space in the canonical unitary realization (1.2) is the Pontryagin space $\mathfrak{H} = \mathfrak{D}(S)$ with reproducing kernel kernel $D_S(w, z)$. The operators A, B, C, D in the colligation and their adjoints are given by

$$A \begin{pmatrix} h \\ k \end{pmatrix} (z) = \begin{pmatrix} \dfrac{h(z) - h(0)}{z} \\ zk(z) - \tilde{S}(z)h(0) \end{pmatrix} \qquad (Bf)(z) = \begin{pmatrix} \dfrac{S(z) - S(0)}{z} f \\ [1_{\mathfrak{F}} - \tilde{S}(z)\tilde{S}(0)^*]f \end{pmatrix} \left.\begin{array}{c} \\ \\ \\ \end{array}\right\}$$

$$C \begin{pmatrix} h \\ k \end{pmatrix} = h(0) \qquad\qquad Df = S(0)f, \qquad\qquad\qquad (2.2)$$

and

$$A^* \begin{pmatrix} h \\ k \end{pmatrix} (z) = \begin{pmatrix} zh(z) - S(z)k(0) \\ \dfrac{k(z) - k(0)}{z} \end{pmatrix}, \quad (C^*g)(z) = \begin{pmatrix} [1_{\mathfrak{G}} - S(z)S(0)^*]g \\ \dfrac{\tilde{S}(z) - \tilde{S}(0)}{z} g \end{pmatrix},$$

$$B^* \begin{pmatrix} h \\ k \end{pmatrix} = k(0), \qquad\qquad D^*g = S(0)^*g,$$

for all typical elements of $\mathfrak{D}(S)$, $\mathfrak{F}$, $\mathfrak{G}$ and all $z \in \Omega(S) \cap \Omega(\tilde{S})$. The theory of complementation of de Branges yields a representation of $\mathfrak{D}(S)$ in terms of the operator

$$\Gamma = \begin{pmatrix} 1 & \Lambda^* \\ \Lambda & 1 \end{pmatrix}.$$

on $\mathfrak{H}(S) \oplus \mathfrak{H}(\tilde{S})$. As shown in [2, Theorem 3.4.3], $\mathfrak{D}(S)$ is contained continuously in $\mathfrak{H}(S) \oplus \mathfrak{H}(\tilde{S})$ and $\operatorname{ran}\Gamma$ is dense in $\mathfrak{D}(S)$, with

$$\langle \Gamma u, \Gamma v \rangle_{\mathfrak{D}(S)} = \langle \Gamma u, v \rangle_{\mathfrak{H}(S) \oplus \mathfrak{H}(\tilde{S})}, \qquad u, v \in \mathfrak{H}(S) \oplus \mathfrak{H}(\tilde{S}). \qquad (2.3)$$

The mapping

$$\Pi_1 \begin{pmatrix} h \\ k \end{pmatrix} = h, \qquad\qquad (2.4)$$

is a coisometry from $\mathfrak{D}(S)$ onto $\mathfrak{H}(S)$ with adjoint

$$\Pi_1^* h = \begin{pmatrix} h \\ \Lambda h \end{pmatrix}.$$

Thus $\mathfrak{D}(S) = \operatorname{ran}\Pi_1^* \oplus \ker \Pi_1 = (\text{graph of } \Lambda) \oplus \ker \Pi_1$ and we note that the second summand is a Hilbert space as Π_1 is contraction.

We also consider **coisometric realizations** of S,

$$S(z) = H + zG(1 - zT)^{-1}F, \qquad z \in \Omega(S), \tag{2.5}$$

where the right side is the characteristic function of a coisometric colligation of bounded operators

$$V = \begin{pmatrix} T & F \\ G & H \end{pmatrix} : \begin{pmatrix} \mathfrak{K} \\ \mathfrak{F} \end{pmatrix} \rightarrow \begin{pmatrix} \mathfrak{K} \\ \mathfrak{G} \end{pmatrix} \tag{2.6}$$

in which $\mathfrak{K}$ is a Pontryagin space with negative index $\geq \kappa$. If this colligation is **closely outer connected**, that is,

$$\overline{\mathrm{span}}\,\{\mathrm{ran}\,(1_{\mathfrak{K}} - zT^*)^{-1}G^* : z \in \Omega(S)\} = \mathfrak{K},$$

then $\mathrm{ind}_-\,\mathfrak{K} = \kappa$ and the colligation is determined up to isomorphism. Among the isomorphic coisometric closely outer connected realizations of $S(z)$ is the special **canonical coisometric realization** described in [2, Theorem 2.2.1]. The state space $\mathfrak{K}$ in this realization is the reproducing kernel Pontryagin space $\mathfrak{H}(S)$ with kernel $K_S(w, z)$, and the operators and their adjoints are given by

$$\left. \begin{array}{rclrcl} (Th)(z) & = & \dfrac{h(z) - h(0)}{z}, & (Ff)(z) & = & \dfrac{S(z) - S(0)}{z}f, \\[2mm] Gh & = & h(0) & Hf & = & S(0)f, \end{array} \right\} \tag{2.7}$$

$$(T^*h)(z) = zh(z) - S(z)\tilde{h}(0), \quad (G^*g)(z) = K_S(0, z)g,$$

$$F^*h = \tilde{h}(0), \qquad\qquad H^*g = S(0)^*g.$$

By [2, Theorem 2.4.1], any coisometric closely outer connected colligation V of the form (2.6) has a dilation to a unitary closely connected colligation

$$U_d = \begin{pmatrix} A_0 & B_0 & C_0 \\ 0 & T & F \\ 0 & G & H \end{pmatrix} : \begin{pmatrix} \mathfrak{H}_0 \\ \mathfrak{H} \\ \mathfrak{F} \end{pmatrix} \rightarrow \begin{pmatrix} \mathfrak{H}_0 \\ \mathfrak{H} \\ \mathfrak{G} \end{pmatrix}, \tag{2.8}$$

in which $\mathfrak{H}_0$ is the Hilbert space $\mathfrak{L}l^2(\mathfrak{C})$, $\mathfrak{C}$ being any Hilbert space such that $\mathfrak{H} \oplus \mathfrak{F} = \mathrm{ran}\,V^* \oplus \mathfrak{C}$, and for $(c_0, c_1, c_2, \ldots) \in \mathfrak{H}_0$,

$$A_0^*\,(c_0, c_1, c_2, \ldots) = (c_1, c_2, c_3, \ldots),$$

$$\begin{pmatrix} B_0^* \\ C_0^* \end{pmatrix}(c_0, c_1, c_2, \ldots) = c_0.$$

The state space of U_d is $\mathfrak{H}_0 \oplus \mathfrak{H}$, and the characteristic function of U_d also coincides with $S(z)$. Hence the main operator of U_d and the main operator of any unitary closely connected realization U of the form (1.2) are unitarily equivalent. In particular, if V itself is unitary, then $\mathfrak{H}_0 = \{0\}$ and the operators T in (2.6) and A in (1.2) are unitarily equivalent. In [2, Theorems 3.2.3, 3.2.5, and 3.4.2] various equivalent conditions for V to be unitary are given. Among such conditions, we mention:

(1) If $S(z)f \in \mathfrak{H}(S)$ for some $f \in \mathfrak{F}$, then $f = 0$.

(2) The main operator in some unitary closely connected colligation for $S(z)$ has no invariant Hilbert subspace of the state space on which it acts as a shift.

(3) The mapping Π_1 in (2.4) is unitary.

Remark 2.3. In the same situation, each of the following conditions is sufficient for the operator V in (2.6) to be unitary:

(i) if A is the main operator in some unitary closely connected realization of $S(z)$, then $\rho(A) \cap \mathbb{T} \neq \emptyset$, where $\mathbb{T}$ is the unit circle;

(ii) dim $\mathfrak{F} < \infty$, dim $\mathfrak{G} < \infty$, and the operator B_0 in (2.8) is $B_0 = 0$. (By [2, Theorem 3.2.3], the equality $B_0 = 0$ is equivalent to the statement: "if $S(z)f \in \mathfrak{H}(S)$ for some $f \in \mathfrak{F}$, then $S(z)f \equiv 0$.")

Indeed, if (i) holds, then $\rho(A_0) \cap \mathbb{T} \neq \emptyset$. Since A_0 is a shift operator, this is impossible unless $\mathfrak{H}_0 = \{0\}$. As to (ii), since $0 \in \rho(U)$ and

$$U - \begin{pmatrix} A & 0 \\ 0 & 0 \end{pmatrix} = \begin{pmatrix} 0 & B \\ C & D \end{pmatrix}$$

has finite rank by the first two conditions, we have $0 \in \tilde{\rho}(A)$, that is, either 0 is a regular point of A or it is normal eigenvalue of A. By the third condition, then also $0 \in \tilde{\rho}(A_0)$, and again this is impossible unless $\mathfrak{H}_0 = \{0\}$ since A_0 is a shift operator.

3. Main theorems

A bounded operator J on a Kreĭn space $\mathfrak{H}$ will be called a **signature operator** if it is both selfadjoint and unitary in the inner product of $\mathfrak{H}$. Thus if $J \in \mathfrak{L}(\mathfrak{H})$ is a signature operator, then

$$J = J^{-1} = J^*.$$

An operator $A \in \mathfrak{L}(\mathfrak{H})$ is said to be J-**selfadjoint** if JA is selfadjoint.

If $J \in \mathfrak{L}(\mathfrak{H})$ is a signature operator, then $\mathfrak{H} = \mathfrak{H}_1 \oplus \mathfrak{H}_2$, where $\mathfrak{H}_1$ and $\mathfrak{H}_2$ are Kreĭn subspaces of $\mathfrak{H}$ and

$$J = \begin{pmatrix} 1_{\mathfrak{H}_1} & 0 \\ 0 & -1_{\mathfrak{H}_2} \end{pmatrix}$$

relative to the decomposition $\mathfrak{H} = \mathfrak{H}_1 \oplus \mathfrak{H}_2$. This is easily proved by showing that the operators $P_1 = \frac{1}{2}(1_{\mathfrak{H}} + J)$ and $P_2 = \frac{1}{2}(1_{\mathfrak{H}} - J)$ are selfadjoint and idempotent and hence projections. The general form of a J-selfadjoint operator on $\mathfrak{H}$ is then given by

$$A = \begin{pmatrix} A_1 & B \\ -B^* & A_2 \end{pmatrix},$$

where $A_1 \in \mathfrak{L}(\mathfrak{H}_1)$ and $A_2 \in \mathfrak{L}(\mathfrak{H}_2)$ are selfadjoint operators and B is any operator in $\mathfrak{L}(\mathfrak{H}_2, \mathfrak{H}_1)$. A J-selfadjoint operator $A \in \mathfrak{L}(\mathfrak{H})$ is similar to a selfadjoint operator $T \in \mathfrak{L}(\mathfrak{K})$, where $\mathfrak{K}$ is a Kreĭn space which has indices that are in general different from the indices of $\mathfrak{H}$. To see this, let $\mathfrak{K}$ be $\mathfrak{H}$ as a vector space, in the inner product such that if $E : \mathfrak{H} \to \mathfrak{K}$ is the identity mapping, then

$$\langle Ef, Eg \rangle_{\mathfrak{K}} = \langle Jf, g \rangle_{\mathfrak{H}}, \qquad f, g \in \mathfrak{H}. \tag{3.1}$$

A fundamental decomposition $\mathfrak{K} = \mathfrak{K}_+ \oplus \mathfrak{K}_-$ can be exhibited in terms of any fundamental decompositions of $\mathfrak{H}_1$ and $\mathfrak{H}_2$, say $\mathfrak{H}_j = \mathfrak{H}_j^+ \oplus \mathfrak{H}_j^-$, $j = 1, 2$, namely

$$\mathfrak{K}_+ = \mathfrak{H}_1^+ \oplus \mathfrak{H}_2^-,$$
$$\mathfrak{K}_- = \mathfrak{H}_1^- \oplus \mathfrak{H}_2^+.$$

By (3.1), $J = E^*E$. It follows that the operator $T = EAE^{-1} \in \mathfrak{L}(\mathfrak{K})$ is selfadjoint:

$$T^* = E^{*-1}A^*E^* = E^{*-1}A^*JE^{-1} = E^{*-1}JAE^{-1} = EAE^{-1} = T.$$

In other language, A viewed as an operator on $\mathfrak{K}$ is selfadjoint.

Theorem 3.1. *Let* $S \in \mathbf{S}_\kappa(\mathfrak{F}, \mathfrak{G})$ *have the unitary closely connected realization* (1.1)–(1.2). *Then* S *satisfies* (1.3) *if and only if there is a signature operator* $J \in \mathfrak{L}(\mathfrak{H})$ *such that*

$$\begin{pmatrix} J & 0 \\ 0 & X \end{pmatrix} U = U^* \begin{pmatrix} J & 0 \\ 0 & X^* \end{pmatrix} \tag{3.2}$$

that is, (i) A *is* J*-selfadjoint,* (ii) $XC = B^*J$, (iii) $XD = D^*X^*$. *In this case, the signature operator* $J \in \mathfrak{L}(\mathfrak{H})$ *satisfying* (3.2) *is unique.*

Proof. Assume that S satisfies (1.3). Both of the colligations

$$\begin{pmatrix} 1 & 0 \\ 0 & X \end{pmatrix} \begin{pmatrix} A & B \\ C & D \end{pmatrix} \begin{pmatrix} 1 & 0 \\ 0 & X \end{pmatrix} : \begin{pmatrix} \mathfrak{H} \\ \mathfrak{G} \end{pmatrix} \to \begin{pmatrix} \mathfrak{H} \\ \mathfrak{F} \end{pmatrix}$$

and

$$\begin{pmatrix} A^* & C^* \\ B^* & D^* \end{pmatrix} : \begin{pmatrix} \mathfrak{H} \\ \mathfrak{G} \end{pmatrix} \to \begin{pmatrix} \mathfrak{H} \\ \mathfrak{F} \end{pmatrix}$$

are unitary and closely connected. By (1.3), their characteristic functions coincide:

$$X(D + zC(1 - zA)^{-1}B)X = D^* + zB^*(1 - zA^*)^{-1}C^*.$$

Since these properties determine a colligation up to isomorphism, there is a unitary operator $J \in \mathfrak{L}(\mathfrak{H})$ such that

$$\begin{pmatrix} J & 0 \\ 0 & 1 \end{pmatrix} \begin{pmatrix} 1 & 0 \\ 0 & X \end{pmatrix} \begin{pmatrix} A & B \\ C & D \end{pmatrix} \begin{pmatrix} 1 & 0 \\ 0 & X \end{pmatrix} \begin{pmatrix} J^* & 0 \\ 0 & 1 \end{pmatrix} = \begin{pmatrix} A^* & C^* \\ B^* & D^* \end{pmatrix}.$$

Taking adjoints and moving some block matrices to the right, we find that

$$\begin{pmatrix} A^* & C^* \\ B^* & D^* \end{pmatrix} = \begin{pmatrix} J^* & 0 \\ 0 & 1 \end{pmatrix} \begin{pmatrix} 1 & 0 \\ 0 & X \end{pmatrix} \begin{pmatrix} A & B \\ C & D \end{pmatrix} \begin{pmatrix} 1 & 0 \\ 0 & X \end{pmatrix} \begin{pmatrix} J & 0 \\ 0 & 1 \end{pmatrix}.$$

The last two identities are the same except that the roles of J and J^* are interchanged. Comparing the relations, we obtain

$$\begin{aligned} JAJ^* = J^*AJ &= A^*, \\ JBX = J^*BX &= C^*, \\ XCJ^* = XCJ &= B^*. \end{aligned}$$

It follows that $J^*(1 - zA)^{-1}J = J(1 - zA)^{-1}J^*$ and

$$\begin{aligned}
J^*(1 - zA)^{-1}B &= \left[J^*(1 - zA)^{-1}J\right]\left[J^*BX\right]X^* \\
&= \left[J(1 - zA)^{-1}J^*\right]\left[JBX\right]X^* \\
&= J(1 - zA)^{-1}B.
\end{aligned}$$

Similarly, $J(1 - zA^*)^{-1}C^* = J^*(1 - zA^*)^{-1}C^*$. Since the colligation U is closely connected, $J = J^*$. Thus J is a signature operator. The identity (3.2) holds by construction. The steps in this argument are reversible, and so the first statement follows. The uniqueness of J is proved from the close connectedness of U by an argument similar to the reasoning above.   □

The arguments in the proof of Theorem 3.1 cannot be adapted to coisometric closely outer connected realizations. The adjoint of such a colligation need not be coisometric, and it need not be closely outer connected; thus the theorem on uniqueness up to isomorphism cannot be invoked. Of course, Theorem 3.1 applies to a coisometric closely outer connected realization V which is unitary, but a little bit more can be said in this situation:

Theorem 3.2. *Let $S \in \mathbf{S}_\kappa(\mathfrak{F}, \mathfrak{G})$ have the coisometric closely outer connected realization (2.5)–(2.6).*

(1) *Assume that S satisfies (1.3) and that V is unitary. Then there is a unique signature operator $J \in \mathfrak{L}(\mathfrak{H})$ such that*

$$\begin{pmatrix} J & 0 \\ 0 & X \end{pmatrix} V = V^* \begin{pmatrix} J & 0 \\ 0 & X^* \end{pmatrix}, \tag{3.3}$$

that is, (i) T is J-selfadjoint, (ii) $XG = F^*J$, (iii) $XH = H^*X^*$.

(2) *Assume that (3.3) holds for some signature operator $J \in \mathfrak{L}(\mathfrak{H})$ and a unitary operator $X \in \mathfrak{L}(\mathfrak{G}, \mathfrak{F})$. Then V is isometric (hence unitary), closely inner connected, and S satisfies (1.3).*

In part (2), closely inner connected means that $\overline{\operatorname{span}}\,\{\operatorname{ran}(1_\mathfrak{R} - zT)^{-1}F : z \in \Omega(S)\} = \mathfrak{R}$.

Proof. (1) The hypotheses imply that V is unitary and closely connected, so (1) follows from the necessity part of Theorem 3.1.

(2) Since V is coisometric,

$$T^*T + G^*G = T^*JJT + JFXX^*F^*J^* = J\left[TT^* + FF^*\right]J^* = 1_\mathfrak{R},$$

and similarly $T^*F + G^*H = 0$ and $F^*F + H^*H = 1_\mathfrak{F}$. Hence V is isometric. The close outer connectedness of V and the equality

$$(1_\mathfrak{R} - zT)^{-1}F = J(1_\mathfrak{R} - zT^*)^{-1}G^*X^*$$

imply that V is closely inner connected. S satisfies (1.3) by Theorem 3.1.   □

Remark 3.3. Theorem 3.2(1) requires a hypothesis that V is unitary. We note a sufficient condition for this to hold in terms of any unitary closely connected realization (1.1)–(1.2) and signature operator J on the state space $\mathfrak{H}$ such that the main operator A is J-selfadjoint. Namely, assume that the space $(\mathfrak{H}, \langle J \cdot, \cdot \rangle_{\mathfrak{H}})$ is a Pontryagin space (for example, this holds when $J = 1$, as in Example 4.1). Then $\sigma(A) \cap (\mathbb{C} \setminus \mathbb{R})$ is a finite set. Thus condition (i) in Remark 2.3 is met, and hence V is unitary.

Part (2) of the next result and the related Example 4.1 in the next section are indefinite generalizations of constructions in [7, pp. 127–128] and [11, Corollary 3].

Theorem 3.4. *Let $S \in \mathbf{S}_\kappa(\mathfrak{F}, \mathfrak{G})$ have the unitary closely connected realization (1.1)–(1.2), and assume that U is a Julia colligation.*

(1) *If A is J-selfadjoint relative to some signature operator $J \in \mathfrak{L}(\mathfrak{H})$, then S satisfies the symmetry condition (1.3) for some unitary operator $X \in \mathfrak{L}(\mathfrak{G}, \mathfrak{F})$.*

(2) *If A is selfadjoint (that is, (1) holds with $J = 1$), then the operator X in (1) can be chosen so that*

$$S(z) = X^*(z + XS(0))(1 + zXS(0))^{-1} = (1 + zS(0)X)^{-1}(S(0) + zX^*). \quad (3.4)$$

Moreover, $S(0)XS(z) = S(z)XS(0)$.

Proof. (1) We apply [2, Theorem 4.5.4] with the two Julia colligations

$$\begin{pmatrix} T_1 & F_1 \\ G_1 & H_1 \end{pmatrix} = \begin{pmatrix} A & B \\ C & D \end{pmatrix} : \begin{pmatrix} \mathfrak{H} \\ \mathfrak{F} \end{pmatrix} \to \begin{pmatrix} \mathfrak{H} \\ \mathfrak{G} \end{pmatrix},$$

$$\begin{pmatrix} T_2 & F_2 \\ G_2 & H_2 \end{pmatrix} = \begin{pmatrix} A^* & C^* \\ B^* & D^* \end{pmatrix} : \begin{pmatrix} \mathfrak{H} \\ \mathfrak{G} \end{pmatrix} \to \begin{pmatrix} \mathfrak{H} \\ \mathfrak{F} \end{pmatrix}.$$

Since $T_1 = A$ is J-selfadjoint, it is unitarily equivalent to $T_2 = A^*$ by means of the signature operator J: $A = J^{-1}A^*J$. Thus by [2, Theorem 4.5.4] and its proof, there are unitary operators $\varphi \colon \mathfrak{F} \to \mathfrak{G}$ and $\psi \colon \mathfrak{G} \to \mathfrak{F}$ such that $S(z) = \psi^{-1}\tilde{S}(z)\varphi$, $B = JC^*\varphi$, $C^* = JB\psi$, and $\varphi D^* = D\psi$. In particular, $B = J^2 B\psi\varphi = B\psi\varphi$, and hence $B(1_{\mathfrak{F}} - \psi\varphi) = 0$. Since B is injective, $\varphi = \psi^{-1}$. Thus $S(z)$ satisfies (1.3) with $X = \psi$.

(2) We use the same choice $X = \psi = \varphi^{-1}$ as in the proof of part (1), but now with $J = 1$. Then $B = C^*X^*$. The symmetry condition $\tilde{S}(z) = XS(z)X$ with $z = 0$ implies that $S(0)X = DX$ is selfadjoint. From the equalities $A^* = A$ and

$$\begin{pmatrix} A & B \\ C & D \end{pmatrix} \begin{pmatrix} A^* & C^* \\ B^* & D^* \end{pmatrix} = \begin{pmatrix} 1_{\mathfrak{D}(S)} & 0 \\ 0 & 1_{\mathfrak{G}} \end{pmatrix},$$

$$\begin{pmatrix} A^* & C^* \\ B^* & D^* \end{pmatrix} \begin{pmatrix} A & B \\ C & D \end{pmatrix} = \begin{pmatrix} 1_{\mathfrak{D}(S)} & 0 \\ 0 & 1_{\mathfrak{F}} \end{pmatrix},$$

we readily obtain

$$D^*S(z) = D^*D + zD^*C(1 - zA)^{-1}B$$
$$= D^*D - zB^*A(1 - zA)^{-1}B$$

$$= D^*D - zB^*(1 - zA)^{-1}AB$$
$$= D^*D - zB^*(1 - zA)^{-1}A^*B$$
$$= D^*D + zB^*(1 - zA)^{-1}C^*D$$
$$= \tilde{S}(z)D$$
$$= XS(z)XD,$$

$$DD^* + zC(1 - zA)^{-1}BD^* = 1 - CC^* - zC(1 - zA)^{-1}AC^*$$
$$= 1 - C(1 + z(1 - zA)^{-1}A)C^*$$
$$= 1 - C(1 - zA)^{-1}C^*,$$

and consequently, since $DX = X^*D^*$, $XD = D^*X^*$, and $B = C^*X^*$, we have

$$DXS(z) = X^*D^*S(z) = S(z)XD$$

and

$$S(z)(1 + zXD) = S(z) + zS(z)D^*X^*$$
$$= S(z) + z(DD^* + zC(1 - zA)^{-1}BD^*)X^*$$
$$= [D + zC(1 - zA)^{-1}B] + z(I - C(1 - zA)^{-1}C^*)X^*$$
$$= D + zX^* + zC(1 - zA)^{-1}(B - C^*X^*)$$
$$= D + zX^*$$
$$= X^*(XD + z).$$

Thus $S(0)XS(z) = S(z)XS(0)$, $S(z) = X^*(z + XD)(1 + zXD)^{-1}$, which yields the assertions, except for the second equality in (3.4), but this follows from the first. $\qquad\square$

The following result on the scalar case extends [11, Corollary 1] to generalized Schur functions.

Corollary 3.5. *Let* $S(z) = \sum_{n=0}^{\infty} S_n z^n$ *be a scalar-valued function in* $\mathbf{S}_\kappa(\mathfrak{F}, \mathfrak{G})$, $\mathfrak{F} = \mathfrak{G} = \mathbb{C}$. *The following assertions are equivalent:*

(1) *The main operator A in any unitary closely connected realization (1.1)–(1.2) is J-selfadjoint for some signature operator $J \in \mathfrak{L}(\mathfrak{H})$.*

(2) *There is a complex number λ of modulus one such that λS_n is real for every $n = 0, 1, 2, \ldots$.*

Proof. The assertion is trivial if $S(z)$ is a constant of absolute value one (which can only occur when $\kappa = 0$). Excluding this case, we conclude that U is a Julia colligation; in fact, we can assume that (1.1)–(1.2) is the canonical unitary realization, and we verify the Julia property by the explicit formulas for the operators in this colligation.

If (1) holds, then by Theorem 3.4 there is a complex number λ of modulus one such that $\tilde{S}(z) = \lambda^2 S(z)$, and this implies (2). The other direction follows from Theorem 3.1. $\qquad\qquad\qquad\qquad\qquad\qquad\qquad\qquad\qquad\qquad\qquad\qquad\qquad\qquad\quad\square$

When the realization (1.1)–(1.2) of a given function $S \in \mathbf{S}_\kappa(\mathfrak{F}, \mathfrak{G})$ in Theorem 3.1 uses the canonical unitary colligation, we can exhibit the signature operator J whose existence is asserted when the symmetry condition (1.3) holds.

Theorem 3.6. *In Theorem* 3.1 *assume that* U *is the canonical unitary colligation. If the symmetry condition* (1.3) *is satisfied, then the unique signature operator* $J = J_{\mathfrak{D}(S)}$ *on* $\mathfrak{D}(S)$ *satisfying* (3.2) *is given by* $J = \hat{J}|_{\mathfrak{D}(S)}$, *where* $\hat{J}$ *is the signature operator on* $\mathfrak{H}(S) \oplus \mathfrak{H}(\tilde{S})$ *defined by*

$$\hat{J} = \begin{pmatrix} 0 & \mathcal{X}^* \\ \mathcal{X} & 0 \end{pmatrix}.$$

Proof. By Lemma 2.1, $\hat{J}$ is a unitary operator on $\mathfrak{H}(S) \oplus \mathfrak{H}(\tilde{S})$. It is clearly selfadjoint and hence a signature operator. It will be shown that $J = \hat{J}|_{\mathfrak{D}(S)}$ is a bounded operator on $\mathfrak{D}(S)$ into itself, and that it is a signature operator satisfying (3.2). Then the theorem follows from the uniqueness assertion in Theorem 3.1.

We use the notation for the canonical unitary colligation given in Section 2. Let $\mathbf{R}$ be the linear relation in $\mathfrak{D}(S) \oplus \mathfrak{D}(S)$ spanned by pairs of the form

$$(\Gamma u, \Gamma \hat{J}u), \qquad u \in \mathfrak{H}(S) \oplus \mathfrak{H}(\tilde{S}).$$

By Lemma 2.2, $\hat{J}\Gamma = \Gamma \hat{J}$. Hence from (2.3) we obtain, for $u, v \in \mathfrak{H}(S) \oplus \mathfrak{H}(\tilde{S})$,

$$\langle \Gamma u, \Gamma v \rangle_{\mathfrak{D}(S)} = \langle \Gamma u, v \rangle_{\mathfrak{H}(S) \oplus \mathfrak{H}(\tilde{S})} = \langle \hat{J}\Gamma u, \hat{J}v \rangle_{\mathfrak{H}(S) \oplus \mathfrak{H}(\tilde{S})}$$
$$= \langle \Gamma \hat{J}u, \hat{J}v \rangle_{\mathfrak{H}(S) \oplus \mathfrak{H}(\tilde{S})} = \langle \Gamma \hat{J}u, \Gamma \hat{J}v \rangle_{\mathfrak{D}(S)}.$$

It follows that $\mathbf{R}$ is isometric. Since its domain and its range are dense in $\mathfrak{D}(S)$, we have by [2, Theorem 1.4.2] that the closure of $\mathbf{R}$ in $\mathfrak{D}(S) \oplus \mathfrak{D}(S)$ is the graph of a unitary operator W from $\mathfrak{D}(S)$ onto itself. By construction,

$$W\Gamma u = \Gamma \hat{J}u = \hat{J}\Gamma u, \qquad u \in \mathfrak{H}(S) \oplus \mathfrak{H}(\tilde{S}),$$

and thus $W|_{\operatorname{ran}\Gamma} = \hat{J}|_{\operatorname{ran}\Gamma}$. But this readily implies that $W = \hat{J}|_{\mathfrak{D}(S)} = J$. Indeed, let $h \oplus k$ be any element of $\mathfrak{D}(S)$, and let u_n be a sequence in $\mathfrak{H}(S) \oplus \mathfrak{H}(\tilde{S})$ such that in $\mathfrak{D}(S)$,

$$\begin{pmatrix} h_n \\ k_n \end{pmatrix} := \Gamma u_n \to \begin{pmatrix} h \\ k \end{pmatrix}, \qquad n \to \infty.$$

Then, pointwise in $\mathfrak{G} \oplus \mathfrak{F}$,

$$\begin{pmatrix} h_n(z) \\ k_n(z) \end{pmatrix} \to \begin{pmatrix} h(z) \\ k(z) \end{pmatrix}, \qquad n \to \infty,$$

and by the continuity of W and X, this implies that

$$W\begin{pmatrix} h \\ k \end{pmatrix}(z) = \lim_{n\to\infty} W\begin{pmatrix} h_n \\ k_n \end{pmatrix}(z) = \lim_{n\to\infty} \hat{J}\begin{pmatrix} h_n \\ k_n \end{pmatrix}(z)$$

$$= \lim_{n\to\infty} \begin{pmatrix} X^*k_n(z) \\ Xh_n(z) \end{pmatrix} = \begin{pmatrix} X^*k(z) \\ Xh(z) \end{pmatrix} = \hat{J}\begin{pmatrix} h \\ k \end{pmatrix}(z).$$

It follows that $J = \hat{J}|_{\mathfrak{D}(S)}$ is a bounded operator and in fact unitary on $\mathfrak{D}(S)$ into itself. To see that it is a signature operator, we must show that it is selfadjoint. In fact, for any $u, v \in \mathfrak{H}(S) \oplus \mathfrak{H}(\tilde{S})$ we have

$$\langle J\Gamma u, \Gamma v\rangle_{\mathfrak{D}(S)} = \langle \Gamma \hat{J} u, \Gamma v\rangle_{\mathfrak{D}(S)} = \langle \hat{J} u, \Gamma v\rangle_{\mathfrak{H}(S)\oplus\mathfrak{H}(\tilde{S})}$$

$$= \langle u, \hat{J}\Gamma v\rangle_{\mathfrak{H}(S)\oplus\mathfrak{H}(\tilde{S})} = \langle \Gamma u, J\Gamma v\rangle_{\mathfrak{D}(S)}.$$

The selfadjointness of J follows from the continuity of J and the denseness of $\operatorname{ran}\Gamma$. Thus J is a signature operator on $\mathfrak{D}(S)$.

From the explicit formulas for the canonical unitary colligation in Section 2, we get

$$(JA^*\begin{pmatrix} h \\ k \end{pmatrix})(z) = \begin{pmatrix} X^*\dfrac{k(z) - k(0)}{z} \\ X(zh(z) - S(z)k(0)) \end{pmatrix}.$$

On the other hand,

$$(AJ\begin{pmatrix} h \\ k \end{pmatrix})(z) = (A\begin{pmatrix} \mathcal{X}^*k \\ \mathcal{X}h \end{pmatrix})(z)$$

$$= \begin{pmatrix} \dfrac{X^*k(z) - X^*k(0)}{z} \\ zXh(z) - \tilde{S}(z)X^*k(0) \end{pmatrix}$$

$$= \begin{pmatrix} X^*\dfrac{k(z) - k(0)}{z} \\ zXh(z) - XS(z)XX^*k(0) \end{pmatrix}$$

$$= \begin{pmatrix} X^*\dfrac{k(z) - k(0)}{z} \\ X(zh(z) - S(z)k(0)) \end{pmatrix},$$

so $JA^* = AJ$, and hence $(JA)^* = JA$. Finally, by the formulas for B and C^*,

$$(JBXg)(z) = \begin{pmatrix} X^*(1 - \tilde{S}(z)\tilde{S}(0)^*)Xg \\ X\dfrac{S(z) - S(0)}{z}Xg \end{pmatrix}$$

$$= \begin{pmatrix} (1_\mathfrak{G} - S(z)S(0)^*)g \\ \dfrac{\tilde{S}(z) - \tilde{S}(0)}{z} g \end{pmatrix}$$

$$= (C^*g)(z),$$

which shows that $JBX = C^*$. The identity $XD = D^*X^*$ is automatic by the symmetry condition. It follows that (3.2) holds with this choice of J, and so the result follows from the uniqueness part of Theorem 3.1. $\quad\square$

Corollary 3.7. *In Theorem 3.2(1) assume that V is the canonical coisometric colligation. If the symmetry condition (1.3) is satisfied, then the unique signature operator $J = J_{\mathfrak{H}(S)}$ on $\mathfrak{H}(S)$ satisfying (3.3) is given by*

$$J = \mathcal{X}^*\Lambda = \Pi_1 J_{\mathfrak{D}(S)} \Pi_1^*,$$

where Λ and Π_1 are as in Section 2 and (2.4), and $J_{\mathfrak{D}(S)}$ is the operator from Theorem 3.6.

Recall from Section 2 that Π_1 is unitary whenever V is unitary.

Proof. Theorem 3.2(2) implies that V is an isometric closely inner connected realization of $S(z)$, which is unitary. The uniqueness theorem for isometric closely inner connected realizations implies that the canonical isometric realization is isomorphic to V and hence also unitary. Thus the canonical coisometric colligation and the canonical isometric colligation for $S(z)$ are both unitary, and so by [2, Theorem 3.4.2(C)], the mapping Λ is unitary. On account of Lemmas 2.1 and 2.2, the operator $W := \Lambda^*\mathcal{X} = \mathcal{X}^*\Lambda$ is a signature operator on $\mathfrak{H}(S)$. From the formulas for Π_1 and $J_{\mathfrak{D}(S)}$ we obtain $W = \Pi_1 J_{\mathfrak{D}(S)} \Pi_1^*$. We show that $W = J_{\mathfrak{H}(S)}$. From the explicit formulas for the operators, we find for $h \in \mathfrak{H}(S)$,

$$(\Pi_1 A J_{\mathfrak{D}(S)} \Pi_1^* h)(z) = \left(\Pi_1 A \begin{pmatrix} \mathcal{X}^*\Lambda h \\ \mathcal{X}h \end{pmatrix}\right)(z) =$$

$$= \frac{X^*(\Lambda h)(z) - X^*(\Lambda h)(0)}{z} = (T\mathcal{X}^*\Lambda h)(z) = (TWh)(z),$$

that is, $\Pi_1 A J_{\mathfrak{D}(S)} \Pi_1^* = TW$. Since, by Theorem 3.6, A is $J_{\mathfrak{D}(S)}$-selfadjoint on $\mathfrak{D}(S)$, the equality implies that T is W-selfadjoint on $\mathfrak{H}(S)$. It also follows directly from the formulas for the operators that for $g \in \mathfrak{G}$,

$$(\mathcal{X}FXg)(z) = X\frac{S(z) - S(0)}{z}Xg = \frac{\tilde{S}(z) - \tilde{S}(0)}{z}g = (\Lambda G^*g)(z).$$

Hence, since Λ is unitary, $WFX = G^*$, or equivalently, $XG = F^*W$. So by the uniqueness statement of Theorem 3.2, $W = J_{\mathfrak{H}(S)}$. $\quad\square$

4. Examples

As a first example we start with a function of the form (3.4) and show that it is a generalized Schur function which satisfies the symmetry condition (1.3).

Example 4.1. Let $\mathfrak{F}$ and $\mathfrak{G}$ be Pontryagin spaces having the same negative index. Let $X \in \mathfrak{L}(\mathfrak{G}, \mathfrak{F})$ be unitary, and let $D \in \mathfrak{L}(\mathfrak{F}, \mathfrak{G})$ be an operator such that XD is selfadjoint and $\mathrm{ind}_- (1 - DD^*) < \infty$. Set

$$\kappa = \mathrm{ind}_- (1 - DD^*) = \mathrm{ind}_- (1 - D^*D)$$

and

$$S(z) = X^*(z + XD)(1 + zXD)^{-1} = (1 + zDX)^{-1}(D + zX^*). \qquad (4.1)$$

(A related example appears in [7, pp. 127–128]. There the function $S(z) = (1 + zX^*D^*)^{-1}(D + zX^*)$ is considered in a Hilbert space situation, with no symmetry condition but with other technical hypotheses. The main one is the identity $1 + X^*D^*DX = DD^* + X^*X$. The formula for $S(z)$ coincides with (4.1) and the identity holds in our case because X is unitary and XD is selfadjoint.) Then $S \in S_\kappa(\mathfrak{F}, \mathfrak{G})$ and satisfies (1.3), its canonical coisometric realization is unitary, and the main operator in this realization is selfadjoint. To see this, first check the identity $\tilde{S}(z) = XS(z)X$ by straightforward algebra. Next note that

$$K_S(w, z) = (1 + zDX)^{-1}(1 - DD^*)(1 + wDX)^{*-1}.$$

In particular, $S \in S_\kappa(\mathfrak{F}, \mathfrak{G})$. Factor

$$1 - DD^* = EE^*, \qquad E \in \mathfrak{L}(\mathfrak{K}, \mathfrak{G}), \ \ker E = \{0\}.$$

Then the mapping

$$u \to (1 + zDX)^{-1}Eu$$

is an isomorphism from $\mathfrak{K}$ onto $\mathfrak{H}(S)$. To show that the coisometric colligation V is unitary, we use a criterion in Section 2: namely, we show that the only $f \in \mathfrak{F}$ such that $S(z)f \in \mathfrak{H}(S)$ is $f = 0$. In fact, if $S(z)f \in \mathfrak{H}(S)$, there is a vector $u \in \mathfrak{K}$ such that

$$(1 + zDX)^{-1}(zX^* + D)f = S(z)f = (1 + zDX)^{-1}Eu,$$

hence $(zX^* + D)f = Eu$ for all z in a neighborhood of the origin, and so $X^*f = 0$ and $f = 0$. Thus V is unitary.

To see that $T: h(z) \to [h(z) - h(0)]/z$ is selfadjoint on $\mathfrak{H}(S)$, consider an element of $\mathfrak{H}(S)$ of the form

$$h(z) = (1 + zDX)^{-1}EE^*g, \qquad g \in \mathfrak{G}.$$

We easily check that DX commutes with $1 - DD^*$, so

$$\frac{h(z) - h(0)}{z} = (1 + zDX)^{-1}(-DX)(1 - DD^*)g$$

$$= (1 + zDX)^{-1}(1 - DD^*)(-DX)g = (1 + zDX)^{-1}E(-E^*DXg),$$

that is, the unitarily equivalent operator R on $\mathfrak{K}$ acts on a dense set by the rule

$$R\colon E^*g \to -E^*DXg, \qquad g \in \mathfrak{G}.$$

For any $g_1, g_2 \in \mathfrak{G}$,

$$\langle RE^*g_1, E^*g_2\rangle_{\mathfrak{K}} = \langle -EE^*DXg_1, g_2\rangle_{\mathfrak{G}} = \langle g_1, -EE^*DXg_2\rangle_{\mathfrak{G}} = \langle E^*g_1, RE^*g_2\rangle_{\mathfrak{K}}.$$

It follows that R is selfadjoint, and therefore T is selfadjoint.

We show that this colligation is a Julia colligation if and only if $\ker(1 - DD^*) = \{0\}$. This follows from the equality

$$(Ff)(z) = (1 + zDX)^{-1}(1 - DD^*)X^*f.$$

It implies that f belongs to $\ker F$ if and only if $(1 - DD^*)X^*f = 0$. Therefore, $\ker F = \{0\}$ if and only if $\ker(1 - DD^*) = \{0\}$.

Finally, we remark that $S(z)$ has unitary values at all points on the circle $|z| = 1$ where it is holomorphic. For straightforward algebra shows that if $|z| = 1$ then

$$(1 + zDX)^{-1}(D + zX^*)(D^* + \bar{z}X)(1 + \bar{z}X^*D^*)^{-1} = 1_{\mathfrak{G}}$$

and

$$(1 + \bar{z}D^*X^*)^{-1}(\bar{z}X + D^*)(zX^* + D)(1 + zXD)^{-1} = 1_{\mathfrak{F}}$$

whenever the inverses exist.

We present a common setting for the next two examples. The following result is, in part, adapted from [3, Theorem 4.1].

Theorem 4.2. *Let $\mathfrak{G}, \mathfrak{K}$ be Pontryagin spaces, $A \in \mathfrak{L}(\mathfrak{K})$, $C \in \mathfrak{L}(\mathfrak{K}, \mathfrak{G})$, and assume that*

$$A^*A + C^*C = 1_{\mathfrak{K}} \qquad and \qquad \bigcap_{0}^{\infty} \ker CA^j = \{0\}. \tag{4.2}$$

Assume also that there is a complex number γ, $|\gamma| = 1$, such that $1 - \gamma A$ is invertible and set

$$B = (\bar{\gamma} - A)(1 - \bar{\gamma}A^*)^{-1}C^* \qquad and \qquad D = 1_{\mathfrak{G}} - C(1 - \bar{\gamma}A^*)^{-1}C^*.$$

Then

$$U := \begin{pmatrix} A & B \\ C & D \end{pmatrix} \colon \begin{pmatrix} \mathfrak{K} \\ \mathfrak{G} \end{pmatrix} \to \begin{pmatrix} \mathfrak{K} \\ \mathfrak{G} \end{pmatrix}$$

is a closely outer connected unitary colligation. Its characteristic function is given by

$$S(z) = 1_{\mathfrak{G}} - (1 - z\bar{\gamma})C(1 - zA)^{-1}(1 - \bar{\gamma}A^*)^{-1}C^* \tag{4.3}$$

and belongs to $\mathbf{S}_\kappa(\mathfrak{G}, \mathfrak{G})$, where $\kappa = \operatorname{ind}_- \mathfrak{K}$. Moreover,

(1) *if T, F, G, and H stand for the operators in the canonical coisometric colligation for $S(z)$ (see (2.7)) then*

$$\begin{pmatrix} W & 0 \\ 0 & 1_{\mathfrak{G}} \end{pmatrix} \begin{pmatrix} A & B \\ C & D \end{pmatrix} = \begin{pmatrix} T & F \\ G & H \end{pmatrix} \begin{pmatrix} W & 0 \\ 0 & 1_{\mathfrak{G}} \end{pmatrix}, \tag{4.4}$$

where the mapping

$$W : k \to C(1 - zA)^{-1}k, \quad k \in \mathfrak{K}, \tag{4.5}$$

is an isomorphism from $\mathfrak{K}$ onto $\mathfrak{H}(S)$, and

(2) *if A_u, B_u, C_u, and D_u stand for the operators in the canonical unitary collation for $S(z)$ (as in (2.2)) then*

$$\begin{pmatrix} W_u & 0 \\ 0 & 1_{\mathfrak{G}} \end{pmatrix} \begin{pmatrix} A & B \\ C & D \end{pmatrix} = \begin{pmatrix} A_u & B_u \\ C_u & D_u \end{pmatrix} \begin{pmatrix} W_u & 0 \\ 0 & 1_{\mathfrak{G}} \end{pmatrix}, \tag{4.6}$$

where the mapping

$$W_u : k \to \begin{pmatrix} C(1 - zA)^{-1}k \\ B^*(1 - zA^*)^{-1}k \end{pmatrix}, \quad k \in \mathfrak{K},$$

is an isomorphism from $\mathfrak{K}$ onto $\mathfrak{D}(S)$.

Proof. Define $S(z)$ by (4.3). We first verify that the identity

$$K_S(w, z) = C(1 - zA)^{-1}(1 - \bar{w}A^*)^{-1}C^* \tag{4.7}$$

holds for w, z in a neighborhood of the origin. In fact,

$$1 - S(z)S(w)^* = 1_{\mathfrak{G}} - \left[1_{\mathfrak{G}} - (1 - z\bar{\gamma})C(1 - zA)^{-1}(1 - \bar{\gamma}A^*)^{-1}C^* \right] \cdot$$
$$\cdot \left[1_{\mathfrak{G}} - (1 - \bar{w}\gamma)C(1 - \gamma A)^{-1}(1 - \bar{w}A^*)^{-1}C^* \right]$$
$$= (1 - \bar{w}\gamma)C(1 - \gamma A)^{-1}(1 - \bar{w}A^*)^{-1}C^*$$
$$\quad + (1 - z\bar{\gamma})C(1 - zA)^{-1}(1 - \bar{\gamma}A^*)^{-1}C^*$$
$$\quad - (1 - z\bar{\gamma})(1 - \bar{w}\gamma)C(1 - zA)^{-1}(1 - \bar{\gamma}A^*)^{-1} \cdot$$
$$\quad \cdot (1_{\mathfrak{K}} - A^*A)(1 - \gamma A)^{-1}(1 - \bar{w}A^*)^{-1}C^*$$
$$= C(1 - zA)^{-1}(1 - \bar{\gamma}A^*)^{-1}\left[(1 - \bar{w}\gamma)(1 - \bar{\gamma}A^*)(1 - zA) \right.$$
$$\quad \left. + (1 - z\bar{\gamma})(1 - \bar{w}A^*)(1 - \gamma A) - (1 - z\bar{\gamma})(1 - \bar{w}\gamma)(1_{\mathfrak{K}} - A^*A) \right] \cdot$$
$$\quad \cdot (1 - \gamma A)^{-1}(1 - \bar{w}A^*)^{-1}C^*$$

The part in brackets simplifies to $(1 - z\bar{w})(1 - \bar{\gamma}A^*)(1 - \gamma A)$, yielding (4.7). From the second equality in (4.2) it now follows that $S \in \mathbf{S}_\kappa(\mathfrak{G}, \mathfrak{G})$. Thus the space $\mathfrak{H}(S)$ is well defined. The mapping W, which is also characterized by

$$(W(1 - \bar{w}A^*)^{-1}C^*u)(z) = K_S(w, z)u, \quad u \in \mathfrak{G},$$

is an isometry with dense domain in $\mathfrak{K}$ (by the second equality in (4.2)) and dense range in $\mathfrak{H}(S)$ and hence an isomorphism between these two spaces. The first equality in (4.2) and the definitions of the operators B and D together with some standard calculations readily yield that U is isometric. By the second equality, U

is closely outer connected. From the definitions of T, F, G, and H and straight-forward calculations we get (4.4) and so U is also coisometric. From the definition of $S(z)$ we obtain

$$\begin{aligned} S(z) - D &= 1_{\mathfrak{G}} - (1 - z\bar{\gamma})C(1 - zA)^{-1}(1 - \bar{\gamma}A^*)^{-1}C^* \\ &\quad - (1_{\mathfrak{G}} - C(1 - \bar{\gamma}A^*)^{-1}C^* \\ &= C(1 - zA)^{-1}\Big[-1 + z\bar{\gamma} + 1 - zA\Big](1 - \bar{\gamma}A^*)^{-1}C^* \\ &= z\,C(1 - zA)^{-1}(\bar{\gamma} - A)(1 - \bar{\gamma}A^*)^{-1}C^* \\ &= z\,C(1 - zA)^{-1}B, \end{aligned}$$

that is, $S(z)$ is the characteristic function of the colligation U. The equality (4.6) follows from the definition of W_u and the explicit formulas (2.2) for A_u, B_u, C_u, and D_u. The rest of property (2) follows from the relation

$$(W_u((1 - \bar{w}A^*)^{-1}C^*u + (1 - \bar{w}A)^{-1}Bv))(z) = D_S(w, z)\begin{pmatrix} u \\ v \end{pmatrix}, \quad u, v \in \mathfrak{G};$$

see also [2, Theorem 2.1.2]. □

We now ask for conditions under which the function S constructed in Theorem 4.2 satisfies the symmetry condition (1.3). By Theorem 3.2 we have to produce a signature operator $J = J_{\mathfrak{H}(S)}$ on $\mathfrak{H}(S)$ and a unitary operator $X \in \mathfrak{L}(\mathfrak{G})$ such that (i) $T^*J = JT$, (ii) $XG = F^*J$, and (iii) $XH = H^*X^*$. We express these conditions in terms of the equivalent signature operator

$$J_{\mathfrak{K}} = W^{-1}J_{\mathfrak{H}(S)}W. \tag{4.8}$$

Formula (4.4) implies that (i)–(iii) hold if and only if

(i') $A^*J_{\mathfrak{K}} = J_{\mathfrak{K}}A$,

(ii') $C^*X^* = J_{\mathfrak{K}}(\bar{\gamma} - A)(1 - \bar{\gamma}A^*)^{-1}C^* = J_{\mathfrak{K}}B$,

(iii') $X\big[1_{\mathfrak{G}} - C(1 - \bar{\gamma}A^*)^{-1}C^*\big] = \big[1_{\mathfrak{G}} - C(1 - \gamma A)^{-1}C^*\big]X^*$, or equivalently, $XD = D^*X^*$.

Similarly, according to Theorem 3.1 and in the notation of Theorem 4.2, we also have to produce a signature operator $J = J_{\mathfrak{D}(S)}$ on $\mathfrak{D}(S)$ and a unitary operator $X \in \mathfrak{L}(\mathfrak{G})$ such that $A_u^*J = JA_u$, $XC_u = B_u^*J$, and $XD_u = D_u^*X^*$. Formula (4.6) implies that these conditions hold if and only if the conditions (i')–(iii') are valid, where now

$$J_{\mathfrak{K}} = W_u^{-1}J_{\mathfrak{D}(S)}W_u. \tag{4.9}$$

Thus assuming the symmetry condition on $S(z)$ in Theorem 4.2, the signature operator $J_{\mathfrak{K}}$ in (4.8) as well as in (4.9) is uniquely determined by the properties (i')–(iii') together with (iv') $J_{\mathfrak{K}}^2 = 1_{\mathfrak{K}}$ and (v') $J_{\mathfrak{K}} = J_{\mathfrak{K}}^*$. To calculate this operator, we multiply $A^*A + C^*C = 1_\Re$ (the first equation in (4.2)) by $J_\Re$ and use (i'), (iv') and (ii'), and we obtain the equation

$$J_\Re - AJ_\Re A = BXC =: Y.$$

By [6, Theorem 3.2 on p. 23, with $A = B$ and $P(\lambda, \mu) = 1 - \lambda\mu$], this equation has the unique solution

$$
\begin{aligned}
J_\Re &= -\frac{1}{4\pi^2} \int_{\Gamma_A} \int_{\Gamma_A} \frac{(A - \lambda)^{-1} Y (A - \mu)^{-1}}{1 - \lambda\mu} d\lambda d\mu \\
&= -\frac{1}{2\pi i} \int_{\Gamma_A} (A - \lambda)^{-1} Y (1 - \lambda A)^{-1} d\lambda,
\end{aligned}
\tag{4.10}
$$

where Γ_A is a suitable contour which surrounds the spectrum $\sigma(A)$ of A for which the expression in the integrals are defined. Such a contour exists if $1 - \lambda\mu \neq 0$ for all points λ and $\mu \in \sigma(A)$.

In the next two examples A and C act on finite dimensional spaces and the provision on A holds: $\lambda\mu \neq 1$ for any two eigenvalues λ and μ of A. In both examples we calculate the corresponding $J_\Re$.

Example 4.3. We apply Theorem 4.2 to the case where $\Re = \mathbb{C}^n$ and $\mathfrak{G} = \mathbb{C}$ in their standard Euclidean metric and with operators replaced by matrices. We take for A the shift and for C the transpose of the first unit vector:

$$
A = \begin{pmatrix}
0 & 1 & 0 & \cdots & 0 \\
0 & 0 & 1 & \cdots & 0 \\
 & & \cdots & & \\
0 & 0 & 0 & \cdots & 1 \\
0 & 0 & 0 & \cdots & 0
\end{pmatrix}, \qquad
C = \begin{pmatrix} 1 & 0 & 0 & \cdots & 0 & 0 \end{pmatrix}.
$$

Then (4.2) is satisfied and with arbitrary $\gamma \in \mathbb{T}$, the unit circle, we find $S(z) = \bar\gamma^n z^n$. Hence condition (1.3) is satisfied with $X = \gamma^n$. We calculate the corresponding signature operator $J_\Re$. First notice that

$$
Y = BXC = \begin{pmatrix}
0 & 0 & 0 & \cdots & 0 \\
0 & 0 & 0 & \cdots & 0 \\
 & & \cdots & & \\
0 & 0 & 0 & \cdots & 0 \\
1 & 0 & 0 & \cdots & 0
\end{pmatrix}
$$

(so Y is independent of γ). We find that the i, j-th entry in the $n \times n$ matrix $(A - \lambda)^{-1} Y (1 - \lambda A)^{-1}$ is given by

$$
((A - \lambda)^{-1} Y (1 - \lambda A)^{-1})_{ij} = -\frac{1}{\lambda^{n+2-i-j}}, \qquad i, j = 1, 2, \ldots, n.
$$

The integral representation (4.10) for $J_{\mathfrak{K}}$ readily yields

$$
J_{\mathfrak{K}} = \begin{pmatrix} 0 & 0 & 0 & \cdots & 1 \\ 0 & 0 & 0 & \cdots & 0 \\ & & \cdots & & \\ 0 & 1 & 0 & \cdots & 0 \\ 1 & 0 & 0 & \cdots & 0 \end{pmatrix}.
$$

We apply the last part of Theorem 4.2 to find the signature operators $J_{\mathfrak{H}(S)}$ on $\mathfrak{H}(S)$ and $J_{\mathfrak{D}(S)}$ on $\mathfrak{D}(S)$. From

$$
(Wk)(z) = C(1 - zA)^{-1} \begin{pmatrix} k_1 \\ k_2 \\ \vdots \\ k_n \end{pmatrix} = k_1 + k_2 z + \cdots + k_n z^{n-1}, \quad k = \begin{pmatrix} k_1 \\ k_2 \\ \vdots \\ k_n \end{pmatrix} \in \mathfrak{K},
$$

and since W is an isomorphism from $\mathfrak{K}$ onto $\mathfrak{H}(S)$, we find that $\mathfrak{H}(S)$ is the linear space of polynomials $p(z)$ of degree $< n$ in the inner product of the Hardy space H^2, and the elements $1, z, \ldots, z^{n-1}$ form an orthonormal basis of $\mathfrak{H}(S)$. Evidently this can also be obtained directly from the definition of the space $\mathfrak{H}(S)$ as a reproducing kernel Hilbert space (or other well known equivalent characterizations of this space). The isomorphic copy $J_{\mathfrak{H}(S)}$ on $\mathfrak{H}(S)$ of $J_{\mathfrak{K}}$ under W follows from (4.8):

$$
J_{\mathfrak{H}(S)} p(z) = z^{n-1} p(1/z).
$$

Similarly, either via the definition of W_u or via the definition of the space $\mathfrak{D}(S)$ as a reproducing kernel space, we obtain that

$$
\mathfrak{D}(S) = \left\{ \begin{pmatrix} p(z) \\ \gamma^n z^{n-1} p(1/z) \end{pmatrix} : p(z) \text{ is a polynomial of degree } < n \right\}
$$

in the inner product that makes W_u an isomorphism, and the elements

$$
\begin{pmatrix} 1 \\ \gamma^n z^{n-1} \end{pmatrix}, \begin{pmatrix} z \\ \gamma^n z^{n-2} \end{pmatrix}, \ldots, \begin{pmatrix} z^j \\ \gamma^n z^{n-j-1} \end{pmatrix}, \ldots, \begin{pmatrix} z^{n-1} \\ \gamma^n \end{pmatrix}
$$

form an orthonormal basis of $\mathfrak{D}(S)$. By (4.9), the isomorphic copy $J_{\mathfrak{D}(S)}$ on $\mathfrak{D}(S)$ of $J_{\mathfrak{K}}$ under W_u is given by

$$
J_{\mathfrak{D}(S)} \begin{pmatrix} p(z) \\ \gamma^n z^{n-1} p(1/z) \end{pmatrix} = \begin{pmatrix} z^{n-1} p(1/z) \\ \gamma^n p(z) \end{pmatrix},
$$

which is in agreement with the formula for $J_{\mathfrak{D}(S)}$ given in Theorem 3.6.

Example 4.4. Let $w_1, w_2, \ldots, w_n$ be n distinct real points such that $w_i w_j \neq 1$ for all $i, j = 1, 2, \ldots, n$, and denote by $\mathbb{P}$ the $n \times n$ matrix

$$
\mathbb{P} = \left(\frac{1}{1 - w_i w_j} \right)_{i,j=1}^n.
$$

We apply Theorem 4.2 to the case where $K = \mathbb{C}^n$ provided with the inner product $\langle x, y \rangle_{\mathfrak{K}} := (\mathbb{P}x, y)$ in which $(\cdot, \cdot)$ is the standard Euclidean inner product and $\mathfrak{G} = \mathbb{C}$

equipped with the standard Euclidean inner product. As in the previous example we replace operators by matrices and take

$$A = \operatorname{diag}(w_1, w_2, \ldots, w_n), \quad C = (1, 1, \ldots, 1).$$

Then (4.2) is satisfied:

$$A^*A + C^*C = \mathbb{P}^{-1}(A\mathbb{P}A + C^T C) = 1,$$

where C^T stands for the transpose of C, and the intersection in (4.2) is trivial because the $n \times n$ Vandermonde matrix $(w_j^i)_{i,j=0}^{n-1}$ is invertible. Let $\gamma \in \mathbb{T}$ be arbitrary. By (4.3),

$$S(z) = 1 - (1 - z\bar\gamma)C(1 - zA)^{-1}\mathbb{P}^{-1}(1 - \bar\gamma A)^{-1}C^T$$

$$= 1 - (1 - z\bar\gamma)\left(\frac{1}{1 - zw_1} \quad \frac{1}{1 - zw_2} \quad \cdots \quad \frac{1}{1 - zw_n}\right)\mathbb{P}^{-1}\begin{pmatrix}\dfrac{1}{1 - \bar\gamma w_1}\\[4pt]\dfrac{1}{1 - \bar\gamma w_2}\\[2pt]\vdots\\[2pt]\dfrac{1}{1 - \bar\gamma w_n}\end{pmatrix}$$

$$= \frac{p(z)}{(1 - zw_1)(1 - zw_2)\cdots(1 - zw_n)},$$

in which $p(z)$ is a polynomial of degree at most n. From $S(w_i) = 0$, $i = 1, 2, \ldots, n$, and $S(\gamma) = 1$ we infer

$$S(z) = \varepsilon\prod_{i=1}^n \frac{z - w_i}{1 - zw_i}, \qquad \varepsilon = \prod_{i=1}^n \frac{1 - \gamma w_i}{\gamma - w_i},$$

and so, since $|\varepsilon| = 1$, (1.3) holds with $X = \bar\varepsilon$. We determine B:

$$\begin{aligned}B &= (\bar\gamma - A)(1 - \bar\gamma\mathbb{P}^{-1}A\mathbb{P})^{-1}\mathbb{P}^{-1}C^T\\[4pt]&= (\bar\gamma - A)\mathbb{P}^{-1}(1 - \bar\gamma A)^{-1}C^T\\[4pt]&= (\bar\gamma - A)\mathbb{P}^{-1}\begin{pmatrix}\dfrac{1}{1 - \bar\gamma w_1}\\[4pt]\dfrac{1}{1 - \bar\gamma w_2}\\[2pt]\vdots\\[2pt]\dfrac{1}{1 - \bar\gamma w_n}\end{pmatrix}\\[4pt]&= (\bar\gamma - A)\left(\frac{q_j(\bar\gamma)}{(1 - \bar\gamma w_1)(1 - \bar\gamma w_2)\cdots(1 - \bar\gamma w_n)}\right)_{j=1}^n,\end{aligned}$$

where $q_j(z)$ is a polynomial of degree $< n$ such that for $i, j = 1, 2, \ldots, n$,

$$q_j(w_i) = 0, \quad i \neq j; \quad q_j(w_j) = (1 - w_j w_1)(1 - w_j w_2)\ldots(1 - w_j w_n).$$

It follows that

$$q_j(z) = \frac{\Pi_{i=1}^n(1 - w_j w_i)}{\Pi_{i=1,i\neq j}^n(w_j - w_i)}\Pi_{i=1,i\neq j}^n(z - w_i)$$

$$= \frac{\varepsilon}{S'(w_j)}\Pi_{i=1,i\neq j}^n(z - w_i)$$

and

$$\frac{q_j(z)}{\Pi_{i=1}^n(1 - z w_i)} = \frac{S(z)}{(z - w_j)S'(w_j)}, \quad j = 1, 2, \ldots, n. \tag{4.11}$$

Thus, as $S(\bar{\gamma}) = \varepsilon^2$,

$$B = \varepsilon^2 \mathrm{diag}\left(\frac{1}{S'(w_1)}, \frac{1}{S'(w_2)}, \ldots, \frac{1}{S'(w_n)}\right)C^T$$

and

$$Y = BXC = \varepsilon\,\mathrm{diag}\left(\frac{1}{S'(w_1)}, \frac{1}{S'(w_2)}, \ldots, \frac{1}{S'(w_n)}\right)C^T C.$$

Since $C^T C$ is the $n \times n$ matrix with all entries equal to 1, $Y_{ij} = \dfrac{\varepsilon}{S'(w_i)}$ and

$$((A - \lambda)^{-1}Y(1 - \lambda A)^{-1})_{ij} = \frac{Y_{ij}}{(w_i - \lambda)(1 - \lambda w_j)}, \quad i, j = 1, 2, \ldots, n.$$

The integral formula (4.10) for $J_\mathfrak{K}$ gives

$$(J_\mathfrak{K})_{ij} = -\frac{1}{2\pi i}\int_{\Gamma_A}\frac{Y_{ij}}{(w_i - \lambda)(1 - \lambda w_j)}\,d\lambda = \frac{Y_{ij}}{1 - w_i w_j}.$$

Hence

$$J_\mathfrak{K} = \left(\frac{\varepsilon}{(1 - w_i w_j)S'(w_i)}\right)_{i,j=1}^n$$

$$= \varepsilon\,\mathrm{diag}\left(\frac{1}{S'(w_1)}, \frac{1}{S'(w_2)}, \ldots, \frac{1}{S'(w_n)}\right)\mathbb{P}.$$

In the setting of the reproducing kernel space $\mathfrak{H}(S)$ the signature operator $J_{\mathfrak{H}(S)}$ in (4.8) also takes a simple form. To see this, first note that the functions

$$\frac{1}{1 - z w_1}, \frac{1}{1 - z w_2}, \ldots, \frac{1}{1 - z w_n}$$

form an algebraic basis of $\mathfrak{H}(S)$. With W defined by (4.5) and using $J_\mathfrak{K} = J_\mathfrak{K}^{-1}$ and (4.11) we find that for $u_j \in \mathbb{C}$, $j = 1, 2, \ldots, n$,

$$J_{\mathfrak{H}(S)}\sum_{j=1}^n\frac{u_j}{1 - z w_j} = WJ_\mathfrak{K}W^{-1}\sum_{j=1}^n\frac{u_j}{1 - z w_j}$$

$$= WJ_\mathfrak{K}^{-1}\begin{pmatrix}u_1\\u_2\\\vdots\\u_n\end{pmatrix}$$

$$
= \ \bar{\varepsilon}\,(\mathbb{P}^{-1}(1-zA)^{-1}C^T)^T
\begin{pmatrix}
u_1 S'(w_1) \\
u_2 S'(w_2) \\
\vdots \\
u_n S'(w_n)
\end{pmatrix}
$$

$$
= \ \bar{\varepsilon} \sum_{j=1}^{n} \frac{q_j(z)}{\Pi_{i=1}^{n}(1-zw_i)} u_j S'(w_j)
$$

$$
= \ \bar{\varepsilon}\, S(z) \sum_{j=1}^{n} \frac{u_j S'(w_j)}{(z-w_j)S'(w_j)}
$$

$$
= \ \bar{\varepsilon}\, S(z) \sum_{j=1}^{n} \frac{u_j}{z-w_j},
$$

that is, $J_{\mathfrak{H}(S)} \dfrac{1}{1-zw_j} = \bar{\varepsilon}\, \dfrac{S(z)}{z-w_j}$, $j = 1, 2, \ldots, n$. This result also follows from Corollary 3.7, which implies that $J_{\mathfrak{K}} = \varepsilon\Lambda$, where by (2.1) and since $S(w_j) = 0$ and $\tilde{S}(z) = \bar{\varepsilon}^2 S(z)$ (on account of (1.3)), $\Lambda \dfrac{1}{1-zw_j} = \bar{\varepsilon}^2 \dfrac{S(z)}{z-w_j}$, $j = 1, 2, \ldots, n$. Similar calculations show that

$$
\mathfrak{D}(S) = \{W_u u : u \in \mathfrak{K}\}
$$

in the inner product that makes

$$
W_u u =
\begin{pmatrix}
\sum_{j=1}^{n} \dfrac{u_j}{1-zw_j} \\
\varepsilon \sum_{j=1}^{n} \dfrac{(\mathbb{P}u)_j}{S'(w_j)(1-zw_j)}
\end{pmatrix},
\quad u = (u_j)_{j=1}^{n} \in \mathfrak{K},
$$

an isomorphism from $\mathfrak{K}$ onto $\mathfrak{D}(S)$, and

$$
J_{\mathfrak{D}(S)} = \begin{pmatrix} 0 & \varepsilon \\ \bar{\varepsilon} & 0 \end{pmatrix}.
$$

The equality for $J_{\mathfrak{D}}$ also follows from Theorem 3.6.

In the remaining part of this section we study the hypotheses (4.2) in Theorem 4.2 for operators A and C acting in infinite dimensional spaces in more detail. Theorem 4.2 depends on the inner product of $\mathfrak{K}$. Sometimes this inner product can be chosen so that the hypotheses of the theorem are satisfied. We give conditions on A and C which assure that this can be done. It is convenient to work with an equivalent form of the problem, namely the operator equation (4.12) below; an apropriate inner product exists whenever this equation has an invertible selfadjoint solution P. By $\rho_\infty(A)$ we denote the connected component of $\rho(A)$ which contains the point ∞.

Theorem 4.5. *Let $\mathfrak{H}$ and $\mathfrak{G}$ be Hilbert spaces. Assume $A \in \mathfrak{L}(\mathfrak{H})$ and $1-\lambda\bar\mu \neq 0$ for all $\lambda, \mu \in \sigma(A)$. Then for every $C \in \mathfrak{L}(\mathfrak{H}, \mathfrak{G})$, there is a unique solution $P \in \mathfrak{L}(\mathfrak{H})$*

of the equation

$$P - A^*PA = C^*C, \tag{4.12}$$

P is selfadjoint, and

$$\bigcap_{j=0}^{\infty} \ker CA^j \subset \ker P. \tag{4.13}$$

If $1 \in \rho_\infty(A)$, then

$$\bigcap_{j=0}^{\infty} \ker CA^j = \ker P. \tag{4.14}$$

Proof. The existence and uniqueness of P follow from [6, p. 23]:

$$P = -\frac{1}{4\pi^2} \int_{\Gamma_{A^*}} \int_{\Gamma_A} \frac{(A^* - \lambda)^{-1}C^*C(A - \mu)^{-1}}{1 - \lambda\mu} \, d\lambda d\mu, \tag{4.15}$$

where $\Gamma_{A^*} \subset \rho(A^*)$ and $\Gamma_A \subset \rho(A)$ are Jordan contours around $\sigma(A^*)$ and $\sigma(A)$, respectively. The solution P is selfadjoint because P^* is also a solution of (4.12). Since for large $|\mu|$ we have

$$C(A - \mu)^{-1} = -\sum_{j=0}^{\infty} \frac{1}{\mu^{j+1}} CA^j,$$

(4.13) holds.

We now assume that $1 \in \rho_\infty(A)$ and show that (4.14) holds. Denote the Hilbert space inner product on $\mathfrak{H}$ by $(\cdot, \cdot)$. Assume $\ker P \neq \{0\}$. Then it is the isotropic part of $\mathfrak{H}$ equipped with the P-inner product $\langle \cdot, \cdot \rangle := (P\cdot, \cdot)$. From (4.12) it follows that A is a P-contraction: $\langle Ax, Ax \rangle \leq \langle x, x \rangle$. Hence the operator

$$S = (A + 1_{\mathfrak{H}})(A - 1_{\mathfrak{H}})^{-1}$$

is P-dissipative: $\operatorname{Re} \langle Sx, x \rangle \leq 0$.

Let $x \in \ker P$ and $y \in \mathfrak{H}$. Since $[u, v] := \operatorname{Re} \langle Su, v \rangle + \operatorname{Re} \langle Sv, u \rangle$ is a nonpositive inner product on the real span of x, y, we have

$$[x, y]^2 \leq [x, x][y, y]$$

and from $[x, x] = 0$ it follows that $[x, y] = 0$, that is, $\operatorname{Re} \langle Sx, y \rangle + \operatorname{Re} \langle Sy, x \rangle = 0$. Since $x \in \ker P$, we have $\langle Sy, x \rangle = 0$ and hence $\operatorname{Re} \langle Sx, y \rangle = 0$. If we replace y by iy we obtain $\operatorname{Im} \langle Sx, y \rangle = 0$, and $\langle Sx, y \rangle = 0$ for arbitrary $y \in \mathfrak{H}$. Therefore $S \ker P \subset \ker P$. The assumption $1 \in \rho_\infty(A)$ implies $1 \in \rho_\infty(S)$. Indeed, by construction $1 \in \rho(S)$ and because $\rho_\infty(A)$ is connected and open, there is a continuous function $\lambda : [0, 1) \to \mathbb{C}$ such that $\lambda([0, 1)) \subset \rho_\infty(A)$, $\lambda(0) = 1$, and $\lambda(t) \to \infty$ as $t \to 1$. The function $\mu : (0, 1) \to \rho(S)$ with

$$\mu(t) := (\lambda(t) + 1)(\lambda(t) - 1)^{-1}, \quad t \in (0, 1),$$

has the properties: $\mu(t) \to 1$ as $t \to 1$ and $\mu(t) \to \infty$ as $t \to 0$. Consequently, $1 \in \rho_\infty(S)$. Now we observe that for each $t \in (0, 1]$, $\mu(t)$ is a point of regular type of $S|_{\ker P}$ [1] (we set $\mu(1) = 1$). Since $\mu(t) \to \infty$ as $t \to 0$ and for $|\mu(t)|$ large enough we have $\mu(t) \in \rho(S|_{\ker P})$, $\mu((0,1]) \subset \rho(S|_{\ker P})$; in particular, $1 \in \rho(S|_{\ker P})$. The identity $A = (S + 1_{\mathfrak{H}})(S - 1_{\mathfrak{H}})^{-1}$ implies

$$A \ker P \subset \ker P. \tag{4.16}$$

Hence by (4.12), we have $Cx = 0$, $x \in \ker P$, and by (4.16), $\ker P \subset \bigcap_{j=0}^{\infty} \ker C A^j$. Together with (4.13) this implies (4.14). □

Examples 4.6 and 4.7 show what can go wrong when the hypotheses of Theorem 4.5 are not satisfied.

Example 4.6. Consider the sequence spaces $\mathfrak{H} = \mathfrak{L}l^2(-\infty, \infty)$, $\mathfrak{H}_0 = \mathfrak{L}l^2(-\infty, 0)$, and $\mathfrak{H}_1 = \mathfrak{L}l^2(1, \infty)$; here, for example, the space in the middle is the space of square summable sequences of complex numbers x_n where the index n runs to the left from 0 to $-\infty$. Define A on $\mathfrak{H}$ by $A = 2U$, where U is the shift operator to the right, and define C on $\mathfrak{H} = \mathfrak{H}_0 \oplus \mathfrak{H}_1$ by $C = \operatorname{diag}(2Q_0, \sqrt{3})$, where Q_0 is the orthogonal projection onto $\mathfrak{H}_0 \ominus \mathfrak{L}l^2(-\infty, -1)$; notice that $\operatorname{ran} U^*|_{\mathfrak{H}_0} = \mathfrak{L}l^2(-\infty, -1)$. Then the unique solution of (4.12) is $P = \operatorname{diag}\{0, -1_{\mathfrak{H}_1}\}$. In this case $\ker P = \mathfrak{H}_0$ and $\bigcap_{j=0}^{\infty} \ker C A^j = \{0\}$.

Example 4.7. Let $\mathfrak{H} = \mathbb{C}^2$ and consider

$$A = \begin{pmatrix} 0 & i \\ -i & 0 \end{pmatrix}.$$

If $C = 0$ then $P = \pm 1_{\mathfrak{H}}$ are solutions of (4.12), so in this case there is no uniqueness. If $C = (\ 1 \quad 1\ )$ then (4.12) has no solutions; in this case we have $\ker C \cap \ker C A = \{0\}$.

Corollary 4.8. *Let $\mathfrak{H}$ and $\mathfrak{G}$ be Hilbert spaces. Assume $A \in \mathfrak{L}(\mathfrak{H})$ is compact and $1 - \lambda\bar{\mu} \neq 0$ for all $\lambda, \mu \in \sigma(A)$, and let $C \in \mathfrak{L}(\mathfrak{H}, \mathfrak{G})$ be arbitrary. Then the solution $P \in \mathfrak{L}(\mathfrak{H})$ of (4.12) is boundedly invertible if and only if*

$$\text{(i)} \bigcap_{j=0}^{\infty} \ker C A^j = \{0\}, \quad \text{(ii)} \dim \ker C < \infty, \quad \text{and} \quad \text{(iii)} \operatorname{ran} C = \overline{\operatorname{ran}} C.$$

If these conditions hold then the space $\mathfrak{H}$ equipped with the P-inner product $(P\cdot, \cdot)$ is a Pontryagin space.

Proof. Assume $0 \in \rho(P)$. Then (i) holds by (4.13). By (4.12), C^*C is a compact perturbation of P, which implies (ii) and (iii). Assume that (i)–(iii) hold. Then $P = C^*C - A^*PA$ is a compact perturbation of C^*C, which is nonnegative and 0

---

[1] A complex number λ is called a **point of regular type** of the closed operator T on a Hilbert space if there is a positive number $c = c(\lambda)$ such that for all $x \in \operatorname{dom} T$, $\|(T - \lambda)x\| \geq c\|x\|$; for equivalent formulations, see, for example, [10, p. 37] and [4, Section 2.6, (6.14)].

is a normal point of C^*C [2]. Consequently, $0 \in \rho(P)$ and $\sigma(P) \cap (-\infty, 0)$ is a finite set which therefore consists of eigenvalues of P. This implies the last statement of the corollary. □

Corollary 4.9. *Let $\mathfrak{H}$ and $\mathfrak{G}$ be Krein spaces. Assume $A \in \mathfrak{L}(\mathfrak{H})$ and $1 - \lambda\bar{\mu} \neq 0$ for all $\lambda, \mu \in \sigma(A)$. Then for every $C \in \mathfrak{L}(\mathfrak{H}, \mathfrak{G})$ there is a unique solution $P \in \mathfrak{L}(\mathfrak{H})$ of the equation* (4.12)*, this P is selfadjoint, and the inclusion* (4.13) *holds. If $1 \in \rho_\infty(A)$ and the subspace $\overline{\operatorname{ran}} C$ is uniformly positive, then equality in* (4.13) *prevails.*

Proof. The first part of the corollary can be proved in the same way as Theorem 4.5 which concerns the Hilbert space case; we omit the details. The last statement follows from Theorem 4.5 by reducing the statement to a Hilbert space setting: Fix a fundamental symmetry $J_\mathfrak{H}$ on $\mathfrak{H}$ and a fundamental symmetry $J_\mathfrak{G}$ on $\mathfrak{G}$ such that $J_\mathfrak{G}|_{\overline{\operatorname{ran}} C}$ is the identity on $\overline{\operatorname{ran}} C$. Then (4.12) can be written as

$$P - J_\mathfrak{H} A^\times J_\mathfrak{H} P A = J_\mathfrak{H} C^\times J_\mathfrak{G} C,$$

where, for example, $C^\times$ stands for the adjoint of C in the Hilbert space inner products of $\mathfrak{H}$ and $\mathfrak{G}$ induced by the symmetries. Multiply this equation by $J_\mathfrak{H}$ from the left. Then

$$J_\mathfrak{H} P - A^\times (J_\mathfrak{H} P) A = C^\times C,$$

because $C^\times J_\mathfrak{G} C = C^\times C$. Now Theorem 4.5 can be applied. □

Corollary 4.10. *Let $\mathfrak{H}$ and $\mathfrak{G}$ be Krein spaces. Assume $A \in \mathfrak{L}(\mathfrak{H})$ and $1 - \lambda\bar{\mu} \neq 0$ for all $\lambda, \mu \in \sigma(A)$. Assume $C \in \mathfrak{L}(\mathfrak{H}, \mathfrak{G})$ has the properties: $\operatorname{ran} C = \overline{\operatorname{ran}} C$ is uniformly positive and $\dim \ker C = n < \infty$. Assume further that the unique solution $P \in \mathfrak{L}(\mathfrak{H})$ of* (4.12) *satisfies*

$$\ker P = \bigcap_{j=0}^{\infty} \ker C A^j.$$

Then P is boundedly invertible if and only if $\ker P = \{0\}$.

Proof. The existence and uniqueness and the selfadjointness of P follows from Corollary 4.9. The arguments in the proof of Corollary 4.9 show that we may as well assume that $\mathfrak{H}$ and $\mathfrak{G}$ are Hilbert spaces. The only if part is trivial. We prove the if part and assume $\ker P = \{0\}$.

Set $\mathfrak{L} = \bigcap_{j=0}^{\infty} \ker C A^j$ and $\mathfrak{L}_k = \bigcap_{j=0}^{k} \ker C A^j$, $k = 0, 1, 2, \dots$. By assumption, $\mathfrak{L} = \{0\}$. Clearly, $\mathfrak{L}_{k+1} \subset \mathfrak{L}_k$ and $\dim \mathfrak{L}_k \leq \dim \ker C = n < \infty$. We claim that

---

[2] A complex number λ is called a **normal point** of the closed linear operator T on a Hilbert space $\mathfrak{H}$ if λ belongs to $\rho(T)$ or is a normal eigenvalue. The latter means that there is a direct sum decomposition $\mathfrak{H} = \mathfrak{F} + \mathfrak{R}$ of $\mathfrak{H}$ into T-invariant subspaces $\mathfrak{F}$ and $\mathfrak{R}$ such that $\dim \mathfrak{F} < \infty$, $\sigma(T|_\mathfrak{F}) = \{\lambda\}$, and $\lambda \in \rho(T|_\mathfrak{R})$; for details, see [9, Section I.2] and also [4, Section 1.1] and [10, p. 37].

$\mathfrak{L}_n = \bigcap_{j=0}^n \ker C A^j = \{0\}$. When $n = 0$ this is obvious: $\mathfrak{L}_0 = \ker C = \{0\}$. If $n > 0$ this follows from $\mathfrak{L}_0 = \ker C \neq \{0\}$ and the implication

$$\mathfrak{L}_k \neq \{0\}, \quad k = 0, 1, \ldots, n-1 \Rightarrow \dim \mathfrak{L}_{k+1} < \dim \mathfrak{L}_k.$$

The implication can be proved by contradiction: if on the contrary $\mathfrak{L}_k = \mathfrak{L}_{k+1}$, then $\mathfrak{L}_k$ is A-invariant and $\mathfrak{L}_k \subset \mathfrak{L} = \{0\}$, which contradicts $\mathfrak{L}_k \neq \{0\}$.

Consider the operator

$$D = \sum_{j=0}^n A^{*j} C^* C A^j.$$

By assumption, D is nonnegative, and $\ker D = \mathfrak{L}_n = \{0\}$. We claim that D is a uniformly positive operator, that is, there is an $\alpha > 0$ such that for all $x \in \mathfrak{H}$, $(Dx, x) \geq \alpha(x, x)$. To prove the claim we first observe that since $\operatorname{ran} C$ is closed, by, for example, [5, Theorem VI.1.10], $\operatorname{ran} C^*$ is closed. From this and the decomposition $\mathfrak{H} = \ker C^* \oplus \operatorname{ran} C$ it follows that $\operatorname{ran} C^* C = \operatorname{ran} C^*$ is closed and hence $\mathfrak{H} = \ker C^* C \oplus \operatorname{ran} C^* C$. Assume D is not uniformly positive. Then there is a sequence $x_k = x_k^1 + x_k^2$, $x_k^1 \in \ker C^* C$, $x_k^2 \in \operatorname{ran} C^* C$, such that $\|x_k\| = 1$ and $Dx_k \to 0$ as $k \to \infty$, that is, for $j = 0, 1, \ldots, n$, $C A^j x_k \to 0$. Since $\ker C^* C = \ker C$ and $\dim \ker C < \infty$ we can assume without loss of generality that x_k^1 has a limit x_0, say. We distinguish two cases:

(i) $\|x_0\| \neq 1$. Then without loss of generality we can assume that the sequence x_k^2 is bounded and bounded away from 0. For $j = 0$ we have $C^* C x_k = C^* C x_k^2 \to 0$ and $C^* C|_{\operatorname{ran} C^* C}$ is boundedly invertible and hence $x_k^2 \to 0$, but this cannot be true as the sequence x_k^2 is bounded away from 0.

(ii) $\|x_0\| = 1$. Then $\|x_k^1\| \to 1$ and hence $x_k^2 \to 0$. Therefore

$$C A^j x_k \to C A^j x_0 = 0, \quad j = 0, 1, \ldots, n,$$

which implies $x_0 \in \mathfrak{L}_n = \{0\}$, contradicting $\|x_0\| = 1$.

Hence $D \geq \alpha$ for some $\alpha > 0$.

Now consider the operator $B = A^{n+1}$. By (4.12),

$$P - B^* P B = P - A^{*n+1} P A^{n+1} = \sum_{j=0}^n A^{*j} C^* C A^j = D \geq \alpha. \tag{4.17}$$

In particular, B is P-contractive. Because $\mathbb{T} \subset \rho(A)$, $\mathbb{T} \subset \rho(B)$, and this implies that there exist spectral B-invariant subspaces $\mathfrak{H}_+$ and $\mathfrak{H}_-$ of $\mathfrak{H}$ such that $\mathfrak{H} = \mathfrak{H}_+ + \mathfrak{H}_-$, direct sum, $|\sigma(B|_{\mathfrak{H}_+})| < 1$, $|\sigma(B|_{\mathfrak{H}_-})| > 1$, and the corresponding Riesz projections $x \mapsto x^\pm$, $x = x^+ + x^-$, $x^\pm \in \mathfrak{H}_\pm$, are bounded on $\mathfrak{H}$. We show that $\mathfrak{H}_+$ and $\mathfrak{H}_-$ are uniformly P-positive and uniformly P-negative, respectively. Denote the P-inner product by $\langle x, y \rangle = (Px, y)$.

(a) Let $x \in \mathfrak{H}_+$. Then $B^k x \in \mathfrak{H}_+$ and $|\sigma(B|_{\mathfrak{H}_+})| < 1$ implies $\langle x, x \rangle \geq \langle B^k x, B^k x \rangle \to 0$. Hence $\mathfrak{H}_+$ is a P-nonnegative subspace. From (4.17) it now follows that

$$\langle x, x \rangle \geq \langle x, x \rangle - \langle Bx, Bx \rangle \geq \alpha(x, x),$$

that is, $\mathfrak{H}_+$ is uniformly P-positive.

(b) Let $x \in \mathfrak{H}_-$. Consider the operator $B_1 = (B|_{\mathfrak{H}_-})^{-1}$. Then $B_1^k x \in \mathfrak{H}_-$ and $|\sigma(B_1|_{\mathfrak{H}_-})| < 1$ implies $\langle x, x \rangle \leq \langle B_1^k x, B_1^k x \rangle \to 0$. Hence $\mathfrak{H}_-$ is a P-nonpositive subspace. From (4.17) it follows that with $y = B_1 x$

$$-\langle x, x \rangle \geq \langle y, y \rangle - \langle By, By \rangle \geq \alpha(y, y) \geq \frac{\alpha}{\|B_1^{-1}\|^2}(x, x),$$

that is, $\mathfrak{H}_-$ is uniformly P-negative.

Finally, we show that $0 \in \rho(P)$. Assume that this is not the case. Then there is a sequence $x_k = x_k^+ + x_k^-$, $x_k^\pm \in \mathfrak{H}_\pm$, such that $\|x_k\| = 1$ and $Px_k \to 0$. As the Riesz projections $x \mapsto x^\pm$ are bounded operators, the sequences $\langle x_k^+, x_k^+ \rangle$ and $\langle x_k^+, x_k^- \rangle$ are bounded, and therefore we can assume without loss of generality that they have limits $\alpha^\pm$ and β, say. Then

$$\langle x_k, x_k^+ \rangle = \langle x_k^+, x_k^+ \rangle + \langle x_k^-, x_k^+ \rangle \to \alpha^+ + \bar{\beta},$$
$$\langle x_k, x_k^- \rangle = \langle x_k^+, x_k^- \rangle + \langle x_k^-, x_k^- \rangle \to \beta + \alpha^-$$

and since these limits are zero, $\beta = -\alpha^+ = -\alpha^-$. From $\alpha^+ \geq 0$ and $\alpha^- \leq 0$, it follows that $\beta = \alpha^+ = \alpha^- = 0$. Thus $\langle x_k^\pm, x_k^\pm \rangle \to 0$, which implies $\|x_k^\pm\| \to 0$ and this contradicts the assumption that $\|x_k\| = 1$. Conclusion: P is boundedly invertible. $\qquad\square$

Acknowledgements. T. Ya. Azizov was supported by the Netherlands Organization for Scientific Research NWO (NB 61–432 and 047-008-008) and by the Russian Foundation for Basic Research RFBR 99–01–00391. A. Dijksma was supported by a Harry T. Dozor fellowship at the Ben-Gurion University of the Negev, Beer-Sheva, Israel. J. Rovnyak was supported by the National Science Foundation grant DMS–9801016.

References

[1] D. Alpay, J. A. Ball, I. Gohberg, and L. Rodman, *Realization and factorization for rational matrix functions with symmetries*, Extension and interpolation of linear operators and matrix functions, Oper. Theory Adv. Appl., vol. 47, Birkhäuser, Basel, 1990, pp. 1–60.

[2] D. Alpay, A. Dijksma, J. Rovnyak, and H. S. V. de Snoo, *Schur functions, operator colligations, and reproducing kernel Pontryagin spaces*, Oper. Theory Adv. Appl., vol. 96, Birkhäuser, Basel, 1997.

[3] D. Alpay and H. Dym, *On a new class of reproducing kernel spaces and a new generalization of the Iohvidov laws*, Linear Algebra Appl. **178** (1993), 109–183.

[4] T. Ya. Azizov and I. S. Iokhvidov, *Foundations of the theory of linear operators in spaces with an indefinite metric*, "Nauka", Moscow, 1986; English transl.: Wiley, New York, 1989.

[5] J. B. Conway, *A course in functional analysis*, Graduate Texts in Mathematics, vol. 96, Springer-Verlag, New York, Berlin, 1985.

[6] Ju. L. Daleckii and M. G. Krein *Stability of solutions of differential equations in Banach space*, Translations Amer. Math. Soc., vol. 43, Amer. Math. Soc., Providence RI, 1974.

[7] L. de Branges, *The expansion theorem for Hilbert spaces of entire functions*, Entire Functions and Related Parts of Analysis (Proc. Sympos. Pure Math., La Jolla, Calif., 1966), Amer. Math. Soc., Providence, R.I., 1968, pp. 79–148.

[8] P. A. Fuhrmann, *On J-symmetric restricted shifts*, Proc. Amer. Math. Soc. **51** (1975), 421–426.

[9] I. Gohberg and M. G. Krein, *Introduction to the theory of linear nonselfadjoint operators*, "Nauka", Moscow, 1965; English transl.: Amer. Math. Soc., Providence RI, fourth printing, 1988.

[10] I. S. Iokhvidov, M. G. Krein, and H. Langer, *Introduction to the spectral theory of operators in spaces with an indefinite metric*, Mathematical Research, vol. 9, Akademie-Verlag, Berlin, 1982.

[11] A. Lubin, *J-symmetric canonical models*, Acta Sci. Math. (Szeged) **38** (1976), no. 1-2, 121–126.

Department of Mathematics
Ben-Gurion University of the Negev
P. O. Box 653
84105 Beer-Sheva, Israel

Department of Mathematics
Voronezh State University
394693 Voronezh, Russia

Department of Mathematics
University of Groningen
P. O. Box 800
9700 AV Groningen, The Netherlands

Department of Mathematics
University of Virginia
Charlottesville, Virginia 22903-3199
USA

1991 Mathematics Subject Classification.
Primary 47A48, 47A50; Secondary 47A45, 47A62

Received May 29, 2000

Operator Theory:
Advances and Applications, Vol. 130, 83–106

Logarithmic Residues of Fredholm Operator Valued Functions and Sums of Finite Rank Projections

Harm Bart, Torsten Ehrhardt and Bernd Silbermann

Dedicated to Peter Lancaster on the occasion of his seventieth birthday

Abstract. Logarithmic residues of analytic Fredholm operator valued functions are identified as sums of finite rank projections. The set of all such sums is closed and the restriction of the trace to it is continuous; its connected components are determined by the (integer) values of the trace. Two finite rank bounded linear operators can be written as the left and right logarithmic residues of a single Fredholm operator valued function if and only if they belong to the same connected component, i.e., if and only if they are sums of finite rank projections having the same trace.

1. Introduction

Let B be a complex Banach algebra with unit element e. A logarithmic residue in B is a contour integral of a logarithmic derivative of an analytic B-valued function F. There is a left version and there is a right version of this notion. The left version corresponds to the left logarithmic derivative $F'(\lambda)F(\lambda)^{-1}$, the right version to right logarithmic derivative $F(\lambda)^{-1}F'(\lambda)$.

The first to consider integrals of this type – in a vector valued context – was L. Mittenthal [M]. His goal was to generalize the spectral theory of a single Banach algebra element (i.e. the case where $F(\lambda) = \lambda e - b$, $b \in B$). He gave sufficient conditions for a logarithmic residue to be an idempotent. The conditions are very restrictive.

Logarithmic residues also appear in the paper [GS1] by I.C. Gohberg and E.I. Sigal. The setting there is that $B = \mathcal{B}(X)$, the Banach algebra of all bounded operators on a complex Banach space, and F is Fredholm operator valued. For such functions Gohberg and Sigal introduced the concept of algebraic multiplicity. It turns out that the algebraic multiplicity of F with respect to a given contour is equal to the trace of the corresponding (left/right) logarithmic residues (see also [GKL] and [GGK]).

Further progress was made in [BES2-4]. In these papers, logarithmic residues are studied from different angles and perspectives. The problems dealt with are of the following type.

1. *If a logarithmic residue vanishes, does it follows that F takes invertible values inside the (integration) contour?* This question was first posed in [B1]. The answer turns out to depend very much on the underlying Banach algebra. For certain important classes it is positive, for other (equally relevant) classes it is negative.

2. *What kind of elements are logarithmic residues?* Here a strong connection with (sums of) idempotents appears (cf. also [BES1]). As for the problem posed under 1, the answer depends on the Banach algebra under consideration too.

3. *How about left versus right logarithmic residues?* In all situations where a definite answer could be obtained, the set of left logarithmic residues coincides with the set of right logarithmic residues. In some situations it was possible to identify the pairs of left and right logarithmic residues associated with one single function F (and the same integration contour).

4. *What can be said about the topological properties of the set of logarithmic residues?* In some cases it was possible, for instance, to identify the connected components of this set.

The papers [BES2] and [BES4] touch briefly on the situation where F is Fredholm operator valued and the underlying Banach space X is possibly infinite dimensional. Here we undertake a more detailed and systematic investigation of this case. An outline of the present paper reads as follows.

Section 2 is of a preliminary nature. It mainly consists of definitions and notations.

In Section 3 we discuss the problem described under 1 and 2 above. The results can already be found in [BES2] and [BES4]. There the proofs employ ideas from Systems Theory; here a factorization technique inspired by [GS1] and [T] is used. The new proofs are more transparent than the ones in [BES2] and [BES4].

Theorem 3.4 tells us that the sets of left, respectively right, logarithmic residues of Fredholm operator valued functions both coincide with that of the sums of finite rank projections (= finite rank idempotent operators). We denote these (coinciding) sets by $\mathcal{P}(X)$, where X is the underlying Banach space. Section 4 studies the topological properties of $\mathcal{P}(X)$. One of the results obtained is that $\mathcal{P}(X)$ is closed and that the trace – regardless of whether X has finite or infinite dimension – is continuous on $\mathcal{P}(X)$. This appears as a corollary to a more general theorem about (norm) closed sets of finite rank operators on which the trace is a continuous function. The sets in question are determined by rank/trace conditions. We also describe the connected components of $\mathcal{P}(X)$.

Section 5 deals with the issue of left versus right logarithmic residues. In particular, it identifies the pairs of operators that can be written as the left and right logarithmic residue of a single Fredholm operator valued function (associated with the same integration contour). As it turns out, there is an intriguing connection with the description of the connected components of $\mathcal{P}(X)$ given in Section 4. The results generalize those obtained for the matrix case in [BES4].

In (the very short) Section 6, we briefly indicate the topics that will be treated in two forthcoming publications.

2. Preliminaries

Throughout this section, B will be a complex Banach algebra with unit element e. If F is a B-valued function with domain Δ, then F^{-1} stands for the function given by $F^{-1}(\lambda) = F(\lambda)^{-1}$ with domain the set of all $\lambda \in \Delta$ such that $F(\lambda)$ is invertible. If Δ is an open subset of the complex plane $\mathbb{C}$ and $F : \Delta \to B$ is analytic, then so is F^{-1} on its domain. The derivative of F will be denoted by F'. The *left*, respectively *right, logarithmic derivative* of F is the function given by $F'(\lambda)F^{-1}(\lambda)$, respectively $F^{-1}(\lambda)F'(\lambda)$, with the same domain as F^{-1}.

Logarithmic residues are contour integrals of logarithmic derivatives. To make this notion more precise, we shall employ bounded Cauchy domains in $\mathbb{C}$ and their (positively oriented) boundaries. For a discussion of these notions, see, for instance [TL].

Let D be a bounded Cauchy domain. The (positively oriented) boundary of D will be denoted by ∂D. We write $\mathcal{A}_\partial(D; B)$ for the set of all B-valued functions F with the following properties: F is defined and analytic on an open neighborhood of the closure $\overline{D}(= D \cup \partial D)$ and F takes invertible values on all of ∂D (hence F^{-1} is analytic on a neighborhood of ∂D). For $F \in \mathcal{A}_\partial(D; B)$, one can define

$$LR_{left}(F; D) = \frac{1}{2\pi i} \int_{\partial D} F'(\lambda)F^{-1}(\lambda)d\lambda, \tag{2.1}$$

$$LR_{right}(F; D) = \frac{1}{2\pi i} \int_{\partial D} F^{-1}(\lambda)F'(\lambda)d\lambda. \tag{2.2}$$

The elements of the form (2.1) or (2.2) are called *logarithmic residues* in B. More specifically, we call $LR_{left}(F; D)$ the *left* and $LR_{right}(F; D)$ the *right logarithmic residue of F with respect to D*.

It is convenient to also introduce a local version of these concepts. Given a complex number λ_0, we let $\mathcal{A}(\lambda_0; B)$ be the set of all B-valued functions F with the following properties: F is defined and analytic on an open neighborhood of λ_0 and F takes invertible values on a deleted neighborhood of λ_0. For $F \in \mathcal{A}(\lambda_0; B)$, one can introduce

$$LR_{left}(F; \lambda_0) = \frac{1}{2\pi i} \int_{|\lambda-\lambda_0|=\rho} F'(\lambda)F^{-1}(\lambda)d\lambda, \tag{2.3}$$

$$LR_{right}(F; \lambda_0) = \frac{1}{2\pi i} \int_{|\lambda-\lambda_0|=\rho} F^{-1}(\lambda)F'(\lambda)d\lambda, \tag{2.4}$$

where ρ is positive and sufficiently small. The orientation of the integration contour $|\lambda - \lambda_0| = \rho$ is, of course, taken positive, that is counterclockwise. Note that the right hand sides of (2.3) and (2.4) do not depend on the choice of ρ. In fact (2.3), respectively (2.4), is equal to the coefficient of $(\lambda - \lambda_0)^{-1}$ in the Laurent expansion at λ_0 of the left, respectively the right, logarithmic derivative of F at λ_0. Obviously, $LR_{left}(F; \lambda_0)$, respectively $LR_{right}(F; \lambda_0)$, is a left, respectively right, logarithmic residue in the sense of the definitions given in the preceding paragraphs (take for D the disc with radius ρ centered at λ_0). We call $LR_{left}(F; \lambda_0)$ the *left* and $LR_{right}(F; \lambda_0)$ the *right logarithmic residue of F at λ_0*.

In certain cases, the study of logarithmic residues with respect to bounded Cauchy domains can be reduced to the study of logarithmic residues with respect to points. The typical situation is as follows. Let D be a bounded Cauchy domain, let $F \in \mathcal{A}_\partial(D; B)$ and suppose F takes invertible values on all of D, except in a finite number of distinct points $\lambda_1, \ldots, \lambda_n \in D$. Then, for $j = 1, \ldots, n$, the function F belongs to $\mathcal{A}(\lambda_j; B)$ and

$$LR_{left}(F; D) = \sum_{j=1}^n LR_{left}(F; \lambda_j),$$

$$LR_{right}(F; D) = \sum_{j=1}^n LR_{right}(F; \lambda_j).$$

This occurs, in particular, when F^{-1} is meromorphic on D, a state of affairs that will be encountered in the subsequent sections.

In the remainder of this paper, we specialize to the case when $B = \mathcal{B}(X)$, the Banach algebra of bounded linear operators on a complex Banach space X. The identity element in $\mathcal{B}(X)$ is of course I_X, the identity operator on X. Instead of I_X we usually write I.

More specifically, this paper deals with the situation where the function F under consideration is Fredholm operator valued. The underlying Banach space is possibly infinite dimensional.

3. Logarithmic residues and idempotents

It is convenient to recall a few definitions and facts from the literature (cf., e.g. [GGK]). Let X be a complex Banach space. For $T \in \mathcal{B}(X)$, we write Ker T for the null space and Im T for the range space of T. If Ker T is finite dimensional and Im T has finite codimension in X, then T is called (a) *Fredholm (operator)*. These conditions imply that Im T is closed, so this need not be required in the definition (as is sometimes done in the literature). The set of all Fredholm operators on X will be denoted by $\Phi(X)$. It is well-known that $\Phi(X)$ is an open subset of $\mathcal{B}(X)$ which is closed under multiplication and also closed under addition of a compact operator. In particular, the sum of a Fredholm operator and a finite rank bounded linear operator is again a Fredholm operator.

For a finite rank operator T, the rank and trace will be denoted by rank T and trace T, respectively.

Let D be a bounded Cauchy domain in $\mathbb{C}$ and let $F \in \mathcal{A}_\partial(D; \mathcal{B}(X))$. We say that F is *Fredholm (operator valued) on D* if $F(\lambda)$ is a Fredholm operator for all $\lambda \in D$. Since $F \in \mathcal{A}_\partial(D; \mathcal{B}(X))$, the values of F on ∂D are invertible. Hence, if F is Fredholm on D, then F takes invertible values on all of D, except in a finite number of points where F^{-1} has poles. The coefficients of the principal parts of these poles have finite rank; in other words, F^{-1} is *finite meromorphic* on D. It follows that the left and right logarithmic residues with respect to D have finite rank. Their traces coincide and are equal to the algebraic multiplicity of F relative to the contour ∂D (see [GS1], [GGK] and [BKL]).

A *projection* on X is an idempotent in $\mathcal{B}(X)$. Such an operator has finite rank if and only if it is the sum of rank one projections on X. If P is a finite rank projection, then trace $P = $ rank P.

The following factorization result is partly contained in and partly inspired by [T] chapter II and the references given there, in particular [GS2]. For the convenience of the reader we present the proposition with a full proof.

Proposition 3.1. *Let X be a complex Banach space, let Δ be a non-empty open subset of $\mathbb{C}$ and let $F : \Delta \rightarrow \mathcal{B}(X)$ be analytic. Suppose F takes invertible values on all of Δ except in a finite number of points where the values of F are Fredholm. Then there exist finite rank projections $P_1, \ldots, P_n$, $Q_1, \ldots, Q_n$ on X, analytic functions $G, H : \Delta \rightarrow \mathcal{B}(X)$ and complex numbers $\lambda, \ldots, \lambda_n \in \Delta$ (possibly non-distinct) such that*

(i) *G and H take invertible values on all of Δ,*

(ii) *F can be factorized as*

$$F(\lambda) = \left(\prod_{j=1}^{n} \left(I - P_j + (\lambda - \lambda_j)P_j \right) \right) G(\lambda) \tag{3.1}$$

$$= H(\lambda) \left(\prod_{j=1}^{n} \left(I - Q_j + (\lambda - \lambda_j)Q_j \right) \right), \qquad \lambda \in \Delta, \tag{3.2}$$

(iii) *P_j and Q_j are similar, $j = 1, \ldots, n$.*

In products involving non-commuting factors, the order of the factors corresponds to the order of the indices. So, for example, in the above product involving the projections $P_1, \ldots, P_n$, the first factor is $I - P_1 + (\lambda - \lambda_1)P_1$, the second factor (provided that $n \geq 2$) is $I - P_2 + (\lambda - \lambda_2)P_2$ and the last factor is $I - P_n + (\lambda - \lambda_n)P_n$.

In terms of [GKL] and [GGK], Proposition 3.1 claims the equivalence on Δ of the function F and a product of functions of the form $I - P + (\lambda - \alpha)P$ with P a finite rank projection. Both in (3.1) and (3.2), the equivalence is "one-sided" in the sense that one of the functions establishing the equivalence is absent; more precisely, it is identically equal to the identity operator. Indeed, (3.1) features only

G as a "right equivalence function" and (3.2) features only H as a "left equivalence function". This situation is somewhat special because in general both left and right equivalence functions are needed.

Proof. Suppose F takes non-invertible values at the distinct points $\mu_1, \ldots, \mu_k \in \Delta$. The function F^{-1} is meromorphic on Δ and takes its poles at $\mu_1, \ldots, \mu_k$. The coefficients of the principal parts of F at these poles are all of finite rank (i.e., F^{-1} is finite meromorphic on Δ). This will play an important role in what follows.

Let ν be the sum of the orders of the poles of F^{-1} at $\mu_1, \ldots, \mu_k$. Then $\nu = 0$ if and only if $k = 0$ which corresponds to the trivial case where we can take $G = H = F$. So assume ν and k positive.

Write

$$F(\lambda) = \sum_{j=0}^{\infty} (\lambda - \mu_1)^j A_j, \qquad F^{-1}(\lambda) = \sum_{j=-s}^{\infty} (\lambda - \mu_1)^j B_j,$$

where s is the order of μ_1 as a pole of F^{-1}. Since B_{-s} has finite rank, there exists a finite rank projection P_1 on X such that $\operatorname{Ker} P_1 = \operatorname{Ker} B_{-s}$. Note that $B_{-s}A_0 = 0$, so $\operatorname{Im} A_0 \subset \operatorname{Ker} B_{-s} = \operatorname{Ker} P_1$ and $P_1 F(\mu_1) = P_1 A_0 = 0$.

Introduce

$$F_1(\lambda) = \begin{cases} \left(I - P_1 + \frac{1}{\lambda - \mu_1} P_1\right) F(\lambda), & \lambda \in \Delta; \lambda \neq \mu_1, \\ (I - P_1)F(\mu_1) + P_1 F'(\mu_1), & \lambda = \mu_1. \end{cases}$$

Then F_1 is analytic on $\Delta \setminus \{\mu_1\}$ and $F_1(\lambda) \to F_1(\mu_1)$ when $\lambda \to \mu_1$. Hence F_1 is analytic on all of Δ. It is easily verified that

$$F(\lambda) = \left(I - P_1 + (\lambda - \mu_1)P_1\right) F_1(\lambda), \qquad \lambda \in \Delta.$$

Clearly F_1 takes invertible values on all of Δ, except in the points $\mu_2, \ldots, \mu_k$ and possibly μ_1. For $j = 2, \ldots, n$, the value $F_1(\mu_j)$ of F_1 at μ_j is the product of two Fredholm operators, hence Fredholm again. Also, for these values of j, the order of μ_j as a pole of F_1^{-1} is the same as the order of μ_j as a pole of F^{-1}. At μ_1 the situation is as follows. The value $F_1(\mu_1) = (I - P_1)F(\mu_1) + P_1 F'(\mu_1)$ is the sum of a Fredholm operator and a finite rank operator, hence Fredholm again and possibly even invertible. In the latter case the sum of the orders of the poles of F_1^{-1} at $\mu_2, \ldots, \mu_k$ is obviously smaller than ν. If $F_1(\mu_1)$ is non-invertible, then F_1^{-1} has a pole at μ_1. Since the coefficient of $(\lambda - \mu_1)^{-s}$ of the Laurent expansion of F_1^{-1} at μ_1 is $B_{-s}(I - P_1)$ and $\operatorname{Im}(I - P_1) = \operatorname{Ker} P_1 = \operatorname{Ker} B_{-s}$, the order of μ_1 as a pole of F_1^{-1} is smaller than s, the order of μ_1 as a pole of F^{-1}. Hence the sum of the orders of the pole of F_1^{-1} at $\mu_1, \ldots, \mu_k$ is smaller than ν.

By induction it follows that there exist finite rank projections $P_1, \ldots, P_n$, an analytic function $G : \Delta \to \mathcal{B}(X)$ and complex numbers $\lambda_1, \ldots, \lambda_n \in \Delta$ (possibly non-distinct) such that G takes invertible values on all of Δ and

$$F(\lambda) = \left(\prod_{j=1}^{n} \Big(I - P_j + (\lambda - \lambda_j) P_j \Big) \right) G(\lambda), \qquad \lambda \in \Delta.$$

The rest of the proposition is now an immediate consequence of this observation: *Let $\hat{G} : \Delta \to \mathcal{B}(X)$ be analytic, let $\hat{P} \in \mathcal{B}(X)$ be a projection and let $\alpha \in \Delta$. Suppose $\hat{G}$ takes invertible values on Δ. Then there exist a projection $\hat{Q} \in \mathcal{B}(X)$ and an analytic function $\hat{H} : \Delta \to \mathcal{B}(X)$ such that $\hat{H}$ takes invertible values on Δ, $\hat{Q}$ is similar to $\hat{P}$ and*

$$\Big(I - \hat{P} + (\lambda - \alpha)\hat{P} \Big) \hat{G}(\lambda) = \hat{H}(\lambda) \Big(I - \hat{Q} + (\lambda - \alpha)\hat{Q} \Big), \qquad \lambda \in \Delta$$

For the proof, put $\hat{Q} = \hat{G}^{-1}(\alpha) \hat{P} \hat{G}(\alpha)$ and

$$\hat{H}(\lambda) = \Big(I - \hat{P} + (\lambda - \alpha)\hat{P} \Big) \hat{G}(\lambda) \left(I - \hat{Q} + \frac{1}{\lambda - \alpha} \hat{Q} \right), \qquad \lambda \in \Delta; \ \lambda \neq \alpha.$$

Then $\hat{H}$ takes invertible values on $\Delta \backslash \{\alpha\}$ and

$$\hat{H}^{-1}(\lambda) = \Big(I - \hat{Q} + (\lambda - \alpha)\hat{Q} \Big) \hat{G}^{-1}(\lambda) \left(I - \hat{P} + \frac{1}{\lambda - \alpha} \hat{P} \right), \qquad \lambda \in \Delta; \ \lambda \neq \alpha.$$

Thus $\hat{H}$ and $\hat{H}^{-1}$ are analytic on $\Delta \backslash \{\alpha\}$ and have at most a simple pole at α. It is easy to see that the coefficients of the term $(\lambda - \alpha)^{-1}$ in the Laurent expansion of $\hat{H}$ and $\hat{H}^{-1}$ equal $(I - \hat{P})\hat{G}(\alpha)\hat{Q}$ and $(I - \hat{Q})\hat{G}^{-1}(\alpha)\hat{P}$, respectively. However these two expressions vanish due to the definition of $\hat{Q}$. Hence $\hat{H}$ and $\hat{H}^{-1}$ have a removable singularity at α and can therefore be continued analytically to all of Δ. With this the observation formulated above follows. $\qquad \square$

In connection with the last paragraph of the proof of Proposition 3.1, we observe that $\hat{H}(\alpha) = \hat{G}(\alpha) + (I - \hat{P})\hat{G}'(\alpha)\hat{G}^{-1}(\alpha)\hat{P}\hat{G}(\alpha)$ and $\hat{H}^{-1}(\alpha) = \hat{G}^{-1}(\alpha) - \hat{G}^{-1}(\alpha)(I - \hat{P})\hat{G}'(\alpha)\hat{G}^{-1}(\alpha)\hat{P}$.

Before we proceed, we make an intermediate remark. Let D be a bounded Cauchy domain in $\mathbb{C}$, let $F \in \mathcal{A}_\partial \big(D; \mathcal{B}(X) \big)$ and suppose F is Fredholm on D. Then there exists an open neighborhood Δ of $\overline{D}$ such that F takes invertible values on the set $\Delta \backslash D$ (which includes the boundary ∂D of D on which, by definition, F takes invertible values). Apply Proposition 3.1 and write F in the form

$$F(\lambda) = \left(\prod_{j=1}^{n} \Big(I - P_j + (\lambda - \lambda_j) P_j \Big) \right) G(\lambda), \qquad \lambda \in \Delta,$$

where $\lambda_1, \ldots, \lambda_n \in D$, $P_1, \ldots, P_n$ are finite rank projections on the Banach space X, the function G is analytic on Δ and G takes invertible values there. Assuming that all operators involved commute – a very special situation – we see that

$$LR_{left}(F;D) = LR_{right}(F;D) = \sum_{j=1}^{n} P_j.$$

The following result (directly related to Proposition 3.1) is instrumental for overcoming the difficulties that arise in the non-commutative case.

Proposition 3.2. *Let* $F \in \mathcal{A}_{\partial}(D; \mathcal{B}(X))$, *where* X *is a complex Banach space and* D *is a bounded Cauchy domain in* $\mathbb{C}$. *Suppose* F *is Fredholm on* D *and write* F *on an open neighborhood* Δ *of* $\overline{D}$ *as*

$$F(\lambda) = \left(\prod_{j=1}^{n} \left(I - P_j + (\lambda - \lambda_j)P_j \right) \right) G(\lambda)$$

$$= H(\lambda) \left(\prod_{j=1}^{n} \left(I - Q_j + (\lambda - \lambda_j)Q_j \right) \right), \qquad \lambda \in \Delta,$$

where $\lambda_1, \ldots, \lambda_n \in D$, $P_1, \ldots, P_n$, $Q_1, \ldots, Q_n$ *are finite rank projections on* X, *and* G *and* H *are analytic* $\mathcal{B}(X)$-*valued functions on* Δ *that take invertible values only. Then*

(i) $\mathrm{trace}\ LR_{left}(F;D) = \sum\limits_{j=1}^{n} \mathrm{trace}\ P_j,$

(ii) $\mathrm{trace}\ LR_{right}(F;D) = \sum\limits_{j=1}^{n} \mathrm{trace}\ Q_j,$

(iii) $\mathrm{rank}\ LR_{left}(F;D) \leq \sum\limits_{j=1}^{n} \mathrm{rank}\ P_j,$

(iv) $\mathrm{rank}\ LR_{right}(F;D) \leq \sum\limits_{j=1}^{n} \mathrm{rank}\ Q_j.$

Note that the right hand sides of (i) and (iii) and, similarly, those of (ii) and (iv) coincide.

Proof. We focus on (i) and (iii). The arguments for (ii) and (iv) are of course analogous. We begin with (i).

For $k = 1, \ldots, n$ and $\lambda \in \Delta$, put

$$P_k(\lambda) = I - P_k + (\lambda - \lambda_k)P_k,$$

and for $k = 0, \ldots, n$ and $\lambda \in \Delta$, define

$$F_k(\lambda) = \left(\prod_{j=1}^{k} P_{n+j-k}(\lambda) \right) G(\lambda).$$

Then $F_k \in \mathcal{A}_\partial (D; \mathcal{B}(X))$ and F_k is Fredholm on D. Write

$$\tilde{L}_k = LR_{left}(F_k; D) = \frac{1}{2\pi i} \int_{\partial D} F_k'(\lambda) F_k^{-1}(\lambda) d\lambda.$$

Then $\tilde{L}_0 = 0$ (as G is analytic and takes invertible values on Δ) and $\tilde{L}_n = LR_{left}(F; D)$. Clearly, for $k = 1, \ldots, n$,

$$F_k(\lambda) = \Big(I - P_{n+1-k} + (\lambda - \lambda_{n+1-k}) P_{n+1-k} \Big) F_{k-1}(\lambda),$$

and a straightforward calculation gives

$$F_k'(\lambda) F_k^{-1}(\lambda) = \frac{1}{\lambda - \lambda_{n+1-k}} P_{n+1-k} + P_{n+1-k}(\lambda) F_{k-1}'(\lambda) F_{k-1}^{-1}(\lambda) P_{n+1-k}^{-1}(\lambda).$$

Thus $\tilde{L}_k$ can be expressed as

$$\tilde{L}_k = P_{n+1-k} + (I - P_{n+1-k}) \tilde{L}_{k-1} (I - P_{n+1-k}) + P_{n+1-k} \tilde{L}_{k-1} P_{n+1-k} +$$

$$+ (I - P_{n+1-k}) \left(\frac{1}{2\pi i} \int_{\partial D} \frac{1}{\lambda - \lambda_{n+1-k}} F_{k-1}'(\lambda) F_{k-1}^{-1}(\lambda) d\lambda \right) P_{n+1-k} +$$

$$+ P_{n+1-k} \left(\frac{1}{2\pi i} \int_{\partial D} (\lambda - \lambda_{n+1-k}) F_{k-1}'(\lambda) F_{k-1}^{-1}(\lambda) d\lambda \right) (I - P_{n+1-k}).$$

Recall that $\tilde{L}_0, \ldots, \tilde{L}_n$ have finite rank. Taking traces and using the commutativity property of the trace, we get

$$\text{trace } \tilde{L}_k = \text{trace } P_{n+1-k} + \text{trace } \tilde{L}_{k-1}(I - P_{n+1-k}) + \text{trace } \tilde{L}_{k-1} P_{n+1-k}.$$

Note in this context that the traces of the last two terms in the above expression relating $\tilde{L}_k$ and $\tilde{L}_{k-1}$ are zero (because $(I - P_{n+1-k}) P_{n+1-k} = P_{n+1-k}(I - P_{n+1-k}) = 0$). From the additivity of the trace, we now obtain

$$\text{trace } \tilde{L}_k = \text{trace } P_{n+1-k} + \text{trace } \tilde{L}_{k-1}$$

and (i) follows by induction.

Next we turn to (iii). Note that $P_k^{-1}(\lambda) = I - P_k + (\lambda - \lambda_k)^{-1} P_k$. For $k = 0, \ldots, n$, write

$$\hat{L}_k = \frac{1}{2\pi i} \int_{\partial D} F'(\lambda) G^{-1}(\lambda) \left(\prod_{j=1}^{k} P_{n+1-j}^{-1}(\lambda) \right) d\lambda,$$

Then $\hat{L}_0 = 0$ (as F' and G^{-1} are analytic on Δ) and $\hat{L}_n = LR_{left}(F; D)$. We shall prove that

$$\text{rank } \hat{L}_k \leq \sum_{j=1}^{k} \text{rank } P_{n+1-j}$$

which for $k = n$ is just (iii). For $k = 0$, the inequality is trivial. So assume $1 \leq k \leq n$. Then

$$\hat{L}_k = \hat{L}_{k-1}(I - P_{n+1-k})$$

$$+ \left(\frac{1}{2\pi i} \int_{\partial D} \frac{1}{\lambda - \lambda_{n+1-k}} F'(\lambda) G^{-1}(\lambda) \left(\prod_{j=1}^{k-1} P_{n+1-j}^{-1}(\lambda) \right) d\lambda \right) P_{n+1-k},$$

and hence rank $\hat{L}_k \leq$ rank $\hat{L}_{k-1} +$ rank P_{n+1-k}. The desired result now follows by induction. This proves (iii). $\square$

Corollary 3.3. *Let* $F \in \mathcal{A}_\partial(D; \mathcal{B}(X))$, *where* X *is a complex Banach space and* D *is a bounded Cauchy domain in* $\mathbb{C}$. *Assume* F *is Fredholm on* D. *Then* $LR_{left}(F; D)$ *and* $LR_{right}(F; D)$ *are finite rank operators with non-negative integer traces. Further, the following statements are equivalent:*

(i) $LR_{left}(F; D) = 0$;

(ii) $LR_{right}(F; D) = 0$;

(iii) trace $LR_{left}(F; D) = 0$;

(iv) trace $LR_{right}(F; D) = 0$;

(v) F *takes invertible values on* D.

For left logarithmic residues, this result was already obtained in [BES2]. The equivalence of (i), (iii) and (v) was proved there using the State Space Method in Analysis (cf. [BGK1],[BGK2] and [B2]). This method involves ideas from Systems Theory. It was also indicated in [BES2] that a quick proof can be obtained by using the material on multiplicities of Fredholm operator valued functions developed by I.C. Gohberg and E.I. Sigal (cf. [GGK], and the references given there, in particular [GS1] and [BKL]; see also the third paragraph of this section above). Our argument here is more transparent and will be based on Propositions 3.1 and 3.2.

Proof. We know already that $LR_{left}(F; D)$ and $LR_{right}(F; D)$ have finite rank. Indeed, this follows from the fact that F^{-1} is finite meromorphic on D with a finite number of poles.

Let Δ be an open neighborhood of $\overline{D}$ such that F takes invertible values on $\Delta \backslash \overline{D}$. Apply Proposition 3.1 to write F as in Proposition 3.2. Recall that the right hand sides of Proposition 3.2 (i) and (iii) coincide and the same holds for the right hand sides of Proposition 3.2 (ii) and (iv). Proposition 3.2 now gives that the traces of $LR_{left}(F; D)$ and $LR_{right}(F; D)$ are non-negative integers.

Obviously (v) $\Rightarrow$ (i) $\Rightarrow$ (iii) and (v) $\Rightarrow$ (ii) $\Rightarrow$ (iv). The implication (iii) $\Rightarrow$ (i) is clear from Proposition 3.2 (i) and (iii). Analogously the implication (iv) $\Rightarrow$ (ii) is obtained from Proposition 3.2 (ii) and (iv). If (iii) holds, then Proposition 3.2 (i) leads to the conclusion that all projections P_j are zero. Hence $F = G$ takes invertible values on D. So (iii) implies (v). Similarly, (v) is implied by (iv), and the proof is complete. $\square$

The next result is a slight refinement of [BES4], Theorem 5.1. One of the main steps in the proof differs from the corresponding step in [BES4] in that it is based on Propositions 3.1 and 3.2 instead of using ideas from Systems Theory.

Theorem 3.4. *Let X be a complex Banach space, let D be a bounded Cauchy domain in $\mathbb{C}$ and let $L \in \mathcal{B}(X)$. The following statements are equivalent:*

(i) *L is a sum of finite rank projections on X;*

(ii) *L is a sum of rank one projections on X;*

(iii) *L has finite rank and the trace of L is an integer larger than or equal to the rank of L;*

(iv) *L is the left logarithmic residue with respect to D of a function $F \in \mathcal{A}_\partial(D; \mathcal{B}(X))$ which is Fredholm on D;*

(v) *L is the right logarithmic residue with respect to D of a function $F \in \mathcal{A}_\partial(D; \mathcal{B}(X))$ which is Fredholm on D.*

With respect to (ii) we note that if L is the sum of rank one projections on X, the number of terms in such a sum is equal to the trace of L. This follows from the additivity of the trace combined with the fact that for a finite rank projection, the trace and the rank coincide. The proof of the implications (i) $\Rightarrow$ (iv) and (i) $\Rightarrow$ (v) will provide additional information about the freedom one has in choosing the function F.

Note that in Theorem 3.4 the Cauchy domain D is given, whereas in [BES4], Theorem 5.1 it is not (cf. the first paragraph in Section 4 below).

Proof. Each finite rank projection can be written as a sum of rank one projections. Hence (i) implies (ii). Obviously (ii) implies (i). Since for finite rank projections, the rank and the trace coincide, the trace is additive and the rank is subadditive, we have that (i) implies (iii). For (iii) $\Rightarrow$ (i) we note that by using standard techniques things can be reduced to the matrix case (see [BES4] for details). For the matrix case, the result was obtained by [HP] and [W]; cf. also [BES4] for a particularly transparent argument. Thus (i), (ii) and (iii) are equivalent. We now focus on proving that (i) implies (iv) and (v) and that (iii) is implied by (iv) or (v).

First we deal with (i) $\Rightarrow$ (iv) and (i) $\Rightarrow$ (v). The proof that we shall give provides additional information about the freedom one has in choosing F. It is a slightly expanded version of the corresponding argument in [BES4].

Suppose (i) holds and write L as

$$L = \sum_{j=1}^n P_j,$$

where $P_1, \ldots, P_n$ are finite rank non-zero projections on X. Choose distinct (but, for the time being, otherwise arbitrary) complex numbers $\lambda_1, \ldots, \lambda_n$. By [E], there exists an entire $\mathcal{B}(X)$-valued function on F such that F takes invertible values on all of $\mathbb{C}$, except in the points $\lambda_1, \ldots, \lambda_n$ where F^{-1} has simple poles while, in addition,

$$LR_{left}(F; \lambda_j) = LR_{right}(F; \lambda_j) = P_j, \qquad j = 1, \ldots, n.$$

If we now assume that $\lambda_1, \ldots, \lambda_n \in D$, we may conclude that $F \in \mathcal{A}_\partial(D; \mathcal{B}(X))$ and

$$L = LR_{left}(F; D) = LR_{right}(F; D) = \sum_{j=1}^n P_j = L.$$

Since all P_j have finite rank, we see from the construction given in [E] that F can be chosen to be Fredholm operator valued on all of $\mathbb{C}$. Indeed, the function in [E] is a product of functions of the type $I - P + f(\lambda)P$ where $P \in \{P_1, \ldots, P_n\}$ – so P is a finite rank projection on X and $I - P$ is Fredholm – and f is an entire scalar function.

Finally, suppose (iv) or (v) holds. We need to establish (iii). In [BES4], the corresponding argument goes by reduction to the matrix case and then employing ideas from Systems Theory. Here we simply note that (iii) can be obtained by applying Propositions 3.1 and 3.2, taking into account the fact – already mentioned before – that the right hand sides of Proposition 3.2 (i) and (iii) and also the right hand sides of Proposition 3.2 (ii) and (iv), coincide. $\square$

4. Topological issues

From Theorem 3.4 it is clear that the set of left logarithmic residues of Fredholm operator valued functions coincides with the set of right logarithmic residues of Fredholm operator valued functions. In fact – independent of the underlying Cauchy domain – these sets are identical to the set of all bounded linear operators on the given Banach space X that can be written as a sum of finite rank projections on X. We shall denote this set by $\mathcal{P}(X)$. Note that $\mathcal{P}(X)$ can also be described as the set of all finite rank operators $T \in \mathcal{B}(X)$ for which trace T is an integer and rank $T \leq$ trace T. The set $\mathcal{P}(X)$ is obviously closed under addition.

In this section we investigate the topological properties of $\mathcal{P}(X)$. For finite dimensional X (matrix case) this was already done in [BES4]. So the point of interest here is the case when X is infinite dimensional. As we shall see, the results of [BES4] carry over – at least to a large extent – to the infinite dimensional setting. The proofs, however, are more involved, one complication being that one cannot rely on the continuity of the trace.

Theorem 4.1. *Let X be a complex Banach space. Then the set $\mathcal{P}(X)$ consisting of all sums of finite rank projections on X is a closed subset of $\mathcal{B}(X)$ and the trace is continuous on $\mathcal{P}(X)$ with respect to the relative topology on $\mathcal{P}(X)$ induced by the norm topology on $\mathcal{B}(X)$.*

To put things in a somewhat wider framework, let $f : \mathbb{C} \to \mathbb{Z}_+$ be any function, where $\mathbb{Z}_+ = \{0, 1, 2, \ldots\}$, and consider the following set of finite rank operators in $\mathcal{B}(X)$:

$$\mathcal{P}(f;X) \;=\; \Big\{\, T \in \mathcal{B}(X) \;:\; T \text{ has finite rank and rank } T \le f(\text{trace } T) \,\Big\}.$$

In order to show the connection of this set with $\mathcal{P}(X)$, let f_0 denote the function whose values $f_0(z)$ are equal to the greatest integer less than or equal to the modulus of z. Observe that $\mathcal{P}(X)$ is the set of all $T \in \mathcal{P}(f_0;X)$ for which trace T is a non-negative integer. In other words, $\mathcal{P}(X)$ is the inverse image in $\mathcal{P}(f_0;X)$ under the trace of the set of non-negative integers. Hence Theorem 4.1 is a direct corollary to the following result.

Another way of concluding Theorem 4.1 from Theorem 4.2 is by observing that $\mathcal{P}(X) = \mathcal{P}(f_1;X)$, where $f_1(z) = z$ for $z \in \mathbb{Z}_+$ and $f_1(z) = 0$ for $z \in \mathbb{C} \setminus \mathbb{Z}_+$.

Theorem 4.2. *Let X be a complex Banach space, and let $f : \mathbb{C} \to \mathbb{Z}_+$ be a function such that*

$$\sup_{z \in \mathbb{C}} \frac{f(z)}{1 + |z|} \;<\; \infty. \tag{4.1}$$

Then the closure $\overline{\mathcal{P}(f;X)}$ of $\mathcal{P}(f;X)$ is contained in the set of all finite rank operators and the trace is continuous on $\overline{\mathcal{P}(f;X)}$ with respect to the relative topology on $\overline{\mathcal{P}(f;X)}$ induced by the norm topology on $\mathcal{B}(X)$. If, in addition, f is an upper semi-continuous function, then $\mathcal{P}(f;X)$ is a closed subset of $\mathcal{B}(X)$.

When X is finite dimensional, one may assume without loss of generality that $f(z)$ is not greater than the dimension of X. Hence the first assertion of the theorem restates the known fact that the trace is always continuous for finite dimensional X. In this case the second assertion is almost trivial, too. Just use the lower semi-continuity of the rank and the continuity of the trace. The proof given below works both for the finite and the infinite dimensional case.

We prepare for the proof with a few general observations. The spectral radius of $T \subset \mathcal{B}(X)$ will be denoted by $r(T)$. Recall that $r(T) \le \|T\|$.

Let $T \in \mathcal{B}(X)$ be of finite rank. Then

$$|\text{trace } T| \le r(T) \cdot \text{rank } T \le \|T\| \cdot \text{rank } T.$$

To see this, let $\lambda_1, \ldots, \lambda_r$ be the non-zero eigenvalues of T counted according to algebraic multiplicity. Then

$$\text{trace } T = \sum_{j=1}^{r} \lambda_j.$$

The desired result now follows from the fact that $r \le \text{rank } T$ and $|\lambda_j| \le r(T)$, $j = 1, \ldots, r$.

Consider a sequence $T_1, T_2, \ldots$ of finite rank bounded linear operators on X. Suppose the sequence converges to $T \in \mathcal{B}(X)$. If the ranks of the operators T_j form a bounded sequence, then T has finite rank too, rank $T \le$ rank T_n for sufficiently large n and trace $T_n \to$ trace T. The first two statements are immediate from the lower semi-continuity of the rank. The argument for the last statement is as follows. Choose K such that rank $T_n \le K$, $n = 1, 2 \ldots$. Again by the lower semi-continuity of the rank we obtain that rank $T \leq K$ too. Now

$$|\text{trace } T_n - \text{trace } T| = |\text{trace } (T_n - T)| \leq \|T_n - T\| \cdot 2K,$$

hence trace $T_n \to$ trace T as $n \to \infty$.

Finally, let P and Q be finite rank projections on X and assume $\|P-Q\| < 1$. Then Im P and Im Q have the same dimension. This is well-known. The proof amounts to observing that $I - P + Q$ and $I - Q + P$ are invertible and that $(I - P + Q)\text{Im } P \subset \text{Im } Q$ and $(I - Q + P)\text{Im } Q \subset \text{Im } P$.

We are now ready for the proof of Theorem 4.2.

Proof. Let $A_1, A_2, \ldots$ be a sequence in $\mathcal{P}(f; X)$, let $A \in \mathcal{B}(X)$ and assume $A_n \to A$. In order to establish the first part of Theorem 4.2, it is sufficient to show that A has finite rank and trace $A_n \to$ trace A.

Clearly A is compact. Thus, denoting the spectrum of A by $\sigma(A)$, the set $\sigma(A)\backslash\{0\}$ consists of isolated points. Choose real numbers ε and δ such that $0 < \varepsilon < \delta < 1$ and $\varepsilon \leq |\lambda| \leq \delta$ implies $\lambda \notin \sigma(A)$. Note that there is much freedom in choosing ε and δ. Later we shall impose one more condition on ε.

Take $R > \max\{\delta, \|A\|\}$ and put

$$D_0 = \left\{ \lambda \in \mathbb{C} \mid |\lambda| < \varepsilon \right\}, \quad D = \left\{ \lambda \in \mathbb{C} \mid \delta < |\lambda| < R \right\}.$$

Then $\sigma(A) \subset D_0 \cup D$. Using standard techniques, one can prove that, for n sufficiently large, $\sigma(A_n) \subset D_0 \cup D$ too. Thus we may assume that this inclusion holds for all n. The same standard techniques yield that $(\lambda I - A_n)^{-1} \to (\lambda I - A)^{-1}$ uniformly on $\mathbb{C}\backslash(D_0 \cup D)$.

The open annulus D is a bounded Cauchy domain in $\mathbb{C}$. Now introduce

$$P = \frac{1}{2\pi i} \int_{\partial D} (\lambda I - A)^{-1}d\lambda, \quad P_n = \frac{1}{2\pi i} \int_{\partial D} (\lambda I - A_n)^{-1}d\lambda.$$

From the spectral theory of compact operators we know that P and P_n are finite rank projections. Also $P_n \to P$ and $\|P_n - P\| < 1$ for n sufficiently large. Thus we may assume that rank $P_n = $ rank P, $n = 1, 2, \ldots$. Since $A_n \to A$ and $P_n \to P$, we have $A_n P_n \to AP$. The operators AP and $A_n P_n$ have finite rank not exceeding rank $P = $ rank P_n. Hence trace $A_n P_n \to$ trace AP.

Recall that $\sigma(A_n)$ is contained in $D_0 \cup D$. So

$$I - P_n = \frac{1}{2\pi i} \int_{\partial D_0} (\lambda I - A_n)^{-1}d\lambda$$

and $\sigma\big(A_n(I - P_n)\big) \subset D_0 \cup \{0\} = D_0$. Thus $r\big(A_n(I - P_n)\big) \leq \varepsilon$ and

$$|\text{trace } A_n(I - P_n)| \leq \varepsilon \cdot \text{rank } A_n(I - P_n) \leq \varepsilon \cdot \text{rank } A_n.$$

Now trace $A_n = $ trace $A_n(I - P_n) + $ trace $A_n P_n$. Hence

$$|\text{trace } A_n| \leq |\text{trace } A_n(I - P_n)| + |\text{trace } A_n P_n| \leq \varepsilon \cdot \text{rank } A_n + |\text{trace } A_n P_n|.$$

Because trace $A_n P_n \to$ trace AP, there exists $K \geq 0$ such that $|\text{trace } A_n P_n| \leq K$.

Hence

$$|\text{trace } A_n| \leq K + \varepsilon \cdot \text{rank } A_n, \quad n = 1, 2, \ldots .$$

Since $A_n \in \mathcal{P}(f; X)$, we have rank $A_n \leq f(\text{trace } A_n) \leq M + M \cdot |\text{trace } A_n|$, where M stands for the left hand side of inequality (4.1). So

$$\text{rank } A_n \leq M + KM + \varepsilon M \cdot \text{rank } A_n, \quad n = 1, 2, \ldots .$$

Recall that there is freedom in choosing ε (and δ). We take advantage of this by selecting ε such that $\varepsilon M \leq \frac{1}{2}$. It then follows that

$$\text{rank } A_n \leq 2(M + MK), \quad n = 1, 2, \ldots .$$

Thus the ranks of the operators A_n form a bounded sequence. We conclude that A has finite rank and trace $A_n \to \text{trace } A$.

To prove the second part of Theorem 4.2 we argue as follows. For n sufficiently large, rank $A \leq \text{rank } A_n$. Also for all n, rank $A_n \leq f(\text{trace } A_n)$. Combining these facts, we get that rank A does not exceed $\liminf f(\text{trace } A_n)$. When f is upper semi-continuous, the right hand side of this inequality does not exceed $f(\text{trace } A)$. So in that case we may conclude that rank $A \leq f(\text{trace } A)$, i.e. $A \in \mathcal{P}(f; X)$. $\square$

For $\tau = 0, 1, 2, \ldots$, let $\mathcal{P}_\tau(X)$ be the set of all $T \in \mathcal{P}(X)$ for which trace $T = \tau$. Then $\mathcal{P}_\tau(X)$ is a closed subset of $\mathcal{B}(X)$. Clearly

$$\mathcal{P}(X) = \bigcup_{\tau=0}^{\infty} \mathcal{P}_\tau(X)$$

and this union is disjoint. From Theorem 3.4 and the comment following it, one sees that a bounded linear operator on X belongs to $\mathcal{P}_\tau(X)$ if and only if it is the sum of τ rank one projections on X. In particular $\mathcal{P}_0(X) = \{0\}$. If $A \in \mathcal{P}_\alpha(X)$ and $B \in \mathcal{P}_\beta(X)$, then $A + B \in \mathcal{P}_{\alpha+\beta}(X)$.

As we shall see in Theorem 4.4 the sets $\mathcal{P}_\tau(X)$ are the connected components of $\mathcal{P}(X)$. First, however, we establish a lemma. The lemma will be used not only in the proof of Theorem 4.4, but also in that of Theorem 5.1.

Lemma 4.3. *Let X be a complex Banach space and let P_1 and P_2 be rank one projections on X. Then there exists a bounded linear operator S on X such that $P_1 = S^{-1} P_2 S$ and S can be chosen to be an exponential.*

The latter means that S has a logarithm, i.e., $S = \exp T$ for some $T \in \mathcal{B}(X)$. Note that $\exp T$ belongs to the connected component of the set of all invertible elements in $\mathcal{B}(X)$ containing the identity operator I.

We begin with some preparations. Let y be a vector in X and let φ be a bounded linear functional on X. The expression $y \otimes \varphi$ will denote the bounded linear operator on X given by $(y \otimes \varphi)x = \varphi(x)y$. Clearly rank $(y \otimes \varphi) \leq 1$, equality holding if and only if $y \neq 0$ and $\varphi \neq 0$. Also $I - y \otimes \varphi$ is invertible with inverse

$$(I - y \otimes \varphi)^{-1} = I + \frac{1}{1 - \varphi(y)} y \otimes \varphi$$

whenever $\varphi(y) \neq 1$.

Proof. Write

$$P_1 = y_1 \otimes \varphi_1, \qquad P_2 = y_2 \otimes \varphi_2$$

with y_1 and y_2 in X and φ_1 and φ_2 bounded linear functionals on X. Since P_1 and P_2 are projections, we have $\varphi_1(y_1) = \varphi_2(y_2) = 1$. Assume first that $\varphi_1(y_2) = \varphi_2(y_1) = 0$. Put $y = y_1 + y_2$, $\varphi = \varphi_1 + \varphi_2$ and $S = I - y \otimes \varphi$. A straightforward computation shows that $SP_1 = P_2 S = -y_2 \otimes \varphi_1$. As $\varphi(y) = 2$, the operator S is invertible. Also the spectrum of S consists of a finite number of points (in fact at most two). So, by a routine argument involving operational calculus, one has that S is an exponential.

Next assume that (at least) one of the scalars $\varphi_1(y_2)$ and $\varphi_2(y_1)$ is non-zero, for instance $\varphi_2(y_1) \neq 0$. Write

$$y_0 = \frac{1}{\varphi_2(y_1)} y_1, \qquad \varphi_0 = \frac{1}{\varphi_2(y_1)} \varphi_2.$$

Then $y_1 \otimes \varphi_0 = y_0 \otimes \varphi_2$ and we denote this operator by P_0. Since $\varphi_0(y_1) = \varphi_2(y_0) = 1$, P_0 is a rank one projection. Introduce

$$S_1 = I - y_1 \otimes (\varphi_0 + \varphi_1), \quad S_2 = I - (y_0 + y_2) \otimes \varphi_2.$$

Once checks without difficulty that $S_1 P_1 = P_0 S_1 = -y_1 \otimes \varphi_1$. As $(\varphi_0 + \varphi_1)(y_1) = 2$, the operator S_1 is invertible. Analogously $S_2 P_0 = P_2 S_2 = -y_2 \otimes \varphi_2$ and S_2 is invertible. Put $S = S_2 S_1$. Then $SP_1 = P_2 S$. Since $S - I$ is of finite rank, the spectrum of S consists of a finite number of points and S is an exponential. □

Theorem 4.4. *Let X be a complex Banach space and let $\mathcal{P}(X)$ and $\mathcal{P}_\tau(X)$ be as above. The following statements are true:*

(i) *The sets $\mathcal{P}(X)$ and $\mathcal{P}_\tau(X)$ are closed subset of $\mathcal{B}(X)$ with empty interior.*

(ii) *The zero operator in X is an isolated point of $\mathcal{P}(X)$ and, in case X has dimension larger than or equal to two (possibly infinite), it is the only such point in $\mathcal{P}(X)$; in fact, when X has dimension at least two and τ is a positive integer, the set $\mathcal{P}_\tau(X)$ does not have isolated points.*

(iii) *For τ and σ non-negative integers, not both zero,*

$$\operatorname{dist}\big(\mathcal{P}_\tau(X),\ \mathcal{P}_\sigma(X)\big) \geq \frac{|\tau - \sigma|}{\tau + \sigma},$$

where the left hand side in this inequality stands for the distance between $\mathcal{P}_\tau(X)$ and $\mathcal{P}_\sigma(X)$.

(iv) *The sets $\mathcal{P}_\tau(X)$ are the connected components of $\mathcal{P}(X)$.*

When X is one-dimensional (scalar case), the set $\mathcal{P}(X)$ consists of isolated points only. More generally this conclusion holds for the set of logarithmic residues (= the set of sums of idempotents) in any commutative Banach algebra (cf. [BES3]).

Proof. Statement (i) is obvious from Theorem 4.1 and the fact that the trace takes integer values on $\mathcal{P}(X)$.

Let $A_1, A_2, \ldots$ be a sequence in $\mathcal{P}(X)$ converging to the zero operator. Since the trace is continuous on $\mathcal{P}(X)$ and integer valued, one has that trace $A_n = 0$ for sufficiently large n. Now rank $A_n \leq$ trace A_n. So $A_n = 0$ for sufficiently large n. Thus the zero operator is an isolated point in $\mathcal{P}(X)$.

Assume X has dimension at least 2 and τ is positive. Take $T \in \mathcal{P}_\tau(X)$. Then T can be written as a sum of τ rank one projections on X. So to prove that T is not an isolated point in $\mathcal{P}_\tau(X)$, it suffices to show that in the present situation the set $\mathcal{P}_1(X)$ consisting of rank one projections has no isolated points. This is easy. Indeed, let P be a rank one projection and let $N : \operatorname{Im} P \to \operatorname{Ker} P$ be a non-zero bounded linear operator. For $\varepsilon \neq 0$, define $P_\varepsilon : X \to X$ by

$$P_\varepsilon x = Px + \varepsilon N(Px), \qquad x \in X.$$

Then P_ε is a projection different from P and $P_\varepsilon \to P$ when $\varepsilon \to 0$. As $\operatorname{Ker} P_\varepsilon = \operatorname{Ker} P$, the rank of P_ε is one, i.e., $P_\varepsilon \in \mathcal{P}_1(X)$. This proves (ii).

For (iii) we argue as follows. Take $A \in \mathcal{P}_\tau(X)$ and $B \in \mathcal{P}_\sigma(X)$. From the remarks made prior to the proof of Theorem 4.2 we know that

$$|\text{trace } (A - B)| \leq \|A - B\| \cdot \operatorname{rank} (A - B). \qquad (4.2)$$

In the situation considered here, this implies

$$|\tau - \sigma| \leq \|A - B\| \cdot (\operatorname{rank} A + \operatorname{rank} B) \leq \|A - B\| \cdot (\tau + \sigma).$$

It follows that $\|A - B\| \geq |\tau - \sigma| \cdot (\tau + \sigma)^{-1}$ and the proof of (iii) is complete.

We know already that $\mathcal{P}_\tau(X)$ is closed in $\mathcal{P}(X)$. It follows from (iii) that

$$\operatorname{dist}\big(\mathcal{P}_\tau(X), \mathcal{P}_\sigma(X)\big) \geq \frac{1}{2\tau - 1}, \qquad \sigma < \tau,$$

$$\operatorname{dist}\big(\mathcal{P}_\tau(X), \mathcal{P}_\sigma(X)\big) \geq \frac{1}{2\tau + 1}, \qquad \sigma > \tau.$$

Hence $\operatorname{dist}\big(\mathcal{P}_\tau(X), \mathcal{P}(X) \backslash \mathcal{P}_\tau(X)\big) > 0$ and $\mathcal{P}_\tau(X)$ is open in $\mathcal{P}(X)$. It remains to show that the sets $\mathcal{P}_\tau(X)$ are connected. This goes as follows. Take P and Q in $\mathcal{P}_\tau(X)$ and write P and Q as a sum of τ rank one projections on X:

$$P = \sum_{j=1}^{\tau} P_j, \qquad Q = \sum_{j=1}^{\tau} Q_j.$$

By Lemma 4.3 there exist $T_1, \ldots, T_\tau$ in $\mathcal{B}(X)$ such that

$$Q_j = \exp(-T_j) P_j \exp(T_j), \qquad j = 1, \ldots, \tau.$$

Introduce $F : \mathbb{C} \to \mathcal{B}(X)$ by

$$F(\lambda) = \sum_{j=1}^{\tau} \exp(-\lambda T_j) P_j \exp(\lambda T_j).$$

Then F is continuous and takes values in $\mathcal{P}_\tau(X)$. The desired result is now clear from $F(0) = P$ and $F(1) = Q$. Indeed, the reasoning even shows that $\mathcal{P}_\tau(X)$ is arcwise connected. □

We elaborate on Theorem 4.4 (iii) with a remark and an example. If $\dim X = d$ is finite (and non-zero), it follows from (4.2) that

$$\operatorname{dist}\left(\mathcal{P}_\tau(X), \mathcal{P}_\sigma(X)\right) \geq \frac{|\tau - \sigma|}{d} \geq \frac{1}{d}. \qquad \sigma \neq \tau.$$

So in the finite dimensional case there is a uniform lower bound (depending on the dimension of the underlying space) for the distance between the different components of $\mathcal{P}(X)$. In the infinite dimensional situation such a uniform bound (generally) does not exist.

Example 4.5. Regardless of whether the dimension of the (non-trivial) Banach space X is finite or infinite, we have

$$\operatorname{dist}\left(\mathcal{P}_\tau(X), \mathcal{P}_\sigma(X)\right) \leq |\tau - \sigma|. \tag{4.3}$$

To see this, let P be a rank one projection on X with norm one and compare $\tau P \in \mathcal{P}_\tau(X)$ and $\sigma P \in \mathcal{P}_\sigma(X)$.

We now deduce a refinement of the inequality (4.3) for the situation where X is the sequence space $\ell_p, 1 \leq p \leq \infty$. Define A and R in $\mathcal{B}(\ell_p)$ by

$$A(x_1, x_2, \dots) = (x_1, \dots, x_\tau, 0, 0, \dots),$$

$$R(x_1, x_2, \dots) = \frac{\sigma - \tau}{\sigma}(x_1, \dots, x_\tau, \ x_{\tau+1}, \dots, x_\sigma, 0, 0, \dots).$$

Here τ and σ are non-negative integers, $\sigma > \tau$. Clearly $A \in \mathcal{P}_\tau(\ell_p)$ and $A + R \in \mathcal{P}_\sigma(\ell_p)$. Hence

$$\operatorname{dist}\left(\mathcal{P}_\tau(\ell_p), \ \mathcal{P}_\sigma(\ell_p)\right) \leq \|R\| = \frac{\sigma - \tau}{\sigma}.$$

Combining the results obtained so far, we get

$$1 - \frac{2\tau}{\tau + \sigma} \leq \operatorname{dist}\left(\mathcal{P}_\tau(\ell_p), \ \mathcal{P}_\sigma(\ell_p)\right) \leq 1 - \frac{\tau}{\sigma}, \qquad \sigma > \tau.$$

For fixed τ, this gives

$$\lim_{\sigma \to \infty} \operatorname{dist}\left(\mathcal{P}_\tau(\ell_p), \ \mathcal{P}_\sigma(\ell_p)\right) = 1.$$

On the other hand we have

$$\lim_{\tau \to \infty} \operatorname{dist}\left(\mathcal{P}_\tau(\ell_p), \ \mathcal{P}_{\tau+\alpha}(\ell_p)\right) = 0$$

for any given positive integer α. □

5. Left versus right logarithmic residues

In this section we address the issue of *left* versus *right* logarithmic residues. There is an intriguing connection with Theorem 4.4 (iii) above.

Theorem 5.1. *Let X be a complex Banach space, let D be a bounded Cauchy domain in $\mathbb{C}$ and let L and R be bounded linear operators on X. The following statements are equivalent:*

(i) *There exists a function $F \in \mathcal{A}_\partial(D; \mathcal{B}(X))$, Fredholm on D, such that L is the left and R is the right logarithmic residue of F with respect to D, i.e.,*

$$L = LR_{left}(F; D) = \frac{1}{2\pi i} \int_{\partial D} F'(\lambda) F^{-1}(\lambda) d\lambda, \qquad (5.1)$$

$$R = LR_{right}(F; D) = \frac{1}{2\pi i} \int_{\partial D} F^{-1}(\lambda) F'(\lambda) d\lambda; \qquad (5.2)$$

(ii) *The operators L and R have finite rank and*

$$\max\{\operatorname{rank} L, \ \operatorname{rank} R\} \leq \operatorname{trace} L = \operatorname{trace} R;$$

(iii) *The operators L and R belong to the same connected component of $\mathcal{P}(X)$.*

Here, as introduced above, $\mathcal{P}(X)$ is the set of logarithmic residues of Fredholm operator function (with underlying space X). The proof, especially the part dealing with the implication (ii) $\Rightarrow$ (i) will provide additional information about the freedom one has in choosing the function F.

Proof. The implication (i) $\Rightarrow$ (ii) is immediate from Propositions 3.1 and 3.2 (cf. the proof of Theorem 3.4). One can reformulate (iii) as: *there exists a non-negative integer τ such that L and R belong to $\mathcal{P}_\tau(X)$.* Thus the equivalence of (ii) and (iii) is clear from Theorem 4.4. It remains to prove that (ii) implies (i).

Suppose (ii) holds. We shall show that (i) holds by an argument similar to that presented in the proof of [BES4], Theorem 6.4. In fact we shall establish the following version of (i). *Let $\tau = \operatorname{trace} L = \operatorname{trace} R$. Then, given the bounded Cauchy domain D and distinct complex numbers $\lambda_1, \ldots, \lambda_\tau$ in D, there exists a function $F : \mathbb{C} \to \mathcal{B}(X)$ with the following properties:*

(a) *F is entire, i.e., analytic on all of $\mathbb{C}$,*

(b) *F takes invertible values on all of $\mathbb{C}$, except in $\lambda_1, \ldots, \lambda_\tau$ where F^{-1} has simple poles;*

(c) *$F \in \mathcal{A}_\partial(D; \mathcal{B}(X))$ and F is Fredholm on D,*

(d) *L is the left and R is the right logarithmic residue of F with respect to D, i.e., (5.1) and (5.2) are satisfied.*

The proof involves an interpolation argument.

Write L and R as sums of τ rank one projections on X:

$$L = \sum_{j=1}^{\tau} P_j, \qquad R = \sum_{j=1}^{\tau} Q_j. \tag{5.3}$$

By Lemma 4.3 there exist $T_1, \ldots, T_\tau$ in $\mathcal{B}(X)$ such that

$$Q_j = \exp(-T_j)\, P_j \exp(T_j), \qquad j = 1, \ldots, \tau. \tag{5.4}$$

Choose scalar polynomials $r_1, \ldots, r_\tau$ with

$$r_k(\lambda_j) = \delta_{kj}, \qquad r_k'(\lambda_j) = 0, \qquad j, k = 1, \ldots, \tau,$$

(δ_{kj} is the Kronecker delta) and put

$$H_k(\lambda) = \exp\left(r_k(\lambda) T_k\right).$$

Then $H_k : \mathbb{C} \to \mathcal{B}(X)$ is analytic and takes invertible values on all of $\mathbb{C}$. Also

$$\begin{aligned}
H_k(\lambda_j) &= I, & k, j &= 1, \ldots, \tau;\ k \neq j, \\
H_j(\lambda_j) &= \exp(T_j), & j &= 1, \ldots, \tau, \\
H_k'(\lambda_j) &= 0, & k, j &= 1, \ldots, \tau.
\end{aligned}$$

Write $H(\lambda) = H_1(\lambda) \cdots H_\tau(\lambda)$. Then $H : \mathbb{C} \to \mathcal{B}(X)$ is analytic and takes invertible values on all of $\mathbb{C}$. Also

$$H(\lambda_j) = \exp(T_j), \quad H'(\lambda_j) = 0, \quad j = 1, \ldots, \tau,$$

Let $G : \mathbb{C} \to \mathcal{B}(X)$ be an analytic function with the following properties: G takes invertible values on all of $\mathbb{C}$, except in $\lambda_1, \ldots, \lambda_\tau$ where G^{-1} has simple poles and

$$LR_{left}(G; \lambda_j) = LR_{right}(G; \lambda_j) = P_j, \qquad j = 1, \ldots, \tau,$$

For an explicit construction of G, see [E]. In the present situation, where $P_1, \ldots, P_\tau$ are of finite rank (even of rank one), the construction in [E] yields a G which takes Fredholm values in the point $\lambda_1, \ldots, \lambda_\tau$. With such a G, we put $F(\lambda) = G(\lambda)H(\lambda)$. Then $F : \mathbb{C} \to \mathcal{B}(X)$ clearly has the properties (a)–(c). It remains to prove that (d) is satisfied too.

Observe that, for ρ positive and sufficiently small,

$$LR_{left}(F; \lambda_j) = LR_{left}(G; \lambda_j) + \frac{1}{2\pi i} \int_{|\lambda - \lambda_j| = \rho} G(\lambda) H'(\lambda) H^{-1}(\lambda) G^{-1}(\lambda) d\lambda.$$

The first term in the right hand side is equal to P_j; the second vanishes because G^{-1} has a simple pole at λ_j and $H'(\lambda_j) = 0$. So

$$LR_{left}(F; \lambda_j) = P_j, \qquad j = 1, \ldots, \tau.$$

Analogously we have

$$LR_{right}(F; \lambda_j) = LR_{right}(H; \lambda_j) + \frac{1}{2\pi i} \int_{|\lambda - \lambda_j| = \rho} H^{-1}(\lambda) G^{-1}(\lambda) G'(\lambda) H(\lambda) d\lambda.$$

The first term in the right hand side vanishes; the second is equal to

$$H^{-1}(\lambda_j) P_j H(\lambda_j).$$

But
$$H(\lambda_j) = \exp(T_j), \ H^{-1}(\lambda_j) = \exp(-T_j),$$
and it follows from (5.4) that
$$LR_{right}(F; \lambda_j) = Q_j, \ j = 1, \ldots, \tau.$$

Now
$$LR_{left}(F; D) = \sum_{j=1}^{\tau} LR_{left}(F; \lambda_j), \qquad LR_{right}(F; D) = \sum_{j=1}^{\tau} LR_{right}(F; \lambda_j),$$
and the desired result is clear from (5.3). □

Earlier in this paper, we referred to the work [GS1] of Gohberg and Sigal on algebraic multiplicities of Fredholm operator valued function (cf. [GGK] and [BKL]). The statements (ii) and (iii) in Theorem 5.1 can be reformulated as: *The operators L and R are logarithmic residues of analytic operator valued functions that are Fredholm on the underlying Cauchy domains and that have the same algebraic multiplicity relative to the boundaries of these domains.*

In connection with Theorem 5.1 we also note the following. Consider one of the connected components of $\mathcal{P}(X)$, i.e. one of the sets $\mathcal{P}_\tau(X)$ where τ is a non-negative integer. If $\tau = 0$, we are dealing with $\{0\}$ and there is nothing interesting to say. Assume τ is positive. Then $\mathcal{P}_\tau(X)$ consists of non-zero finite rank operators which pairwise can be written as left and right logarithmic residues of a single analytic (even entire) $\mathcal{B}(X)$-valued function. Now let ℓ and r be positive integers (in a moment to be identified with ranks) such that both ℓ and r do not exceed τ and the dimension of X. If X is infinite dimensional, this of course means that $\ell, r \leq \tau$. We claim that there exist $L, R \in \mathcal{P}_\tau(X)$ with rank $L = \ell$, rank $R = r$. To see this, write d for the largest of the integers ℓ and r, let α and β be positive real numbers and observe that there exist $d \times d$ matrices A and B for which rank $A = \ell$, trace $A = \alpha$, rank $B = r$ and trace $B - \beta$. The desired result follows by taking $\alpha = \beta = \tau$, defining L and R first on a d-dimensional subspace of X and then extending the definition in an obvious manner ("adding zeros") to the whole space X.

6. Concluding remarks

In this paper, we considered logarithmic residues of functions whose values on the underlying Cauchy domain are Fredholm operators. A special case is formed by functions of the type
$$F(\lambda) = \alpha(\lambda)I + C(\lambda)$$
where C takes compact values and α is a non-vanishing scalar function. In light of the spectral theory for compact operators ($F(\lambda) = \lambda I - T$, T compact) one may wonder what happens when the condition that α does not vanish is dropped. We shall come back to this in the forthcoming paper [BES5]. In another article [BES6] we shall use the (factorization and interpolation) techniques employed above to

study logarithmic residues of Banach algebra valued functions having a simply meromorphic inverse. By this we mean that the inverse is meromorphic and has poles of order one only.

The paper [BES4] is quite strongly related to the present one in that it deals with logarithmic residues in certain matrix algebras. In all situations considered there, the set of left/right logarithmic residues coincides with the set of sums of idempotents in the matrix algebra in questions. For examples of matrix algebras where these sets are not the same, see the forthcoming paper [BES6].

Finally, let us return to Theorem 4.2. There are many questions that can be asked with regard to this theorem. We mention only a few here. Is the boundedness condition (4.1) essential? How important is upper semi-continuity? What can be said about the structure (topological properties, connections with special linear combinations of finite rank projections etc.) of the set $\mathcal{P}(f; X)$? The authors intend to deal with these issues – even in a somewhat wider framework – in another forthcoming paper.

We conclude with the following example, which shows that the boundedness condition (4.1) is indeed essential for the continuity of the trace in the case where X is the sequence space ℓ^p, $1 \leq p \leq \infty$.

Example 6.1. *Suppose that condition (4.1) does not hold. Then there exists a sequence $\{z_n\}_{n=1}^{\infty}$ such that either $|z_n| \to \infty$ and $|z_n|^{-1} r_n \to \infty$ or $z_n \to z$ for some $z \in \mathbb{C}$ and $r_n \to \infty$, where $r_n = f(z_n)$. In the first case we define the diagonal operator*

$$A_n = \mathrm{diag}\,\underbrace{\left(\frac{z_n}{r_n}, \frac{z_n}{r_n}, \ldots, \frac{z_n}{r_n}, 0, 0, \ldots \right)}_{r_n \text{ times}}.$$

Obviously, trace $A_n = z_n$ *and* rank $A_n \leq r_n$. *Because $r_n = f(z_n)$, it follows that* rank $A_n \leq f(\mathrm{trace}\ A_n)$, *i.e., $A_n \in \mathcal{P}(f; X)$. On the other hand, $A_n \to 0$. Hence trace is not continuous on $\mathcal{P}(f; X)$.*

In the case where $z_n \to z$ and $r_n \to \infty$, we construct a counterexample as follows. Choose $a \in \mathbb{C}$ different from z such that $f(z - a) \geq 1$. This is possible because (4.1) does not hold. Put

$$B_n = \mathrm{diag}\,\left(z_n - a, \underbrace{\frac{a}{r_n - 1}, \ldots, \frac{a}{r_n - 1}}_{r_n - 1 \text{ times}}, 0, 0, \ldots \right).$$

Here we have trace $B_n = z_n$ *and* rank $B_n \leq r_n$. *Thus $B_n \in \mathcal{P}(f; X)$. Clearly $B_n \to B$ with $B = \mathrm{diag}\,(z - a, 0, 0, \ldots)$. Now trace $B = z - a$ and rank $B = 1$. Since $f(z - a) \geq 1$, we may conclude that $B \in \mathcal{P}(f; X)$. Note that trace $B_n = z_n$ converges to z and not to $z - a = $ trace B. Consequently, also in this case, the trace is not continuous on $\mathcal{P}(f; X)$.*   □

References

[B1]   Bart, H., Spectral properties of locally holomorphic vector-valued functions, *Pacific J. Math.* **52** (1974), 321–329.

[B2]   Bart, H., Transfer functions and operator theory, *Linear Algebra Appl.* **84** (1986), 33–61.

[BES1] Bart, H., Ehrhardt, T., Silbermann, B., Zero sums of idempotents in Banach algebras, *Integral Equations and Operator Theory* **19** (1994), 125–134.

[BES2] Bart, H., Ehrhardt, T., Silbermann, B., Logarithmic residues in Banach algebras, *Integral Equations and Operator Theory* **19** (1994), 135–152.

[BES3] Bart, H., Ehrhardt, T., Silbermann, B., Logarithmic residues, generalized idempotents and sums of idempotents in Banach algebras, *Integral Equations and Operator Theory* **29** (1997), 155–186.

[BES4] Bart, H., Ehrhardt, T., Silbermann, B., Sums of idempotents and logarithmic residues in matrix algebras, In: *Operator Theory and Analysis, The M.A. Kaashoek Anniversary Volume* (Eds. H. Bart, I. Gohberg, A.C.M. Ran), Operator Theory: Advances and Applications, Vol. 122, Birkhäuser, Basel 2001, 139–168.

[BES5] Bart, H., Ehrhardt, T., Silbermann, B., Logarithmic residues of analytic Banach algebra valued functions possessing a simply meromorphic inverse, forthcoming.

[BES6] Bart, H., Ehrhardt, T., Silbermann, B., Logarithmic residues and sums of idempotents in the Banach algebra generated by the compact operators and the identity, forthcoming.

[BGK1] Bart H., Gohberg I., Kaashoek, M.A., *Minimal Factorization of Matrix and Operator Functions*, Operator Theory: Advances and Applications, Vol. 1, Birkhäuser, Basel 1979.

[BGK2] Bart, H., Gohberg I., Kaashoek, M.A., The state space method in analysis, in: *Proceedings ICIAM 87, Paris-La Vilette* (A.H.P. van der Burgh and R.M.M. Mattheij, eds.), Reidel, 1987, 1–16.

[BKL]   Bart, H., Kaashoek, M.A., Lay, D.C., The integral formula for the reduced algebraic multiplicity of meromorphic operator functions, *Proceedings Edinburgh Mathematical Society* **21** (1978), 65–72.

[E]   Ehrhardt, T., Finite sums of idempotents and logarithmic residues on connected domains, *Integral Equations and Operator Theory* **21** (1995), 238–242.

[GGK]   Gohberg, I., Goldberg, S., Kaashoek, M.A., *Classes of Linear Operators, Vol. 1*, Operator Theory: Advances and Applications, Vol. 49, Birkhäuser, Basel 1990.

[GKL]   Gohberg, I., Kaashoek, M.A., Lay, D.C., Equivalence, linearization and decompositions of holomorphic operator functions, *J. Funct. Anal.* **28** (1978), 102–144.

[GS1]   Gohberg, I.C., Sigal, E.I., An operator generalization of the logarithmic residue theorem and the theorem of Rouché, *Mat. Sbornik* **84 (126)** (1971), 607–629 (in Russian), English Transl. in: *Math. USSR Sbornik* **13** (1971), 603–625.

[GS2]   Gohberg, I.C., Sigal, E.I., Global factorization of meromorphic operator functions and some applications, *Mat. Issled.* **6**, no.1 (1971), 63–82 (in Russian).

[HP]   Hartwig, R.E., Putcha, M.S., When is a matrix a sum of idempotents ?, *Linear and Multilinear Algebra* **26** (1990), 279–286.

[M]   Mittenthal, L., Operator valued analytic functions and generalizations of spectral theory, *Pacific J. Math.* **24** (1968), 119–132.

[T]   Thijsse, G.Ph.A., *Decomposition Theorems for Finite-Meromorphic Operator Functions*, Thesis Vrije Universiteit Amsterdam, Krips Repro, Meppel 1978.

[TL]   Taylor, A.E., Lay, D.C., *Introduction to Functional Analysis*, Second Edition, John Wiley and Sons, New York 1980.

[W]   Wu, P.Y., Sums of idempotent matrices, *Linear Algebra Appl.* **142** (1990), 43–54.

Econometrisch Instituut
Erasmus Universiteit Rotterdam
Postbus 1738
3000 DR Rotterdam
The Netherlands

Fakultät für Mathematik
Technische Universität Chemnitz
09107 Chemnitz
Germany

1991 Mathematics Subject Classification. Primary 30G30; Secondary 47A53

Received May 31, 2000

Operator Theory:
Advances and Applications, Vol. 130, 107–120
© 2001 Birkhäuser Verlag Basel/Switzerland

Positive Linear Maps and the Lyapunov Equation

Rajendra Bhatia and Ludwig Elsner

Dedicated to Peter Lancaster, a good friend and colleague

Abstract. It is well–known that positivity plays an important role in the study of the discrete time and the continuous time Lyapunov equations. We show how general theorems on positive linear maps on matrices may be used in this context. Our method leads to several old, recent, and new bounds on the sensitivity of these equations. Further, it can be applied to related problems and to other matrix equations as well.

1. Introduction

A linear map on the space of $n \times n$ complex matrices (or, more generally, on any C^*–algebra) is said to be **positive** if it preserves the cone of positive (semidefinite) matrices. Because of their importance in the theory of operator algebras, such maps, and their very special subclass of completely positive maps, have been studied extensively in recent years.

The aim of this article is to show how some elementary results on positive maps can be fruitfully employed in the study of the Lyapunov matrix equation, a topic of great interest in differential equations and control theory.

The matrix equation

$$X - F^* X F = Q \tag{1.1}$$

is called the discrete–time Lyapunov equation, or the Stein equation. It is assumed that F is **stable**; i.e., the spectrum of F is contained in the open unit disk. In this case, the equation (1.1) has a unique solution X for every Q. Further, if Q is positive, then so is X.

Closely related to (1.1) is the equation

$$A^* X + X A = W, \tag{1.2}$$

called the continuous–time Lyapunov equation. Here it is assumed that A is **positively stable**; i.e., the spectrum of A is contained in the open right half plane. In this case, the equation (1.2) has a unique solution X for every W. Further, if W is positive, then so is X.

While the positivity statements about these two equations are of great importance and have been known right from the beginning of their study, general facts from the theory of positive linear maps have scarcely been used in the vast literature concerning these equations.

In this article we give an overview emphasizing positivity and show a unified approach for obtaining results of several authors. Specifically, we provide simple proofs of several known error bounds on solutions and derive some new ones.

We point out how the same approach may be used in studying some generalizations of the Lyapunov equation and for obtaining other interesting matrix inequalities.

2. Positive Linear Maps

Let $\mathbb{M}$ be the space of $n \times n$ complex matrices. For brevity we will use the term **positive matrix** to mean a positive semidefinite matrix. A linear map $\Phi : \mathbb{M} \to \mathbb{M}$ is said to be **positive** if $\Phi(A)$ is positive whenever A is positive.

Let $||A||$ denote the norm of an element A of $\mathbb{M}$ viewed as a linear operator on the Euclidean space $\mathbb{C}^n$. (This is called the **spectral norm** in the numerical analysis literature; it is equal to the largest singular value of A). For a linear map Φ on $\mathbb{M}$ let

$$||\Phi|| := \sup_{||A||=1} ||\Phi(A)||. \tag{2.1}$$

The space $\mathbb{M}$ equipped with the norm $|| \cdot ||$ is a C^*–algebra. One of the basic facts about positive linear maps on C^*–algebras with an identity is that they attain their norm at the identity. In other words

$$||\Phi|| = ||\Phi(I)||. \tag{2.2}$$

See [25] or [20, Exercise 10.5.10]. Since $||A||$ and $||\Phi||$ are difficult to compute in general, this relation is of fundamental importance.

It is instructive to see a special instance of this general theorem in a setting more familiar in matrix theory. Le Z be a positive matrix, and let $\Phi(X) = Z \circ X$ be the **Schur product** or the **Hadamard product** of Z and X; i.e., if we write $X = [x_{ij}]$ where x_{ij} are the entries of the matrix X, then $Z \circ X = [z_{ij}\, x_{ij}]$.

By a well–known theorem of Schur Φ is a positive linear map on $\mathbb{M}$, and we have

$$||Z \circ X|| \leq \max_i z_{ii}\, ||X|| \tag{2.3}$$

for all X. See [16, p. 95], [17, p. 343]. This assertion is a special instance of (2.2).

In the study of operator algebras completely positive maps have turned out to be more useful than positive maps. Let $[A_{ij}]_{N \times N}$ denote an $N \times N$ block matrix whose entries A_{ij} are elements of $\mathbb{M}$. We say that a linear map Φ on $\mathbb{M}$ is N–positive if the matrix $[\Phi(A_{ij})]_{N \times N}$ is positive whenever $[A_{ij}]_{N \times N}$ is positive.

(Thus positive linear maps are 1–positive). We say that Φ is **completely positive** if it is N–positive for all N.

The map $\Phi(A) = A^{tr}$ taking a matrix to its transpose is positive but not 2–positive. A basic structure theorem due to Choi [9] says that a linear map Φ on $\mathbb{M}$ is completely positive if and only if it has the form

$$\Phi(A) = \sum_i V_i^* \, A V_i \tag{2.4}$$

for some V_i in $\mathbb{M}$.

It is easy to see from the definition that the Schur product map $\Phi(X) = Z \circ X$ is completely positive for every positive matrix Z. This is so because the $N \times N$ block matrix with all entries Z is again positive.

A linear map Φ on $\mathbb{M}$ is called **unital** if $\Phi(I) - I$, and **trace preserving** if tr $\Phi(A) = $ tr A for all A. A positive, unital, trace preserving map on $\mathbb{M}$ is said to be **doubly stochastic**. See [1, Section 7] for an admirably concise introduction to the properties of such maps.

If Z is a positive matrix with unit diagonal; i.e., $z_{ii} = 1$ for all i, then the Schur product map $\Phi(X) = Z \circ X$ is doubly stochastic.

Let $||| \cdot |||$ be any **unitarily invariant norm** on $\mathbb{M}$, viz. one that satisfies the properties

$$|||UAV||| = |||A||| \text{ for all } A \in \mathbb{M} \text{ and unitary } U, V,$$

and

$$|||E||| = 1 \text{ for all rank–one orthoprojectors } E.$$

Special examples of such norms are the **Schatten p–norms** defined as

$$||A||_p := \left(\sum_{j=1}^{n} s_j^p(A) \right)^{1/p}, \ 1 \leq p \leq \infty,$$

where $s_1(A) \geq s_2(A) \geq \ldots \geq s_n(A)$ are the singular values of A. Thus

$$||A||_\infty = s_1(A) = ||A||$$

is the operator (spectral) norm, and

$$||A||_2 = (\text{tr } A^*A)^{1/2} = \left(\sum_{i,j} |a_{ij}|^2 \right)^{1/2} = ||A||_F$$

the Frobenius (or Hilbert–Schmidt) norm.

A more general class of norms of some interest is that of **weakly unitarily invariant** norms. These are norms τ on $\mathbb{M}$ that satisfy the relation

$$\tau(A) = \tau(UAU^*) \text{ for all } A \in \mathbb{M} \text{ and unitary } U.$$

This family includes all the unitarily invariant norms and several others. For example, the numerical radius

$$w(A) := \sup_{||x||=1} | < x, Ax > | = \sup_{||x||=1} |x^*Ax|$$

is such a norm. Another example is $||A|| + |\mathrm{tr}\, A|$. See [3, Chapter 40] for a detailed exposition of such norms.

The inequality (2.3) has a generalization. If Z is positive, then for every unitarily invariant norm

$$|||Z \circ X||| \leq \max\, z_{ii}\, |||X|||\ \ \text{for all}\ \ X. \tag{2.5}$$

This follows, e.g., from [17, p. 343] and the Fan dominance theorem. More generally, if Φ is any doubly stochastic map on $\mathbb{M}$, then

$$|||\Phi(X)||| \leq |||X|||\ \ \text{for all}\ \ X. \tag{2.6}$$

See [1, Corollary 7.8]. Indeed, (2.6) implies (2.5): If D is the diagonal matrix whose i-th diagonal entry is the square root of z_{ii} then $\Phi(X) = D^{-1}ZD^{-1} \circ X$ is a doubly stochastic map. Now apply (2.6) and (4.8) to $Z \circ X = D\Phi(X)D$ to get (2.5).

More can be said when X is Hermitian. Then we have for every weakly unitarily invariant norm τ

$$\tau(\Phi(X)) \leq \tau(X)\ \ \text{for all Hermitian}\ \ X. \tag{2.7}$$

This follows from [1, Thm. 7.1].

3. The Lyapunov Equations

Given a stable matrix F and a positively stable matrix A, let $\mathcal{S}_F$ and $\mathcal{L}_F$ be the linear maps on $\mathbb{M}$ defined as

$$\mathcal{S}_F(X) = X - F^*XF, \tag{3.1}$$

$$\mathcal{L}_A(X) = A^*X + XA. \tag{3.2}$$

The statements about the equations (1.1) and (1.2) made in Section 1 can be rephrased to say that the maps $\mathcal{S}_F$ and $\mathcal{L}_A$ are invertible and that their inverses are positive maps. As we will see, they are in fact completely positive.

Let us recall some of the familiar arguments that show that $\mathcal{S}_F^{-1}$ and $\mathcal{L}_A^{-1}$ are positive. This will bring out clearly the relevance of the notions introduced in Section 2.

First assume that F is diagonal, $F = \mathrm{diag}\,(\lambda_1, \ldots, \lambda_n)$. Then we can solve (1.1) simply by comparing matrix entries; we have

$$x_{ij} = \frac{q_{ij}}{1 - \bar{\lambda}_i\, \lambda_j}. \tag{3.3}$$

In the same way, if $A = \operatorname{diag}(\mu_1, \ldots, \mu_n)$, then the equation (1.2) can be solved by putting

$$x_{ij} = \frac{w_{ij}}{\bar{\mu}_i + \mu_j}. \tag{3.4}$$

We can write this as $X = M \circ W$, where M is the matrix with entries $m_{ij} = (\bar{\mu}_i + \mu_j)^{-1}$. In the same way we can write the relation (3.3) as $X = L \circ Q$, where L is the matrix with entries $l_{ij} = (1 - \bar{\lambda}_i \lambda_j)^{-1}$. We will show that the matrices M and L are positive.

To each point λ in the open unit disk there corresponds a unique point μ in the open right half plane such that

$$\lambda = \frac{\mu - 1}{\mu + 1}. \tag{3.5}$$

Using this correspondence, we have

$$\frac{1}{1 - \bar{\lambda}_i \lambda_j} = \frac{(\bar{\mu}_i + 1)(\mu_j + 1)}{2(\bar{\mu}_i + \mu_j)}. \tag{3.6}$$

So, we can write $L = \frac{1}{2} D^* M D$, where $D = \operatorname{diag}(\mu_1 + 1, \ldots, \mu_n + 1)$. Thus L is positive if and only if M is positive. The entries of M are

$$m_{ij} = \frac{1}{\bar{\mu}_i + \mu_j} = \int_0^\infty e^{-t(\bar{\mu}_i + \mu_j)} \, dt.$$

This last expression is the inner product between the functions f_i and f_j in the space $L_2([0, \infty))$ defined as $f_i(t) = e^{-t\mu_i}$. Thus M is a Gram matrix and is, therefore, positive.

Thus the maps $\mathcal{S}_F^{-1}$ and $\mathcal{L}_A^{-1}$ are positive when F and A are diagonal. One can now prove that this is also the case when they are similar to diagonal matrices. Finally a continuity argument gives the result for all F and A. (See [17, pp. 301, 347]).

The solution to (1.1) can be written explicitly as

$$X = \sum_{k=0}^\infty F^{*k} Q F^k, \tag{3.7}$$

and that to (1.2) as

$$X = \int_0^\infty e^{-A^* t} W e^{-At} \, dt. \tag{3.8}$$

See [23, p. 443]. The condition that F be stable guarantees the convergence of the series (3.7), and that A be positively stable guarantees the convergence of the integral (3.8). It is then easy to see that these X satisfy the respective equations. One advantage of writing the solutions in this form is that now the positivity of

$\mathcal{S}_F^{-1}$ and $\mathcal{L}_A^{-1}$ is obvious. Another advantage is that all arguments are independent of matrix representations and are valid for operators in Hilbert space.

The advantages of having the solutions to operator equations in different closed forms are well–known. See, in particular, the papers [5, 8] where several forms of the solution to the Sylvester equation $AX - XB = Y$ are described and used for obtaining different inequalities. In this spirit, the solution to (1.1) can be written in the form of an integral:

$$X = \frac{1}{2\pi} \int_0^{2\pi} [(e^{i\theta} - F)^{-1}]^* \, Q[(e^{i\theta} - F)^{-1}] \, d\theta. \tag{3.9}$$

One easy way of deriving this is as follows. If λ is a point in the complex plane with $|\lambda|$ larger than the spectral radius of F, then we have the Neumann series expansion for the resolvent

$$(\lambda - F)^{-1} = \frac{1}{\lambda}(1 - \lambda F)^{-1} = \sum_{k=0}^{\infty} \lambda^{-k-1} \, F^k.$$

So, the right hand side of (3.9) can be written as

$$\frac{1}{2\pi} \int_0^{2\pi} \left[\sum_{j=0}^{\infty} e^{i(j+1)\theta} \, F^{*j} \right] Q \left[\sum_{k=0}^{\infty} e^{-i(k+1)\theta} F^k \right] d\theta \qquad = \dots$$

$$\frac{1}{2\pi} \int_0^{2\pi} \sum_{j,k=0}^{\infty} e^{i(j-k)\theta} \, F^{*j} \, QF^k \, d\theta \quad = \sum_{k=0}^{\infty} F^{*k} \, QF^k.$$

This is exactly the expression in (3.7). We remark that (3.9) is related to an integral representation in [23], attributed to Krein.

In [8] the authors explained how one might solve matrix equations first for scalars, and then write the solution in different ways to facilitate the entry of noncommuting variables. Here is another instance of this heuristic principle at work. For complex numbers, the equation $x - |a|^2 x = q$ with $|a| < 1$, has the solution

$$x = \frac{q}{1 - |a|^2} = \frac{q}{1 - r^2}, \quad \text{where } a = re^{i\varphi}.$$

We look for different expressions for $\frac{1}{1-r^2}$. One of them is an infinite series. This leads to the solution (3.7). To get an integral expression, recall that [25, p. 112]

$$\frac{1}{2\pi} \int_0^{2\pi} \frac{1 - r^2}{1 - 2r \, \cos\theta + r^2} \, d\theta = 1.$$

From this it is easy to see that

$$\frac{q}{1 - |a|^2} = \frac{1}{2\pi} \int_0^{2\pi} \frac{q}{|e^{i\theta} - a|^2} \, d\theta.$$

The integral (3.9) is a noncommutative version of this expression.

We have noted that $\mathcal{S}_F^{-1}$ and $\mathcal{L}_A^{-1}$ are positive linear maps on $\mathbb{M}$. In fact, they are completely positive. Again, one can see this in different ways. The very form of the solutions (3.7) and (3.8) and Choi's characterisation (2.4) immediately give this fact. Another illuminating way of looking at this is the following. Let $\widetilde{F}$ be the $N \times N$ block diagonal matrix with all diagonal entries equal to F. If F is stable then so is $\widetilde{F}$. The N^2 equations

$$X_{ij} - F^* X_{ij} F = Q_{ij} \quad i, j = 1, 2, \ldots, N$$

can be written as a single matrix equation

$$[X_{ij}] - \widetilde{F}^*[X_{ij}]\widetilde{F} = [Q_{ij}],$$

where $[X_{ij}]$ is the $N \times N$ block matrix with entries X_{ij}. The map $(\mathcal{S}_{\widetilde{F}})^{-1}$ is positive. This is equivalent to saying that $\mathcal{S}_F^{-1}$ is N–positive. Since this is true for all N, $\mathcal{S}_F^{-1}$ is completely positive. The same argument shows that $\mathcal{L}_A^{-1}$ is completely positive.

Let K and H be the matrices defines as

$$K = \mathcal{S}_F^{-1}(I), \tag{3.10}$$

$$H = \mathcal{L}_A^{-1}(I), \tag{3.11}$$

and let Σ_F and Γ_A be the linear maps on $\mathbb{M}$ defined as

$$\Sigma_F(Y) = K^{-1/2} \, \mathcal{S}_F^{-1}(Y) K^{-1/2}, \tag{3.12}$$

$$\Gamma_A(Y) = H^{-1/2} \, \mathcal{L}_A^{-1}(Y) H^{-1/2}. \tag{3.13}$$

It has been shown in [2] that when A is normal, Γ_A is a doubly stochastic map. Following the same arguments it can be shown that when F is normal, Σ_F is doubly stochastic. In fact as shown in [2, Thm. 2] the maps Σ_F and Γ_A are doubly stochastic if and only if the matrices F and A, respectively, are normal.

4. The sensitivity of solutions

In numerical computations, the given matrices in (1.1) and (1.2) are subjected to approximations. It is then of interest to obtain error bounds for the solution. Among the several papers on this subject are [2, 5, 6, 8, 10, 12, 13, 14, 19, 21, 26, 27]. The use of properties of positive linear maps in this context seems to have been made first in [2] and then in [6, 26]. Let us explain the basic ideas.

We begin by summarising the general framework in the numerical analysis of linear systems [15]. Consider the equation

$$Lx = w \qquad (4.1)$$

where x and w are vectors in a Banach space X and L is an invertible bounded linear operator on X. Let ΔL and Δw be small changes in L and w respectively, and consider the perturbed equation

$$(L + \Delta L)(x + \Delta x) = w + \Delta w. \qquad (4.2)$$

From (4.1) and (4.2) we get

$$\Delta x = -L^{-1}\Delta L(x + \Delta x) + L^{-1}\Delta w.$$

Hence,

$$||\Delta x|| \leq ||L^{-1}||\,||\Delta L||\,||x + \Delta x|| + ||L^{-1}||\,||\Delta w||,$$

and therefore,

$$\frac{||\Delta x||}{||x + \Delta x||} \leq ||L^{-1}|| \left(||\Delta L|| + \frac{||\Delta w||}{||x + \Delta x||} \right). \qquad (4.3)$$

To achieve some symmetry use the inequality

$$\frac{1}{||x + \Delta x||} \leq \frac{||L + \Delta L||}{||w + \Delta w||}$$

that can be obtained from (4.2), and get from (4.3)

$$\frac{||\Delta x||}{||x + \Delta x||} \leq ||L^{-1}|| \left(||\Delta L|| + \frac{||L + \Delta L||\,||\Delta w||}{||w + \Delta w||} \right), \qquad (4.4)$$

or equivalently,

$$\frac{||\Delta x||}{||x + \Delta x||} \leq ||L^{-1}||\,||L + \Delta L|| \left(\frac{||\Delta L||}{||L + \Delta L||} + \frac{||\Delta w||}{||w + \Delta w||} \right). \qquad (4.5)$$

Interchanging the roles of (4.1) and (4.2) we could also write

$$\frac{||\Delta x||}{||x||} \leq ||(L + \Delta L)^{-1}||\,||L|| \left(\frac{||\Delta L||}{||L||} + \frac{||\Delta w||}{||w||} \right). \qquad (4.6)$$

Now let us apply this to equations (1.1) and (1.2). Both equations are of the form

$$\mathcal{L}(X) = W,$$

where $\mathcal{L}$ is a linear operator on $\mathbb{M}$. Equip $\mathbb{M}$ with any unitarily invariant norm $|||\cdot|||$ and let the norm it induces on linear operators on $\mathbb{M}$ be also denoted by $|||\cdot|||$; i.e.,

$$|||\mathcal{L}||| = \sup_{|||X|||=1} |||\mathcal{L}(X)|||. \qquad (4.7)$$

All unitarily invariant norms on $\mathbb{M}$ satisfy the inequality

$$|||XYZ||| \leq ||X||\,|||Y|||\,||Z|| \quad \text{for all } X, Y, Z. \qquad (4.8)$$

Thus the norm of $\mathcal{L}$ is easy to estimate. We have

$$|||\mathcal{S}_F||| = \sup_{|||X|||=1} |||X - F^* X F||| \leq 1 + ||F||^2, \tag{4.9}$$

$$|||\mathcal{L}_A||| = \sup_{|||X|||=1} |||A^* X + X A||| \leq 2||A||. \tag{4.10}$$

Note also

$$
\begin{aligned}
|||\mathcal{S}_{F+\Delta F} - \mathcal{S}_F||| &= \sup_{|||X|||=1} |||F^* X \Delta F + (\Delta F)^* X F + (\Delta F)^* X \Delta F||| \\
&\leq 2||F||\,||\Delta F|| + ||\Delta F||^2 \\
&= ||\Delta F||\,(2||F|| + ||\Delta F||),
\end{aligned}
\tag{4.11}
$$

and

$$|||\mathcal{L}_{A+\Delta A} - \mathcal{L}_A||| \leq 2\,||\Delta A||. \tag{4.12}$$

Now proceeding as in the derivation of the general bound (4.4) and using the inequalities (4.9) – (4.12) one can easily prove the following theorem.

Theorem 4.1. *For the sensitivity of the solution X of the equation (1.1) we have the bound*

$$\frac{|||\Delta X|||}{|||X + \Delta X|||} \leq |||\mathcal{S}_F^{-1}||| \left\{ \alpha\,\frac{||\Delta F||}{||F + \Delta F||} + \beta\,\frac{|||\Delta Q|||}{|||Q + \Delta Q|||} \right\}, \tag{4.13}$$

where

$$\alpha = 2(||F|| + ||\Delta F||)^2, \quad \beta = 1 + ||F + \Delta F||^2. \tag{4.14}$$

For the sensitivity of the solution X of the equation (1.2) we have the bound

$$\frac{|||\Delta X|||}{|||X + \Delta X|||} \leq 2|||\mathcal{L}_A^{-1}|||\,||A + \Delta A|| \left\{ \frac{||\Delta A||}{||A + \Delta A||} + \frac{|||\Delta W|||}{|||W + \Delta W|||} \right\}. \tag{4.15}$$

To use (4.13) and (4.15) we need effective estimates for $|||\mathcal{S}_F^{-1}|||$ and $|||\mathcal{L}_A^{-1}|||$. In Section 3 we have observed that $\mathcal{S}_F^{-1}$ and $\mathcal{L}_A^{-1}$ are positive linear maps on $\mathbb{M}$. Hence by (2.2) we have

$$||\mathcal{S}_F^{-1}|| = ||\mathcal{S}_F^{-1}(I)||, \tag{4.16}$$

$$||\mathcal{L}_A^{-1}|| = ||\mathcal{L}_A^{-1}(I)||. \tag{4.17}$$

These results are proved in [10, 13] by other arguments.

As explained in [2] a standard duality argument shows that

$$||\mathcal{S}_F^{-1}||_1 = ||\mathcal{S}_{F^*}^{-1}(I)||, \tag{4.18}$$

$$||\mathcal{L}_A^{-1}||_1 = ||\mathcal{L}_{A^*}^{-1}(I)||, \tag{4.19}$$

where $||T||_1$ stands for the trace norm

$$||T||_1 := s_1(T) + \cdots + s_n(T).$$

Thus positivity helps us in reducing the problem of finding a norm like $||\mathcal{S}_F^{-1}||$ to that of finding the norm of a single matrix $K = \mathcal{S}_F^{-1}(I)$. This is simpler, but we do not know a good way of finding it. In a recent paper [27] Tippett et al have shown how to do this in several special cases of interest. (Incidentally, many of the results in this paper that are stated for Schatten p–norms are in fact true for all unitarily invariant norms).

In one special, but important, case we not only have effective expressions for (4.16) and (4.17), we can even extend the result to all unitarily invariant norms.

Theorem 4.2. *Let F be normal. Then for every unitarily invariant norm*

$$|||\mathcal{S}_F^{-1}||| = ||\mathcal{S}_F^{-1}(I)|| = \frac{1}{1 - ||F||^2}. \tag{4.20}$$

Let A be normal. Then for every unitarily invariant norm

$$|||\mathcal{L}_A^{-1}||| = ||\mathcal{L}_A^{-1}(I)|| = \frac{1}{2 \min \Re \lambda_i(A)}, \tag{4.21}$$

where $\lambda_i(A)$ are the eigenvalues of A.

Proof. If F is normal then $||F|| = \max |\lambda_j(F)|$. Use the form of the solution (3.7) and the property (4.8) to get

$$|||\mathcal{S}_F^{-1}||| = \sup_{|||Q|||=1} ||| \sum_{k=0}^{\infty} F^{*k} Q F^k ||| \leq \sum_{k=0}^{\infty} ||F||^{2k} = \frac{1}{1 - ||F||^2}.$$

If u is an eigenvector of norm 1 corresponding to the eigenvalue $\lambda_1(F)$ for which $|\lambda_1(F)| = ||F||$, then choosing $Q = uu^*$ we can see that there is equality in the inequality proved above.

The same argument can be adopted to yield (4.21). But let us give a different proof. Since the norms involved are unitarily invariant, we may choose a basis in which $A = \text{diag}(\mu_1, \ldots, \mu_n)$. Use the form of the solution (3.4) and the inequality (2.5) to see

$$|||X||| \leq \frac{1}{2 \min \Re \mu_i} |||W|||.$$

This is the same as saying $|||\mathcal{L}_A^{-1}||| \leq \frac{1}{2 \min \Re \mu_i}$. To see that there is equality here choose the index j for which $\Re \lambda_j = \min_i \Re \lambda_i$, and then choose W to be the matrix with $w_{jj} = 1$ and all other entries 0. $\qquad \square$

For the Frobenius norm, the result of Theorem 4.2 is known [10, p. 1211].

We have also remarked that for normal F and A, the operators defined in (3.12) and (3.13) are doubly stochastic. Thus we could use the inequalities (2.6) and (2.7) to write more perturbation bounds. When the right hand side in (1.1) and (1.2) is Hermitian (a case that does arise in applications [19]) this will give more special information.

5. Some Schwarz Inequalities

To demonstrate further the efficacy of the ideas explained above we use them to give simple proofs of some inequalities proved recently [18]. They are loosely connected to the Lyapunov equation (1.1) in that expressions like $X - F^*XF$ occur here again.

We say that A is a **contraction** if $\|A\| \leq 1$.

Theorem 5.1 (Jocić). *Let F, G be normal contractions. Then for every unitarily invariant norm and for every X in $\mathbb{M}$ we have*

$$|||(I - F^*F)^{1/2} X (I - G^*G)^{1/2}||| \leq |||X - F^*XG|||. \tag{5.1}$$

Proof. First consider the special case $F = G$. We have to prove

$$|||(I - F^*F)^{1/2} X (I - F^*F)^{1/2}||| \leq |||X - F^*XF|||. \tag{5.2}$$

We may assume without loss of generality that $F = \text{diag}(\lambda_1, \ldots, \lambda_n)$. Then the inequality (5.2) says that the norm of the matrix with entries $[(1 - |\lambda_i|^2)^{1/2} x_{ij}(1 - |\lambda_j|^2)^{1/2}]$ is not bigger than the norm of one with entries $[(1 - \overline{\lambda}_i \lambda_j)x_{ij}]$. This is the same as saying that the Schur product with the matrix Z whose entries are

$$z_{ij} = \frac{(1 - |\lambda_i|^2)^{1/2} (1 - |\lambda_j|^2)^{1/2}}{1 - \overline{\lambda}_i \lambda_j} \tag{5.3}$$

is a norm–reducing operation. Since $z_{ii} = 1$ for all i, this statement would follow from (2.5) if we could show that Z is positive. But $Z = D^*LD$, where D is the diagonal matrix with $d_{ii} = (1 - |\lambda_i|^2)^{1/2}$ and L is the matrix with entries $(1 - \overline{\lambda}_i \lambda_j)^{-1}$. The positivity of this matrix has been our *leitmotiv*.

One can prove the general inequality (5.1) from its special case (5.2) by a familiar argument using 2×2 block matrices. Given F, G apply (5.2) with $\begin{pmatrix} F & 0 \\ 0 & G \end{pmatrix}$ in place of F and $\begin{pmatrix} 0 & X \\ 0 & 0 \end{pmatrix}$ in place of X. $\qquad\square$

Say that a countable family $\{A_r\}$ in $\mathbb{M}$ is **square summable** if $\sum \|A_r\|^2 < \infty$. The following theorem is another Schwarz type inequality.

Theorem 5.2 (Jocić). *Let $\{A_r\}$ and $\{B_r\}$ be two square summable families of commuting normal matrices. Then for every unitarily invariant norm and for every X*

$$\left|\left|\left| \sum_{r=1}^{\infty} A_r^* X B_r \right|\right|\right| \leq \left|\left|\left| \left(\sum_{r=1}^{\infty} A_r A_r^* \right)^{1/2} X \left(\sum_{r=1}^{\infty} B_r B_r^* \right)^{1/2} \right|\right|\right|. \tag{5.4}$$

Proof. Once again first consider the special case $A_r = B_r$ for all r. Since A_r are pairwise commuting normal matrices, we can choose an orthonormal basis in which all of them are diagonal. So let $A_r = \mathrm{diag}\,(\lambda_1^{(r)}, \ldots, \lambda_n^{(r)})$. As in the proof of Theorem 5.1, the problem then reduces to showing that the matrix Z with entries

$$z_{ij} = \sum_{r=1}^{\infty} \overline{\lambda_i^{(r)}}\, \lambda_j^{(r)} \Big/ \left(\sum_{r=1}^{\infty} |\lambda_i^{(r)}|^2\right)^{1/2} \left(\sum_{r=1}^{\infty} |\lambda_j^{(r)}|^2\right)^{1/2}$$

is positive. This is evident since $Z = D^* E D$ where D is the diagonal matrix with entries $d_{ii} = \left(\sum_r \lambda_i^{(r)}\right) \Big/ \left(\sum_r |\lambda_i^{(r)}|^2\right)^{1/2}$ and E is the matrix all of whose entries are equal to 1.

This proves (5.4) in the special case $A_r = B_r$. The general case follows again by considering 2×2 block matrices. □

For more matrix Schwarz inequalities the reader is referred to [4] and to [3, Ch. IX]. For a fuller application of the ideas used in our proofs of Theorems 5.1 and 5.2 see [7].

6. Remarks and conclusions

We have shown how a lot of information about the Lyapunov equations (1.1) and (1.2) can be obtained by appealing to general theorems on positive linear maps. There are other equations to which our methods may be applied. For example the equation

$$A^* X + X A + t A^{*1/2} X\, A^{1/2} = W \tag{6.1}$$

where A is positively stable, $A^{1/2}$ is its positively stable square root, and t is a real number has been of some interest [22]. It is known [7] that when A is positive, then for each positive W this equation has a positive solution X if and only if $t \in (-2, 2]$. Other generalizations of the Lyapunov equation where positivity of the solutions is guaranteed are considered in [11]. Our general method can be used in all these contexts.

Most of our arguments work equally well in finite and infinite–dimensional Hilbert spaces. We have tried to illustrate this by giving different proofs for some of our statements.

The error bounds we have obtained in Section 4 include those obtained in [10, 13, 19] by different arguments. Our approach provides a unification and simplification of some of the ideas and connects it to another area where a lot is known. The bounds (4.13) and (4.15) are simple and they are similar to the bound (4.4) for general linear systems.

At the same time it is likely that by taking into account the special features of the equations (1.1) and (1.2) more special and sharper perturbation bounds may be obtained. See [14] for such an approach.

In Section 5 we have shown how the idea of positivity can be applied to related problems.

Acknowledgements. This paper is a revised and expanded version of an earlier report [6]. The first author thanks Sonderforschungsbereich 343 *Diskrete Strukturen in der Mathematik* at the University of Bielefeld for its support of visits during which this work was completed.

References

[1] Ando, Majorization, doubly stochastic matrices and comparison of eigenvalues, *Linear Algebra Appl.* **118** (1989), 163–248.

[2] Bhatia R., A note on the Lyapunov Equation, *Linear Algebra Appl.* **259** (1997), 71–76.

[3] Bhatia R., *Matrix Analysis*, Springer, New York 1997.

[4] Bhatia R., Davis C., A Cauchy–Schwarz inequality for operators with applications, *Linear Algebra Appl.* **223** (1995), 119–129.

[5] Bhatia R. Davis, C. McIntosh A., Perturbation of spectral subspaces and solution of linear operator equations, *Linear Algebra Appl.* **52/53** (1983), 45–67.

[6] Bhatia R., Elsner L., Sensitivity of the solutions of some matrix equations, *Preprint SFB 343* **97–106** (1997), .

[7] Bhatia R., Parthasarathy K.R., Positive definite functions and operator inequalities, *Bull. London Math. Soc.* **32** (2000), 214–228.

[8] Bhatia R., Rosenthal P., How and why to solve the operator equation $AX - XB = Y$, *Bull. London Math. Soc.* **29** (1997), 1–21.

[9] Choi, M.D., Completely positive linear maps on complex matrices, *Linear Algebra Appl.* **10** (1975), 285–290.

[10] Gahinet P.M., Laub A., Kenney C.S., Hewer G., Sensitivity of the stable discrete–time Lyapunov equation, *IEEE Trans. Automat. Control* **35** (1990), 1209–1217.

[11] Godunov S.K., *Modern Aspects of Linear Algebra*, American Mathematical Society, Providence 1998.

[12] Hammarling S.J., Numerical solution of the stable, non–negative definite Lyapunov equation, *IMA J. Numer. Anal.* **2** (1982), 303–323.

[13] Hewer G., Kenney C., The sensitivity of the stable Lyapunov equation, *SIAM J. Control Optim.* **26** (1988), 321–344.

[14] Higham N.J., Perturbation theory and backward error for $AX - XB = C$, *BIT* **33** (1993), 124–136.

[15] Higham N.J., *Accuracy and stability of numerical algorithms*, Society for Industrial and Applied Mathematics 1996.

[16] Horn R., *The Hadamard product*, Matrix Theory and Applications, Proceedings of Symposia in Applied Math., Vol. 20 American Mathematical Society.1990

[17] Horn R., Johnson C.R., *Topics in Matrix Analysis*, Cambridge University Press 1991.

[18] Jocić D.R., Cauchy–Schwarz and means inequalities for elementary operators into norm ideals, *Proc. Amer. Math. Soc.* **126** (1998), 2705–2711.

[19] Jonckheere E.A., New bound on the sensitivity of the solution of the Lyapunov equation, *Linear Algebra Appl.* **60** (1984), 57–64.

[20] Kadison R.V., Ringrose J.R., *Fundamentals of the Theory of Operator Algebras, Vol. 4*, Birkh"auser, Boston 1992.

[21] Kenney C., Hewer G., Trace norm bounds for stable Lyapunov equations, *Linear Algebra Appl.* **221** (1995), 1–18.

[22] Kwong M.K., On the definiteness of the solutions of certain matrix equations, *Linear Algebra Appl.* **108** (1988), 177–197.

[23] Lancaster P., Explicit solutions of linear matrix equations, *SIAM Review* **12** (1970), 544–566.

[24] Lancaster P., Tismenetsky M., *The Theory of Matrices*, Academic Press, New York 2nd ed. 1985.

[25] Paulsen V., *Completely Bounded Maps and Dilations*, Longman, Harlow 1986.

[26] Rudin W., *Real and Complex Analysis*, McGraw–Hill Book Co., New York 1966.

[27] Tippett M.K., Cohn S.E., Todling R., Marchesin D., Conditioning of the stable, discrete Lyapunov operator, *SIAM J. Matrix Anal. Appl.* **22** (2000), 56–65.

[28] Tippett M.K., Marchesin D., Upper bounds for the solution of the discrete algebraic Lyapunov equation, *Automatica* **35** (1999), 1485–1489.

Rajendra Bhatia
Indian Statistical Institute
Delhi Center, 7, S.J.S. Sansanwal Marg
New Delhi 110016, India

Ludwig Elsner
Fakultät für Mathematik
Universität Bielefeld
33613 Bielefeld, FRG

1991 Mathematics Subject Classification. Primary 15A12; Secondary 47B65

Received September 4, 2000

Operator Theory:
Advances and Applications, Vol. 130, 121–133
© 2001 Birkhäuser Verlag Basel/Switzerland

Full- and Partial-Range Completeness

Paul Binding and Rostyslav Hryniv

Dedicated to Peter Lancaster on the occasion of his 70-th birthday

Abstract. For a selfadjoint definitizable operator in a Krein space we study relations between various completeness properties of its root vectors and regularity of its critical points

1. Introduction

The eigenvalue problem

$$Sx = \lambda Tx, \tag{1.1}$$

where S and T are selfadjoint operators, is called left definite (LD) (resp. right definite (RD)) if S (resp. T) is positive definite, and can be studied via a selfadjoint operator A formally given by $S^{-1}T$ (resp. $T^{-1}S$) on an appropriate Hilbert space H_S (resp. H_T) with inner product generated by S (resp. T). These spaces are sometimes called form domains for S and T. Early work on Sturm-Liouville problems around 1830 concerned RD cases and subsequent completeness theory has implicitly or explicitly used the H_T framework, T being the operator of multiplication by a "weight" function w say. Around 1900, HILBERT and others became interested in problems with indefinite w, and completeness for LD cases was studied in H_S. In the 1970s, forward-backward problems (e.g. of transport and scattering theory) involving indefinite w were analysed for completeness in subspaces specified by the sign of w, and interest reverted to a modified H_T setting.

Example 1.1. Suppose that q and w are continuous functions on $\mathbb{R}$ such that q is nonnegative, $q(x) \to \infty$ as $x \to \infty$ and $xw(x) > 0$ for $x \in \mathbb{R} \setminus \{0\}$. We seek bounded solutions of the forward-backward initial value problem

$$w(x)\frac{\partial u}{\partial t} - \frac{\partial^2 u}{\partial x^2} + q(x)u = 0, \qquad x \in \mathbb{R}, \quad t > 0, \tag{1.2}$$

$$u(0, x) = g^+(x), \qquad x > 0. \tag{1.3}$$

Separation via $u(t, x) = e^{-\lambda t}y(x)$ in (1.2) yields

$$-y'' + qy = \lambda w(x)y,$$

which can be regarded as a spectral problem for the operator A given by

$$Ay = \frac{1}{w(x)}(-y'' + qy).$$

It is easily seen that the operator A is positive and selfadjoint in what is now a Krein space $H_T = L_2(\mathbb{R}, w(x)\,dx)$ (whose topology is that of the Hilbert space $H_{|T|} = L_2(\mathbb{R}, |w(x)|\,dx)$) and that its spectrum is discrete and consists of infinitely many positive eigenvalues $0 < \lambda_1^+ \leq \lambda_2^+ \leq \ldots$ and infinitely many negative eigenvalues $0 > \lambda_1^- \geq \lambda_2^- \geq \ldots$. Let $y_k^\pm$ denote the corresponding eigenfunctions.

Any finite linear combination $\sum c_k e^{-\lambda_k^+ t} y_k^+(x)$ gives a bounded solution of equation (1.2), and in order that (1.3) be satisfied g^+ must belong to the closed linear span of the restrictions $P^+ y_k^+$ of the functions y_k^+ onto $\mathbb{R}^+$. If this is to hold for any g^+, then the system $P^+ y_k^+$, $k \in \mathbb{N}$, must be total in $H_T^+ := L_2(\mathbb{R}^+, w(x)\,dx)$. If, moreover, these functions form a Riesz basis of H_T^+, then problem (1.2)–(1.3) has a unique bounded solution for any such g^+.

If a terminal condition $u(T, x) = g^-(x)$ for $x < 0$ is imposed at $t = T > 0$, then one examines completeness of an analogous system $P^- y_k^-$ in the space $H_T^- := L_2(\mathbb{R}^-, -w(x)\,dx)$. □

Such properties have been investigated mainly under the heading of "partial- (also called half-) range completeness", largely in the context of Sturm-Liouville and more general elliptic equations with indefinite weight functions, see e.g. [Be, CL, FR, KKLZ] and references therein. The "ranges" refer to the subsets of $\mathbb{R}$ where w has given sign. We remark that all works we are aware of use "full-range completeness", i.e., in $H_{|T|}$, as an intermediate step to establish the partial-range results. Most authors in this area have used "completeness" to mean existence of bases (or even Riesz bases), in contrast with the more general usage, i.e., existence of total sets.

In many applications, H_S and H_T become Pontryagin and Krein spaces respectively when S and T are indefinite in (1.1). Full- and partial-range completeness results have been derived for problems of this type via indefinite inner product space theory in, e.g., [CL, DaL, FR, Py] and some of the references above. We shall employ this setting below, and the relevant definitions can be found in [AI, Bo, L3] although most will be given in Section 2. (It should be mentioned that many contributions to finite dimensional indefinite inner product space theory can be found in the work of GOHBERG, LANCASTER and RODMAN [GLR]; see also Section 3). We note that various authors have studied the related problem of finding conditions on the weight function w of Example 1 under which the corresponding operator A has only regular critical points or else a singular critical point at infinity: see, e.g., [CL, CN1, Fl, Vo].

This note began as a stepping stone for a variational principle, but has expanded to include logical relations between most of the above concepts of completeness and regularity of critical points. We have found this a helpful exercise, and although some of the results are known, we hope that their assembly may prove useful. In Section 2 we give some definitions, assumptions and connections with maximal semidefinite subspaces. In preparation for the analysis of singular critical points, we study general (closed) subspaces of Pontryagin spaces (equivalent to degenerate Pontryagin spaces) in Section 3 and in particular we extend the invariant subspace theorem for selfadjoint operators to this setting.

In Section 4 we give a complete set of implications between the various concepts under study for the case where A has critical points only of finite type. We show that some of these implications (but not all) hold for infinite type critical points in Section 5. For example, the existence of both partial-range bases does not guarantee a full-range basis. This raises the question of establishing partial-range completeness directly (without going via the full-range property). Whether this leads to weaker sufficient partial-range completeness conditions, particularly for Sturm-Liouville problems, we leave open.

2. Preliminaries

Let $\mathcal{K}$ be a **Krein space** with an inner product $[\cdot, \cdot]$ and a fundamental decomposition $\mathcal{K} = \mathcal{K}^{+}[+]\mathcal{K}^{-}$. Recall that this means that $\mathcal{K}^{+}$ and $\mathcal{K}^{-}$ are respectively Hilbert and anti-Hilbert spaces under $[\cdot, \cdot]$ and that $\mathcal{K}$ is a **Pontryagin space** if $\mathcal{K}^{-}$ or $\mathcal{K}^{+}$ has finite dimension. We introduce a Hilbert space inner product on $\mathcal{K}$ by setting $(x, x) = [x^{+}, x^{+}] - [x^{-}, x^{-}]$, where $x = x^{+} + x^{-}$, $x^{\pm} \in \mathcal{K}^{\pm}$, and denote the corresponding Hilbert space by $\mathcal{H}$ and the orthogonal projectors in $\mathcal{H}$ onto $\mathcal{K}^{\pm}$ by $P^{\pm}$. All topological properties (closedness of subspaces, boundedness of operators, etc.) will always be understood with respect to the topology of $\mathcal{H}$.

Throughout, **subspaces** of $\mathcal{K}$ are assumed to be closed. A subspace $\mathfrak{L}$ of $\mathcal{K}$ is **neutral** if $[\cdot, \cdot]$ vanishes on $\mathfrak{L}$. $\mathfrak{L}$ is a **degenerate Pontryagin space** if it is the $[\cdot, \cdot]$-orthogonal direct sum of two subspaces, one being finite dimensional and neutral (and called the **isotropic part** of $\mathfrak{L}$), and the other being a Pontryagin space under $[\cdot, \cdot]$. $\mathfrak{L}$ is **nonnegative** if $[x, x] \geq 0$ for each $x \in \mathfrak{L}$, and **maximal nonnegative** if $\mathfrak{L}$ is not properly contained in any nonnegative subspace. **Positive, negative**, etc., subspaces are defined similarly. A positive subspace $\mathfrak{L}$ is **uniformly positive** if there exists a constant $c > 0$ such that $[x, x] \geq c(x, x)$ for all $x \in \mathfrak{L}$. Some connections between these concepts are provided by the following

Lemma 2.1. *Any subspace $\mathfrak{L}$ of a Pontryagin space $\mathcal{K}$ is itself a (perhaps degenerate) Pontryagin space. In particular, $\mathfrak{L}$ admits a decomposition*

$$\mathfrak{L} = \mathfrak{L}^{-}[+]\mathfrak{L}^{0}[+]\mathfrak{L}^{+}, \qquad (2.1)$$

where $\mathfrak{L}^{0}$ is the isotropic part of $\mathfrak{L}$, and $\mathfrak{L}^{-}$ and $\mathfrak{L}^{+}$ are respectively uniformly negative and uniformly positive subspaces of $\mathcal{K}$.

Proof. The decomposition comes from [AI, Theorem 1.6.4], and uniformity from [Bo, Lemma IX.2.1]. □

The following result will also be used frequently.

Lemma 2.2. *Suppose that $\mathfrak{L}$ is a nonnegative subspace. Then P^+ is a homeomorphism of $\mathfrak{L}$ and $P^+\mathfrak{L}$. Moreover, $\mathfrak{L}$ is maximal nonnegative if and only if $P^+\mathfrak{L} = \mathcal{K}^+$. Analogous statements relate nonpositive subspaces, P^-, and $\mathcal{K}^-$.*

Proof. See [Bo, Lemma IV.7.1 and Theorem V.4.2]. □

We shall consider three forms of completeness for systems S of vectors $\{u_n\}_{n=1}^\omega$, $\omega \leq \infty$, in $\mathcal{K}$. The weakest is totality: S is **total** if its cls equals $\mathcal{K}$ (throughout, we abbreviate (closed) linear span to $(c)ls$). By definition, S is a **basis** of $\mathcal{K}$ if any element of $\mathcal{K}$ has a unique expansion as a (perhaps infinite) linear combination of the u_n. The u_n form a **Riesz basis** if they are a (linear) homeomorphic image of a $\mathcal{H}$-orthonormal basis. Recall that any Riesz basis in a Hilbert space is unconditional (i.e., the expansion series for any element converges unconditionally). We remark that a substantial amount of material on totality and bases can be found in [GK, Chapters V and VI].

The above concepts correspond to the "full-range" theory of Section 1. The "partial-range" analogues are defined as follows. S is **positive-range total** if the subspace $\mathfrak{L} = \mathrm{cls}\,\{u_n\}_{n=1}^\omega$ is nonnegative and the system $\{P^+u_n\}_{n=1}^\omega$ is total in $\mathcal{K}^+$. S is a **positive-range (Riesz) basis** in $\mathcal{K}$ if the subspace $\mathfrak{L}^+ = \mathrm{cls}\,\{u_n\}_{n=1}^\omega$ is nonnegative and the system $\{P^+u_n\}_{n=1}^\omega$ forms a (Riesz) basis of $\mathcal{K}^+$. The definitions of **negative-range total** systems and **negative-range (Riesz) bases** are analogous.

Lemma 2.2 implies the following results.

Corollary 2.3. *The system $\{u_n\}_{n=1}^\omega$ is positive-range (resp., negatie-range) total if and only if the subspace $\mathfrak{L} = \mathrm{cls}\,\{u_n\}_{n=1}^\omega$ is maximal nonnegative (resp., maximal nonpositive).*

Corollary 2.4. *The set $\{u_n\}_{n=1}^\omega$ is a positive-range (resp., negative-range) (Riesz) basis if and only if $u_1,...,u_\omega$ is a (Riesz) basis for the subspace $\mathfrak{L} := \mathrm{cls}\{u_1,...,u_\omega\}$ and the latter is maximal nonnegative (resp., maximal nonpositive) in $\mathcal{K}$.*

Throughout we assume A to be a selfadjoint operator in $\mathcal{K}$ with at most countable essential spectrum. By shift and inversion, we may (and shall) assume that A is bounded. The completeness concepts above are transferred to A via its root vectors: we say that A is **positive/negative-range total** if it possesses a positive/negative-range total system of root vectors in $\mathcal{K}$. Similarly A is **positive/negative-range (Riesz) basic** if it possesses a positive/negative-range (Riesz) basis of root vectors in $\mathcal{K}$.

We shall always assume that A is **definitizable**, i.e., $[p(A)x, x] \geq 0$ for some nonconstant polynomial p and all x in $\mathcal{K}$. (An example with many applications is when A is **quasi-uniformly positive**, i.e., when $\mathcal{K}$ is a degenerate Pontryagin space under the form $[A\cdot,\cdot]$, cf. [Bi, CN2].) Thus A possesses a **spectral function** $E(\cdot)$ [L3], which is defined on all **admissible** intervals, i.e., with endpoints not in the set $c(A)$ of **critical points** (consisting of real zeros of p) and takes its values in the set of orthoprojectors of $\mathcal{K}$. For an admissible interval Δ, $E(\Delta)\mathcal{K}$ is an A-invariant subspace and the restriction of A onto this subspace has spectrum in $\overline{\Delta}$. Moreover, if p is positive (negative) on $\overline{\Delta}$, then the corresponding spectral subspace $E(\overline{\Delta})\mathcal{K}$ is uniformly positive (uniformly negative, respectively).

Recall also [L3, Proposition II.2.1] that the nonreal spectrum of a definitizable operator A consists of finitely many pairs of complex conjugate eigenvalues (which are zeros of the definitizing polynomial p). Note that the closed linear span $\mathcal{K}^{nr}$ of the corresponding root subspaces can always be split off by means of the Riesz projector. Since completeness in $\mathfrak{L}^{nr}$ of root vectors of A is automatic, without loss of generality we can assume that $\mathcal{K}^{nr} = \{0\}$, i.e., that the spectrum of A is real.

For a subspace $\mathfrak{L}$, $\varkappa^{+}(\mathfrak{L})$ and $\varkappa^{-}(\mathfrak{L})$ denote the maximal dimensions of nonnegative and nonpositive subspaces in $\mathfrak{L}$ respectively. For a point $\lambda \in \mathbb{R}$, we denote by $\varkappa^{\pm}(\lambda)$ the minima of $\varkappa^{\pm}(E(\Delta)\mathcal{K})$, where Δ runs over all admissible intervals containing λ. A point $\lambda \in \sigma(A)$ has **finite type** if $\varkappa^{+}(\lambda)$ or $\varkappa^{-}(\lambda)$ is finite; in particular, λ is said to be of **positive (negative) type** if $\varkappa^{-}(\lambda) = 0$ ($\varkappa^{+}(\lambda) = 0$, respectively).

A point $\lambda \in \mathbb{R}$ is critical if and only if the spectral subspace $E(\Delta)\mathcal{K}$ is indefinite for any admissible $\Delta \ni \lambda$, i.e., if and only if $\min\{\varkappa^{+}(\lambda), \varkappa^{-}(\lambda)\} > 0$. Next, a critical point $\lambda \in c(A)$ is called **regular** if the set of spectral projectors $E(\Delta)$ for all admissible Δ in some neighbourhood of λ is uniformly bounded and λ is called **singular** otherwise.

Lemma 2.5. (see [L3]) *Suppose that λ is a critical point of A. Denote by $\mathfrak{L}_{\lambda}$ the corresponding root subspace and by $\mathcal{K}_{\lambda}$ the closed linear span of the subspaces $E(\Delta)\mathcal{K}$ for all admissible Δ not containing λ. Then following statements are equivalent:*

(a) *λ is a regular critical point;*

(b) *$\mathfrak{L}_{\lambda}$ is a Krein space;*

(c) *$\mathcal{K}_{\lambda}$ is a Krein space;*

(d) *$\mathcal{K} = \mathfrak{L}_{\lambda}[+]\mathcal{K}_{\lambda}$.*

In the next section, we shall study root subspaces (which are degenerate Pontryagin spaces) corresponding to singular critical points of finite type.

3. Degenerate Pontryagin spaces

Recall that according to PONTRYAGIN's theorem a bounded selfadjoint operator A in a Pontryagin space possesses a maximal semidefinite invariant subspace. This was generalized to our conditions by LANGER [L1] who gave existence of maximal nonnegative and nonpositive invariant subspaces for bounded definitizable operators in Krein spaces. To facilitate the examination of root subspaces of eigenvalues which are (possibly singular) critical points of finite type, we shall treat in this section invariant semidefinite subspaces in a degenerate Pontryagin space $\mathfrak{L}$.

By definition,

$$\mathfrak{L} = \mathfrak{L}^{0}[+]\mathfrak{L}', \tag{3.1}$$

where $\mathfrak{L}^0$ is a neutral subspace of finite dimension $\varkappa^0$ and $\mathfrak{L}'$ is a Pontryagin space (say, with $\varkappa^- := \varkappa^-(\mathfrak{L}') < \infty$). Let A be a bounded symmetric linear operator on $\mathfrak{L}$. We first prove the following generalization of Pontryagin's theorem.

Lemma 3.1. *There exist a maximal nonnegative A-invariant subspace $\mathfrak{L}_A^+$ of $\mathfrak{L}$ of codimension $\varkappa^-$ and a maximal nonpositive A-invariant subspace $\mathfrak{L}_A^-$ of $\mathfrak{L}$ of dimension $\varkappa^- + \varkappa^0$.*

Proof. With respect to decomposition (3.1) A is easily seen to have a matrix representation

$$A = \begin{pmatrix} A^0 & A^{0'} \\ 0 & A'' \end{pmatrix}, \tag{3.2}$$

where A^0 and $A^{0'}$ are arbitrary operators from $\mathfrak{L}^0$ and $\mathfrak{L}'$ to $\mathfrak{L}^0$ and A'' is self-adjoint in $\mathfrak{L}'$.

According to the invariant subspace theorems in Pontryagin space (see above), there is a maximal nonnegative A''-invariant subspace $\mathfrak{L}'_+$ of $\mathfrak{L}'$ of codimension $\varkappa^-$. It follows from (3.2) that $\mathfrak{L}_A^+ := \mathfrak{L}^0[\dotplus]\mathfrak{L}'_+$ is A-invariant, nonnegative and has codimension $\varkappa^-$ in $\mathfrak{L}$. Therefore $\mathfrak{L}_A^+$ is maximal nonnegative as required.

Analogously there exists a maximal nonpositive A''-invariant subspace $\mathfrak{L}'_-$ of $\mathfrak{L}'$. Then $\mathfrak{L}_A^- := \mathfrak{L}^0[\dotplus]\mathfrak{L}'_-$ is A-invariant, nonpositive subspace of dimension $\varkappa^- + \varkappa^0$. The lemma is proved. $\qquad\square$

As is well-known, in finite dimensional Pontryagin spaces one can construct bases in which A and the Gram operator of $[\cdot, \cdot]$ take certain "canonical" forms (which essentially go back to WEIERSTRASS. Cf. [Ma, Section 108], and [Bi] for an infinite dimensional version). Here we follow [GLR] and observe from the proof of Theorem I.3.20 there that maximal semidefinite A-invariant subspaces can be spanned by initial parts of Jordan chains of any canonical Jordan basis of A. In degenerate Pontryagin spaces this is not the case, as the following example demonstrates.

We consider the space $\mathcal{K} = \mathbb{C}^6$ with the degenerate indefinite inner product $[\cdot, \cdot] = (G\cdot, \cdot)$, where $(\cdot, \cdot)$ is the Euclidean scalar product and

$$G = \begin{pmatrix} 0 & 0 & 0 & 0 & 0 & 1 \\ 0 & -1 & 0 & 0 & 1 & 0 \\ 0 & 0 & 0 & 0 & 0 & 0 \\ 0 & 0 & 0 & 0 & 0 & -1 \\ 0 & 1 & 0 & 0 & -1 & 0 \\ 1 & 0 & 0 & -1 & 0 & 0 \end{pmatrix}.$$

The matrix G is unitarily equivalent to the orthogonal sum

$$G \sim 0 \oplus \begin{pmatrix} -1 & 1 \\ 1 & -1 \end{pmatrix} \oplus \begin{pmatrix} 0 & 0 & 1 \\ 0 & 0 & -1 \\ 1 & -1 & 0 \end{pmatrix}$$

and therefore has eigenvalues $0, 0, 0, -2, \pm\sqrt{2}$. It follows that any maximal non-negative subspace of $\mathcal{K}$ has dimension 4.

Denote the standard basis vectors by e_0, e_1, f_0, f_1, f_2, and f_3 and put

$$
A = \begin{pmatrix}
0 & 1 & 0 & 0 & 0 & 0 \\
0 & 0 & 0 & 0 & 0 & 0 \\
0 & 0 & 0 & 1 & 0 & 0 \\
0 & 0 & 0 & 0 & 1 & 0 \\
0 & 0 & 0 & 0 & 0 & 1 \\
0 & 0 & 0 & 0 & 0 & 0
\end{pmatrix}.
$$

Then A is G-symmetric and e_0, e_1 and f_0, f_1, f_2, f_3 are two Jordan chains corresponding to the eigenvalue 0. Therefore any Jordan basis of A consists of two Jordan chains of length 2 and 4.

Lemma 3.2. *There is a unique maximal nonnegative A-invariant subspace $\mathcal{L}^+$ in $\mathcal{K}$; it is spanned by the vectors e_0, f_0, f_1, and $e_1 + f_2$.*

Proof. It is easily verified that the subspace spanned by the vectors indicated is A-invariant and G-neutral and therefore maximal nonnegative.

Suppose that $\mathcal{L}^+$ is any maximal nonnegative A-invariant subspace and that $x_0 = \sum \alpha_k e_k + \sum \beta_k f_k \in \mathcal{L}^+$ with $\beta_3 \neq 0$. Then $Ax_0 = \alpha_1 e_0 + \beta_1 f_0 + \beta_2 f_1 + \beta_3 f_2 \in \mathcal{L}^+$ and $[Ax_0, Ax_0] = -|\beta_3|^2 < 0$, a contradiction. Therefore, $\mathcal{L}^+ \subset \mathcal{L}' = \mathrm{ls}\,\{e_0, e_1, f_0, f_1, f_2\}$. Next, $\mathcal{L}^+$ must contain e_0, f_0, f_1 because $\mathrm{ls}\,\{e_0, f_0, f_1\}$ is an isotropic subspace of $\mathcal{L}'$. Being of dimension 4, $\mathcal{L}^+$ coincides with $\mathrm{ls}\,\{e_0, f_0, f_1, \alpha e_1 + \beta f_2\}$ for some $\alpha, \beta \in \mathbb{C}$. It suffices to note that the vector $\alpha e_1 + \beta f_2$ is nonnegative if and only if $\alpha = \beta$. The proof is complete. $\square$

Lemma 3.3. *There is no Jordan basis of A in $\mathcal{L}^+$ which is extensible to a Jordan basis of A in $\mathcal{K}$.*

Proof. By Lemma 3.2, $e_0, f_0, f_1 + e_0, f_2 + e_1$ form a Jordan basis of $\mathcal{L}^+$. Thus any Jordan basis of $A_+ = A|_{\mathcal{L}^+}$ has one chain x_0, x_1, x_2 of length 3 and one chain y_0 of a single eigenvector. Now $A_+^2 x_2 = x_0 \neq 0$, and in view of $A_+^2 e_0 = A_+^2 f_0 = A_+^2 f_1 = 0$ we deduce that $x_2 \notin \mathrm{ls}\,\{e_0, f_0, f_1\}$. Since $e_1 + f_2 \notin R(A)$, it follows that $x_2 \notin R(A)$ and so the chain x_0, x_1, x_2 does not extend to a chain of length 4. $\square$

4. Finite type critical points

We now revert to the case where A is a self-adjoint operator on the Krein space $\mathcal{K}$ (which is necessarily non-degenerate). Recall that A is bounded and definitizable and has real spectrum. In this section we assume all critical points to be of finite type (as is the case if A is quasi-uniformly positive, for example). Without loss of generality we consider the case where λ_0 is the only critical point (regular or singular), $\varkappa^-(\lambda_0) =: \varkappa < \infty$, and all the other spectral points are of positive type. This situation can always be achieved by restricting A to a spectral subspace $E(\Delta)\mathcal{K}$ with Δ containing no critical points except λ_0. Then $\mathcal{K}$ is in fact a Pontryagin space; now using Pontryagin's invariant subspace theorem (see Section 3) one can easily show that λ_0 is an eigenvalue of A and that the corresponding root subspace $\mathfrak{L}_{\lambda_0}$ satisfies $\varkappa^-(\mathfrak{L}_{\lambda_0}) = \varkappa$.

Proposition 4.1. *Suppose that the spectrum of the operator A is real and of positive type except for the point λ_0, which is a critical point of finite type with $\varkappa^-(\lambda_0) =: \varkappa < \infty$. Then A is negative-range Riesz basic.*

Proof. Let $\mathfrak{L}_{\lambda_0}$ be the root subspace of A corresponding to the eigenvalue λ_0. According to Lemma 2.1 $\mathfrak{L}_{\lambda_0}$ is a (perhaps degenerate) Pontryagin space, so by Pontryagin's theorem (resp. Lemma 3.1) there exists an A-invariant nonpositive subspace $\mathfrak{L}^-$ of $\mathfrak{L}_{\lambda_0}$ of dimension $\varkappa$. Then $\mathfrak{L}^-$ is maximal nonpositive in $\mathcal{K}$ and hence $P^-\mathfrak{L}^- = \mathcal{K}^-$ by Lemma 2.2.

Denote by $v_1, \ldots, v_\varkappa$ any Jordan basis of the restriction $A|_{\mathfrak{L}^-}$. Then by Corollary 2.4 the vectors $P^-v_1, \ldots, P^-v_\varkappa$ form a (Riesz, by finite dimensionality) basis of $\mathcal{K}^-$, and the proof is complete. $\square$

The above proposition shows, in particular, that the root vectors of A are negative-range total. We next study positive-range totality.

Denote by $\lambda_1, \ldots, \lambda_\omega$, $\omega \leq \infty$, the eigenvalues of A not equal to λ_0 and observe that by assumption the corresponding eigenspaces $\mathfrak{L}_n$, $n = 1, \ldots, \omega$, are Hilbert spaces with respect to the norm $[\cdot, \cdot]^{1/2}$. We also put

$$\mathcal{K}_0 := \operatorname{cls}\{\mathfrak{L}_n\}_{n=1}^\omega. \tag{4.1}$$

Lemma 4.2. *Any maximal nonnegative A-invariant subspace contains $\mathcal{K}_0$ of (4.1). Moreover, if $\mathfrak{L}^+$ is a maximal nonnegative A-invariant subspace of $\mathfrak{L}_{\lambda_0}$ (see Lemmata 2.1 and 3.1), then $\mathcal{K}_0[+]\mathfrak{L}^+$ is a maximal nonnegative A-invariant subspace of $\mathcal{K}$.*

Proof. Let $\mathfrak{L}$ be any maximal nonnegative A-invariant subspace. Then its orthogonal complement $\mathfrak{L}^{[\perp]}$ is A-invariant and nonpositive, which means that $\mathfrak{L}^{[\perp]} \subset \mathfrak{L}_{\lambda_0}$ since by assumption λ_0 is the only spectral point of A of nonpositive type. By [L3]

$$\mathcal{K}_0^{[\perp]} := \{x \in \mathcal{K} \mid [x, y] = 0 \ \forall y \in \mathcal{K}_0\} = \mathfrak{L}_{\lambda_0}$$

and therefore

$$\mathcal{K}_0 = \mathfrak{L}_{\lambda_0}^{[\perp]} \subset \left(\mathfrak{L}^{[\perp]}\right)^{[\perp]} = \mathfrak{L},$$

which proves the first claim.

Let $\mathfrak{L}^+$ be any maximal nonnegative A-invariant subspace of $\mathfrak{L}_{\lambda_0}$. Then $\mathcal{K}_0[+]\mathfrak{L}^+$ is evidently nonnegative and A-invariant, and it suffices to prove that it has the required codimension. To this end we observe first that

$$\mathfrak{L}^0 := \mathcal{K}_0 \cap \mathfrak{L}_{\lambda_0}$$

is the isotropic part of $\mathfrak{L}_{\lambda_0}$ and hence the sum $\mathcal{K}_0[+]\mathfrak{L}_{\lambda_0}$ has codimension $\varkappa^0 := \dim \mathfrak{L}^0$ in $\mathcal{K}$. By Lemma 3.1 $\mathfrak{L}^+$ is of codimension $\varkappa^- := \varkappa - \varkappa^0$ in $\mathfrak{L}_{\lambda_0}$, whence the codimension of $\mathcal{K}_0[+]\mathfrak{L}^+$ is at most $\varkappa$ as required. The lemma is proved. $\square$

Proposition 4.3. *Under the assumptions of Proposition 4.1, A is positive-range total.*

Proof. Take a maximal nonnegative A-invariant subspace $\mathcal{L}$ of the form $\mathcal{K}_0[+]\mathcal{L}^+$ of Lemma 4.2. By Corollary 2.3, it suffices to show that the root vectors of A belonging to $\mathcal{L}$ form a total set in $\mathcal{L}$. Since $\mathcal{L}^+$ is a subspace of the root subspace $\mathcal{L}_{\lambda_0}$ and $\mathcal{K}_0$ is the closed linear span of the eigenspaces $\mathcal{L}_n$, the totality of root vectors of A in $\mathcal{L}$ is evident, and the proof is complete. □

To study positive-range bases of A, we need some auxiliary results.

Lemma 4.4. *Under the assumptions of Proposition 4.1, the critical point λ_0 is regular if and only if the subspace $\mathcal{K}_0$ of (4.1) is positive.*

Proof. By Lemma 2.5, λ_0 is regular if and only if $\mathcal{K}_0$ is a Krein space. Since $\mathcal{K}_0$ is evidently nonnegative, it is a Krein space if and only if uniformly positive. Finally we observe that all positive subspaces in a Pontryagin space are also uniformly positive, and the proof is complete. □

Lemma 4.5. *Suppose that A is basic or positive-range basic. Then the subspace $\mathcal{K}_0$ of (4.1) is positive.*

Proof. Assume that the subspace $\mathcal{K}_0$ of (4.1) is not positive; then there exists a neutral vector $x \in \mathcal{K}_0$ and hence $[x, v] = 0$ for all $v \in \mathcal{K}_0$ by nonnegativity of $\mathcal{K}_0$. Denote by P_n the spectral projector $E(\{\lambda_n\})$ onto $\mathcal{L}_n$; then $[P_n x, u] = [x, u] = 0$ for all $u \in \mathcal{L}_n$ and hence $P_n x = 0$ for all $n \in \mathbb{N}$. We shall show that under the assumptions of the lemma this implies $x = 0$.

Suppose that $\{v_k\}_{k=1}^\infty$ is a basis of $\mathcal{K}$ composed of root vectors of A and $x = \sum c_k v_k$. It is easily seen that $P_n x = \sum_{v_k \in \mathcal{L}_n} c_k v_k$ and hence $c_k = 0$ for all k such that $v_k \notin \mathcal{L}_{\lambda_0}$. On the other hand, there exists a sequence x_n in $\mathcal{K}_0$ that converges to x and such that

$$x_n \in \mathcal{L}_1[+]\mathcal{L}_2[+] \cdots [+]\mathcal{L}_n.$$

Since the coordinate functionals $c_k = c_k(x)$ depend continuously on x [GK, p. 307], we see that also $c_k = 0$ for k such that $v_k \in \mathcal{L}_{\lambda_0}$. This finishes the proof for the case of a basis.

Suppose now that $\{v_k\}_{k=1}^\infty$ is a positive-range basis of A spanning a subspace $\mathcal{L}$. By Corollary 2.4 $\mathcal{L}$ is maximal nonnegative in $\mathcal{K}$ and the system $\{v_k\}_{k=1}^\infty$ is a basis of $\mathcal{L}$. Now $\mathcal{K}_0 \subset \mathcal{L}$ by Lemma 4.2, and the proof proceeds as above with $\mathcal{K}$ replaced by $\mathcal{L}$. □

Theorem 4.6. *Under the assumptions of Proposition 4.1, the following are equivalent:*

(a) *A is Riesz basic;*

(b) *A is basic;*

(c) *A is positive-range Riesz basic;*

(d) *A is positive-range basic;*

(e) *A is total;*

(f) λ_0 *is a regular critical point.*

Proof. The implications (a)$\Longrightarrow$(b) and (c)$\Longrightarrow$(d) are obvious, and any of (a)–(d) implies (f) by Lemmata 4.5 and 4.4. For the equivalence of (e) with nondegeneracy of $\mathfrak{L}_{\lambda_0}$, see, e.g., [AI, Theorem 4.2.15]. Equivalence of this with (f) follows from a result of Langer (see [BN] for elaboration). Thus it suffices to establish the implications (f)$\Longrightarrow$(a) and (f)$\Longrightarrow$(c).

Suppose that (f) holds; then by Lemmata 2.5 and 4.4 the subspace $\mathcal{K}_0$ is uniformly positive. Therefore the restriction of A onto its invariant subspace $\mathcal{K}_0$ is similar to a selfadjoint operator in $\mathcal{K}_0$ (with the topology of $\mathcal{H}$) and thus has a system of eigenvectors forming a Riesz basis of $\mathcal{K}_0$.

To prove (a), it remains to find a Riesz basis of $\mathfrak{L}_{\lambda_0}$ composed of root vectors of A. Fix a decomposition (2.1) for the subspace $\mathrm{Ker}(A - \lambda_0 I)$. Then $\mathfrak{L}^+$ has finite codimension in $\mathrm{Ker}(A - \lambda_0 I)$ and hence in $\mathfrak{L}_{\lambda_0}$. Moreover, $\mathfrak{L}'_{\lambda_0} := (\mathfrak{L}^+)^{[\perp]} \cap \mathfrak{L}_{\lambda_0}$ is A-invariant, has finite dimension, and $\mathfrak{L}_{\lambda_0} = \mathfrak{L}^+[+]\mathfrak{L}'_{\lambda_0}$. Now we choose a ($\mathcal{H}$-orthonormal, say) Riesz basis of $\mathfrak{L}^+$ and then add any Jordan basis of A restricted to $\mathfrak{L}'_{\lambda_0}$. Together the chosen system is a Riesz basis of $\mathcal{K}$, and (a) is proved.

To prove (c), we fix a maximal nonnegative A-invariant subspace $\mathfrak{L}$ of the form $\mathcal{K}_0[+]\mathfrak{L}^+$ of Lemma 4.2. By Corollary 2.4 it suffices to find a system of root vectors of A forming a Riesz basis of $\mathfrak{L}$. This is done as in the previous paragraph, replacing $\mathfrak{L}_{\lambda_0}$ by $\mathfrak{L}^+$, and the proof is complete. $\qquad\square$

5. Infinite type critical points

Now we study critical points of infinite type. By restricting ourselves to a spectral subspace as in Section 4, we can assume that $\lambda = 0$ is the only critical point, that it is of infinite type and that A is a positive operator in $\mathcal{K}$. We also assume that $\lambda = 0$ is not an eigenvalue, which corresponds to treatment of the critical point at infinity in most of the literature. Then there are infinitely many eigenvalues of both signs, positive eigenvalues λ_k^+ are of positive type, and negative ones λ_k^- are of negative type. Denote by $v_k^\pm$ the corresponding $\mathcal{K}$-**orthonormalized** eigenvectors. This means that $[v_k^+, v_l^-] = 0$ and $[v_k^\pm, v_l^\pm] = \pm\delta_{k,l}$ for all $k, l \in \mathbb{N}$. We need the following elementary result.

Lemma 5.1. *Suppose the vectors v_k^+ form a $\mathcal{K}$-orthonormalized Riesz basis of the closed linear span $\mathfrak{L}^+$ of the v_k^+. Then $\mathfrak{L}^+$ is uniformly positive.*

Proof. From the definition of a Riesz basis, there is a number $c > 0$ such that, for any

$$x = \sum_{k=1}^{\infty} c_k v_k^+ \in \mathfrak{L}^+, \qquad \text{we have} \qquad c^{-1}\|x\|^2 \le \sum_{k=1}^{\infty} |c_k|^2 \le c\|x\|^2.$$

This shows that $[x, x] = \sum |c_k|^2 \ge c^{-1}\|x\|^2$, i.e., that $\mathfrak{L}^+$ is uniformly positive. $\qquad\square$

Theorem 5.2. *Suppose the vectors $v_k^\pm$ are $\mathcal{K}$-orthonormalized. Then the following are equivalent:*

(a) *the vectors v_k^+ form a positive-range Riesz basis;*

(b) *the vectors v_k^- form a negative-range Riesz basis;*

(c) *$\lambda = 0$ is a regular critical point;*

(d) *the vectors $v_k^\pm$ form a Riesz basis.*

Proof. (a)$\Longrightarrow$(c): Denote by $\mathcal{L}^+$ and $\mathcal{L}^-$ the closed linear spans of the vectors v_k^+ and v_k^- respectively and suppose that (a) holds. Then $\mathcal{L}^+$ is uniformly positive by Lemma 5.1 and maximal nonnegative by Corollary 2.4. It follows [L3] that $(\mathcal{L}^+)^{[\perp]}$ is uniformly negative and hence so is $\mathcal{L}^- \subset (\mathcal{L}^+)^{[\perp]}$. Therefore the sum of $\mathcal{L}^+$ and $\mathcal{L}^-$ is direct and closed; since it is in addition dense in $\mathcal{K}$ by [L3, Corollary 2, p.38], we have $\mathcal{L}^+[+]\mathcal{L}^- = \mathcal{K}$. By Lemma 2.5 $\lambda = 0$ is a regular critical point, and (c) is established.

Similar arguments show that (b) implies (c).

(c)$\Longrightarrow$(d) is essentially established in [CL] (see also [DiL]).

Finally, suppose that (d) holds; then the vectors v_k^+ and v_k^- form Riesz bases of $\mathcal{L}^+$ and $\mathcal{L}^-$ and the latter are uniformly definite by Lemma 5.1. Since $v_k^\pm$ being a Riesz basis implies that $\mathcal{K} = \mathcal{L}^+[+]\mathcal{L}^-$, $\mathcal{L}^+$ and $\mathcal{L}^-$ are maximal nonnegative and maximal nonpositive respectively. This gives (a) and (b) by Corollary 2.4, and the proof is complete.   □

Results similar to Theorem 5.2 have apparently been proved by SHKALIKOV [Sh].

If the completeness conditions are weakened then the above equivalences fail in general. For instance, A is automatically total (cf. [L3, Corollary 2, p.38]) but an example of [L2], where the subspaces we labelled $\mathcal{L}^\pm$ above are not maximal semidefinite, shows that A may be neither positive- nor negative-range total (see Corollary 2.3). Moreover, if A is basic, but the original basis $\{v_k^\pm\}$ is conditional, then it appears possible (although we do not know a specific example) that $\mathcal{L}^+ + \mathcal{L}^- \neq \mathcal{K}$, in which case A would not be both positive- and negative-range basic.

Conversely, if A has both partial-range bases $v_k^\pm$, then it may happen that A is not basic, as demonstrated by the following example. (A similar example was given in [KKLZ] to show lack of an isomorphism needed for a standard approach to partial-range via full-range completeness). Note that after an appropriate scaling $\{P^+v_k^+\}$ and $\{P^-v_k^-\}$ actually become $\mathcal{H}$-orthonormal bases of $\mathcal{K}^+$ and $\mathcal{K}^-$, respectively.

Example 5.3. Let $\mathcal{H} = \oplus\mathbb{C}^2$ with Euclidean metric and let the indefinite inner product of the corresponding Krein space $\mathcal{K}$ be defined by the operator $G = \oplus G_n$, where

$$G_n = \begin{pmatrix} 0 & 1 \\ 1 & 0 \end{pmatrix}.$$

Then $P^{\pm}$ are the orthogonal sums of the 2×2 matrices $P_n^{\pm}$,

$$P_n^{\pm} = \frac{1}{2} \begin{pmatrix} 1 & \pm 1 \\ \pm 1 & 1 \end{pmatrix},$$

and $\mathcal{K}^{\pm}$ are spanned by the orthonormal basis vectors $w_k^{\pm}$, $k \in \mathbb{N}$, which have k-th component equal to $(\pm 2^{-1/2}, 2^{-1/2})^T$ and all the other components zero.

Consider the operator $A = \oplus A_n$ with

$$A_n = \begin{pmatrix} 0 & n^{-1} \\ n^{-3} & 0 \end{pmatrix}.$$

It is positive in $\mathcal{K}$, and has eigenvalues $\lambda_k^{\pm} = \pm k^{-2}$ with corresponding $\mathcal{K}$-ortho-normalized eigenvectors $v_k^{\pm}$ having k-th component $2^{-1/2}(\pm k^{1/2}, k^{-1/2})^T$ and all the other components zero.

A simple calculation shows that $P^{\pm} v_k^{\pm} = 2^{-1}(k^{1/2} + k^{-1/2}) w_k^{\pm}$; therefore A has both partial range bases. Nevertheless it is easily seen that the vector having n-th component $n^{-1}(0, 1)^T$ cannot be expanded as a convergent series in terms of the $v_k^{\pm}$, which therefore do not form a basis of $\mathcal{H}$.

Acknowledgements. Research of PB was partly supported by NSERC of Canada and of RH partly by a PIMS Postdoctoral Fellowship at the University of Calgary.

References

[AI]   Azizov, T. Ya, Iohvidov, I. S., *Linear Operators in Spaces with an Indefinite Metric*, John Wiley & Sons, Chichester-New York-Brisbane-Toronto-Singapore 1989.

[Be]   Beals, R., Indefinite Sturm-Liouville problems and half-range completeness, *J.Differential Equations* **56** (1985), 391–407.

[Bi]   Binding, P., A canonical form for self-adjoint pencils in Hilbert spaces, *Integr. Equat. Oper. Th.* **12** (1989), 324–342.

[BN]   Binding, P., Najman, B., Regularity of finite type critical points for self-adjoint operators in Krein space, *Oper. Th. Adv. Appl.* **80** (1995), 79–89.

[Bo]   Bognár, J., *Indefinite Inner Product Spaces*, Springer-Verlag, Berlin-Heidelberg-New York 1974.

[CL]   Ćurgus, B., Langer, H., A Krein space approach to symmetric ordinary differential operators with an indefinite weight function, *J. Diff. Eqat.* **79** (1989), 31–61.

[CN1] Ćurgus, B., Najman, B., A Krein space approach to elliptic eigenvalue problems with indefinite weights, *Differential Integral Equ.* **7** (1994), 1241–1252.

[CN2] Ćurgus, B., Najman, B., Quasi-uniformly positive operators in Krein spaces, *Oper. Th. Adv. Appl.* **80** (1995), 90–99.

[DaL] Daho, K., Langer, H., Sturm-Liouville operators with an indefinite weight function, *Proc. Roy. Soc. Edinburgh* **87A** (1977), 161–191.

[DiL] Dijksma, A., Langer, H., Operator theory and ordinary differential operators, *Fields Inst. Monographs* **3**, Amer. Math. Soc., Providence, R.I.,1996, 75–139.

[FR]   Faierman, M., Roach, G. F., Linear elliptic eigenvalue problems involving an indefinite weight, *J. Math. Anal. Appl.* **126** (1987), 516–528.

[Fl]   Fleige, A., The "turning point condition" of Beals for indefinite Sturm-Liouville problems, *Math. Nachr.* **172** (1995), 109–112.

[GK]   Gohberg, I., Krein, M. G., *Introduction to the Theory of Linear Non-Selfadjoint Operators in Hilbert Spaces*, Amer. Math. Soc. Transl. of Math. Monographs **18**, Providence, R.I. 1969.

[GLR]   Gohberg, I., Lancaster, P., Rodman, L., *Matrices and Indefinite Scalar Products*, Oper. Th.: Adv. Appl., vol.8, Birkhäuser Verlag, Basel-Boston-Stuttgart 1983.

[KKLZ] Kaper, H. G. Kwong, M. K., Lekkerkerker, C. G., Zettl, A., Full- and partial-range eigenfunction expansions for Sturm-Liouville problems with indefinite weights, *Proc. Roy. Soc. Edinburgh Sect. A* **98** (1984), 69–88.

[L1]   Langer, H., Invariant subspaces of linear operators on a space with indefinite metric, *Soviet Math. Dokl.* **7** (1966), 849–852.

[L2]   Langer, H., On the maximal dual pairs of invariant subspaces of J-self-adjoint operators, *Mat. Zametki* **7** (1970), 443–447.

[L3]   Langer, H. Spectral functions of definitizable operators in Krein spaces, *Lecture Notes in Mathematics* **984**, Springer-Verlag, Berlin, 1982, 1–46.

[Ma]   Malcev, A. I., *Foundations of Linear Algebra*, Freeman 1963.

[Py]   Pyatkov, S., Riesz completeness of the eigenelements and associated elements of linear selfadjoint pencils, *Russian Acad. Sci. Sb. Math.* **81** (1995), 343–361.

[Sh]   Shkalikov, A. A., private communication.

[Vo]   Volkmer, H., Sturm-Liouville problems with indefinite weights and Everitt's inequality, *Proc. Roy. Soc. Edin.* **126** (1996), 1097–1112.

Department of Mathematics and Statistics
University of Calgary
Calgary, AB, Canada T2N 1N4
Institute for Applied Problems
of Mechanics and Mathematics
3b Naukova str.
79601 Lviv, Ukraine

1991 Mathematics Subject Classification. Primary 47B50; Secondary 47A75, 47A70

Received February 18, 2001

Operator Theory:
Advances and Applications, Vol. 130, 135–152
© 2001 Birkhäuser Verlag Basel/Switzerland

Spectral Isomorphisms between Generalized Sturm-Liouville Problems

Paul A. Binding, Patrick J. Browne and Bruce A. Watson

Dedicated to Peter Lancaster on the occasion of his 70-th birthday

Abstract. We characterize all isospectral norming constant preserving maps between certain classes of Sturm-Liouville problems with eigenparameter dependent and constant boundary conditions. In consequence we obtain existence and uniqueness inverse spectral results for Sturm-Liouville problems with eigenparameter dependent boundary conditions.

1. Introduction

We consider regular Sturm-Liouville differential equations

$$-y'' + qy = \lambda y, \quad \lambda \in \mathbb{R} \tag{1.1}$$

on the interval $[0, 1]$, subject to boundary conditions of various types specified by the ratio $\rho = y'/y$. At $x = 0$ we impose

$$\rho(0) \quad = \quad \alpha \tag{1.2}$$

where α is an extended real number. We distinguish the Dirichlet case (D) with $\alpha = \infty$ from the non-Dirichlet case (N) where α is finite. At $x = 1$ we impose one of:

$$\rho(1) \quad = \quad \beta \tag{1.3}$$

(the "constant" case, again either D or N)

$$\rho(1) = a\lambda + b \tag{1.4}$$

(the "affine" case (A), with $a > 0$); and

$$\rho(1) = \frac{a\lambda + b}{c\lambda + d} \tag{1.5}$$

(the "bilinear" case (B), with $ad - bc > 0, c \neq 0$).

We shall label a boundary value problem for (1.1) by two letters referring to the boundary conditions at the left and right ends respectively; for example DA refers to (1.2) with $\alpha = \infty$ and (1.4); NB refers to (1.2) with finite α and (1.5), and so on. Our aim, roughly, is to discuss the strong spectral connections between these various types of problem.

Substantial bibliographies, mostly on the affine case, are contained in [4], [5] and [15], while references to more recent work, especially for the bilinear case and various generalizations, can be found in [6] and [7].

The first direct antecedent of the present investigation seems to be Fulton's work [4] where asymptotics for the affine and bilinear cases were developed, for q continuous and of bounded variation. For example, if the eigenvalues of a bilinear (say NB) problem are labelled $\lambda_0^B < \lambda_1^B < \lambda_2^B < \cdots$ then [4, Case 1] gives a result equivalent to

$$\lambda_n^B = (n-1)^2\pi^2 + 2\alpha - \frac{2a}{c} + \int_0^1 q\, dt + O(n^{-1}) \tag{1.6}$$

as $n \to \infty$. This may be compared with the standard asymptotics (e.g. [11]) for the eigenvalues λ_n^C of a constant NN problem. Indeed, if we use the same q and set $\beta = a/c$ then we find

$$\lambda_{n-1}^C = \lambda_n^B + O(n^{-1})$$

although this relationship was not observed explicitly for many years.

In fact, [1], the stronger spectral connection

$$\lambda_{n-1}^C = \lambda_n^B + O(n^{-2})$$

turns out to be valid for $q \in L^1$. Inverse Sturm-Liouville theory (e.g. [9]) now suggests that one should be able to recover a new q (q^C, say) for a constant NN problem which exactly reproduces the λ_n^B as eigenvalues for the same boundary conditions (1.2) and some (1.3). This, however, is not a rigorous deduction since [9] requires stronger asymptotics than (1.6). Moreover the construction of q^C in [11], say, is quite involved.

Recently another, simpler, construction was discovered for a related problem. To be more precise, it was shown in [3] that from an NB problem one can construct explicitly an NN problem with the same eigenvalues λ_n, say, for $n \geq 1$. On the other hand there could be many such constructions, since it is well known that one spectrum is not in general enough to determine a Sturm-Liouville problem. [3] also contains an inverse result, to the effect that asymptotics similar to (1.6) must come from some problem of NB type with $q \in L^2$, but again the uniqueness question remains.

We shall resolve the above issues by extending the analysis of the spectra in [3] to the "norming constants" $\nu_n = \|y_n\|^2$ where y_n is a (suitably normalized) eigenfunction corresponding to λ_n. Indeed we produce an isomorphism between NB and NN type problems, preserving both spectrum and norming constants (but again with an index shift of one). For this it is necessary to be careful about the domain and range: we map $q \in L^1, \alpha$ and the three free constants in (1.5) (after appropriate scaling) for the NB problem into q, α, β and the two constants λ_0, ν_0 for the NN problem. As a result we show in Section 4 that there is precisely one map with the above preservation property. Moreover, given sequences

$$\lambda_n^B = (n-1)^2\pi^2 + k + o(1),$$

$$\nu_n^B = \frac{1}{2} + o\left(\frac{1}{n}\right),$$

as $n \to \infty$, with k independent of n, there is precisely one NB problem with spectrum and norming constants given by λ_n^B and ν_n^B, respectively. For a different approach to uniqueness (but with no discussion of existence) we refer to [2].

Before proceeding, we remark that NB and NN problems make up only one of several possible pairings. In fact we shall show that each pair (NB,DA), (DB,NA), (DA,NN) and (NA,DN) has strongly related spectral properties and corresponding isomorphisms. By composition one can then pair (NB,NN) and (DB,DN). We depend partly on extensions of [3], as outlined above, and these are detailed in Sections 2 and 3. We also depend on existence and uniqueness results from inverse Sturm-Liouville theory, and since we were unable to find precisely what we needed for the (Dirichlet) case $\alpha = \infty$, we have developed these results as an Appendix in Section 5.

2. Transformation of problems of types NA and NB

In this section we reconsider the transformation of problems of type NA (resp. NB) to problems of type DN (resp. DA) given in [3]. In particular we reformulate the transformations with precise expressions for their domains and ranges, and we show that the resulting maps are invertible and preserve norming constants.

At this stage it is appropriate to recall the operator theoretic approach to NA (resp. NB) problems. We write $\delta = ad - bc$ for both problems, setting $c = 0, d = 1$ for the affine case, so then $\delta = a$. We normalize the bilinear case by setting $c = 1$, so (1.5) is defined by three constants a, d and $\delta = ad - b$. We shall continue to employ the constant $b = \delta - ad$ for ease of notation. Let $L^2[0,1] \oplus \mathbb{C}$ be equipped with the Hilbert space inner product

$$\left\langle \begin{pmatrix} f \\ \eta \end{pmatrix}, \begin{pmatrix} g \\ \zeta \end{pmatrix} \right\rangle_\delta = \int_0^1 f\bar{g} + \frac{\eta\bar{\zeta}}{\delta},$$

and define the operator L by

$$L\begin{pmatrix} y \\ -ay(1) \end{pmatrix} = \begin{pmatrix} -y'' + qy \\ by(1) - dy'(1) \end{pmatrix} \tag{2.1}$$

$$\mathcal{D}(L) = \left\{ \begin{pmatrix} y \\ -ay(1) + cy'(1) \end{pmatrix} \;\middle|\; \begin{array}{l} -y'' + qy \in L^2[0,1] \\ \rho(0) = \alpha, y, y' \in AC[0,1] \end{array} \right\},$$

where as usual $\rho = y'/y$.

For brevity, we introduce the notation

$$\|y\|_a^2 = \left\langle \begin{pmatrix} y \\ -ay(1) \end{pmatrix}, \begin{pmatrix} y \\ -ay(1) \end{pmatrix} \right\rangle_\delta = \int_0^1 y^2 \, dt + a|y(1)|^2$$

$$\left(\text{resp. } \|y\|_{a,\delta}^2 = \left\langle \begin{pmatrix} y \\ (y' - ay)(1) \end{pmatrix}, \begin{pmatrix} y \\ (y' - ay)(1) \end{pmatrix} \right\rangle_\delta \right.$$
$$\left. = \int_0^1 y^2 \, dt + \frac{|y'(1) - ay(1)|^2}{\delta} \right).$$

It is known [15] that L is self-adjoint with compact resolvent in the above Hilbert space and that the eigenvalue problem for (1.1), (1.2) with finite α and (1.4) (resp. (1.5)), which we abbreviate to $(q, \alpha, a, b; NA)$ (resp. $(q, \alpha, a, d, \delta; NB)$) is equivalent to the spectral problem for L.

For $(q, \alpha, a, b) \in L^1 \times \mathbb{R} \times \mathbb{R}^+ \times \mathbb{R}$ (resp. $(q, \alpha, a, d, \delta) \in L^1 \times \mathbb{R}^3 \times \mathbb{R}^+$, denote the spectrum of $(q, \alpha, a, b; NA)$ (resp. $(q, \alpha, a, d, \delta; NB)$) by $\lambda_0 < \lambda_1 < \cdots$, and let y_n be the eigenfunction corresponding to λ_n normalized so that $y_n(0) = 1$. Then we define the norming constants $\nu_n, n \geq 0$, by $\nu_n = \|y_n\|_a^2$ (resp. $\nu_n = \|y_n\|_{a,\delta}^2$). If $(\hat{q}, \beta; DN)$ (resp. $(\hat{q}, a', b'; DA)$) has spectrum $\lambda_1 < \lambda_2 < \cdots$ and u_n is the eigenfunction corresponding to λ_n normalized by $u_n'(0) = 1, n \geq 1$, then we define the norming constants $\xi_n, n \geq 1$, by $\xi_n = (u_n, u_n)_{L^2}$ (resp. $\xi_n = \|u_n\|_{a'}^2$).

Define $E \subset L^1[0,1] \times \mathbb{R} \times \mathbb{R}^+ \times \mathbb{R}$ by

$$E = \{(q, \alpha, a, b) | q \in L^1[0,1], \alpha, b \in \mathbb{R}, a > 0\}$$

(resp. $E \subset L^1[0,1] \times \mathbb{R}^3 \times \mathbb{R}^+$ by

$$E = \{(q, \alpha, a, d, \delta) | q \in L^1[0,1], \alpha, a, d \in \mathbb{R}, \delta > 0\}).$$

We also define $\hat{E} \subset L^1[0,1] \times \mathbb{R} \times \mathbb{R}^+ \times \mathbb{R}$ by

$$\hat{E} = \left\{ (\hat{q}, \beta, \nu_0, \lambda_0) \; \middle| \; \begin{matrix} \hat{q} \in L^1[0,1], \lambda_0, \beta \in \mathbb{R}, \nu_0 > 0 \\ \text{and the equation} \quad -w'' + \hat{q}w = \lambda_0 w \\ \text{has a positive solution with} \quad \frac{w'}{w}(1) = \beta \end{matrix} \right\}$$

(resp. $\hat{E} \subset L^1[0,1] \times \mathbb{R}^+ \times \mathbb{R} \times \mathbb{R}^+ \times \mathbb{R}$ as

$$\hat{E} = \left\{ (\hat{q}, a', b', \nu_0, \lambda_0) \; \middle| \; \begin{matrix} \hat{q} \in L^1[0,1], \lambda_0, b' \in \mathbb{R}, \nu_0, a' > 0, \\ \text{and the equation} \quad -w'' + \hat{q}w = \lambda_0 w \\ \text{has a positive solution with} \quad \frac{w'}{w}(1) = a'\lambda_0 + b' \end{matrix} \right\}\right).$$

The transformation from a problem of type NA (resp. NB) to one of type DN (resp. DA) in [3] can be considered as a mapping $T : E \to \hat{E}$ given by

$$T(q, \alpha, a, b) = (\hat{q}, \beta, \nu_0, \lambda_0)$$

$$(\text{resp. } T(q, \alpha, a, d, \delta) = (\hat{q}, a', b', \nu_0, \lambda_0))$$

where

$$\hat{q} = q - 2z', z = y_0'/y_0 \tag{2.2}$$

$$\beta = -\left(\frac{1}{a} + a\lambda_0 + b\right) \tag{2.3}$$

$$\nu_0 = \|y_0\|_a^2 \tag{2.4}$$

(resp.

$$
\begin{aligned}
a' &= -(\lambda_0 + d)/\delta \qquad (> 0 \quad \text{by} \quad [1]), \\
b' &= -\frac{d(\lambda_0 + d)}{\delta} - \frac{a\lambda_0 + b}{\lambda_0 + d} \\
\nu_0 &= \|y_0\|_{a,\delta}^2).
\end{aligned}
$$

In the sequel we shall frequently use the following observation. Although it is assumed in [3] that $q \in L^2$, in fact all the results hold for $q \in L^1$, with the exception of the inverse result Theorem 4.3. In particular, T maps E into $\hat{E}$ where $T(q, \alpha, a, b) = (\hat{q}, \beta, \nu_0, \lambda_0)$ (resp. $T(q, \alpha, a, d, \delta) = (\hat{q}, a', b', \nu_0, \lambda_0)$) and $(\hat{q}, \beta; DN)$ (resp. $(\hat{q}, a', b'; DA)$) has spectrum $\lambda_1 < \lambda_2 < \cdots$.

Lemma 2.1. *The norming constants ν_n for the original problem in E and ξ_n for the transformed problem in $\hat{E}$ satisfy*

$$
\xi_n = \frac{\nu_n}{\lambda_n - \lambda_0}, \quad n \geq 1. \tag{2.5}
$$

Proof. We calculate

$$
\begin{aligned}
(\lambda_n - \lambda_0)\nu_n &= \left\langle (L - \lambda_0) \begin{pmatrix} y_n \\ -ay_n(1) \end{pmatrix}, \begin{pmatrix} y_n \\ -ay_n(1) \end{pmatrix} \right\rangle_a \\
&= \left\langle \begin{pmatrix} -y_n'' + (q - \lambda_0)y_n \\ by_n(1) - y_n'(1) + a\lambda_0 y_n(1) \end{pmatrix}, \begin{pmatrix} y_n \\ -ay_n(1) \end{pmatrix} \right\rangle_a \\
&= \int_0^1 (-y_n'' + (q - \lambda_0)y_n)\, y_n \, dt - y_n(1)[(b + a\lambda_0)y_n(1) - y_n'(1)] \\
&= \int_0^1 [y_n'^2 + (q - \lambda_0)y_n^2] \, dt + \alpha - y_n^2(1)(a\lambda_0 + b) \tag{2.6}
\end{aligned}
$$

(resp.

$$
(\lambda_n - \lambda_0)\nu_n = \int_0^1 [y_n'^2 + (q - \lambda_0)y_n^2] \, dt \tag{2.7}
$$

$$
+ \alpha - [a(a\lambda_0 + b)y_n^2 - 2(a\lambda_0 + b)y_n y_n' + c(c\lambda_0 + d)y_n'^2](1)/\delta).
$$

Now consider $u_n = (\lambda_0 - \lambda_n)^{-1}(y_n' - y_n z), n \geq 1$, where $z = y_0'/y_0$. From [3], u_n is an eigenfunction corresponding to λ_n of $(\hat{q}, \beta; DN)$ (resp. $(\hat{q}, a', b'; DB)$). It is readily verified that $u_n'(0) = 1$. We note that

$$
\begin{aligned}
(\lambda_0 - \lambda_n)^2 u_n^2 &= y_n'^2 - 2y_n' y_n z + y_n^2 z^2 \\
&= y_n'^2 - (y_n^2 z)' + y_n^2 z' + y_n^2(-z' - \lambda_0 + q) \\
&= y_n'^2 - (y_n^2 z)' + (q - \lambda_0)y_n^2. \tag{2.8}
\end{aligned}
$$

Thus

$$
\begin{aligned}
(\lambda_0 - \lambda_n)^2 \xi_n &= \int_0^1 [y_n'^2 + (q - \lambda_0)y_n^2]\, dt - [y_n^2 z]_0^1 \\
&= \int_0^1 [y_n'^2 + (q - \lambda_0)y_n^2]\, dt - (a\lambda_0 + b)y_n^2(1) + \alpha \qquad (2.9) \\
&= (\lambda_n - \lambda_0)\nu_n
\end{aligned}
$$

(resp.

$$
\begin{aligned}
(\lambda_0 - \lambda_n)^2 \xi_n &= \int_0^1 (y_n'^2 + (q - \lambda_0)y_n^2)\, dt \\
&\quad - [y_n^2 z]_0^1 + a'[z^2 y_n^2 - 2z y_n' y_n + y_n'^2](1) \\
&= \int_0^1 (y_n'^2 + (q - \lambda_0)y_n^2)\, dt \\
&\quad - z(1)y_n^2(1) + \alpha + a'[z^2 y_n^2 - 2z y_n' y_n + y_n'^2](1) \qquad (2.10) \\
&= \int_0^1 (y_n'^2 + (q - \lambda_0)y_n^2)\, dt \\
&\quad + \alpha + [a' y_n'^2 - 2a' z y_n y_n' + (a'z - 1)z y_n^2](1) \\
&= (\lambda_n - \lambda_0)\nu_n)
\end{aligned}
$$

which establishes the result. □

We now define $R(\hat{q}, \beta, \nu_0, \lambda_0) = (q_1, \alpha_1, a_1, b_1)$ for $(\hat{q}, \beta, \nu_0, \lambda_0) \in \hat{E}$ (resp. $R(\hat{q}, a', b', \nu_0, \lambda_0) = (q_1, \alpha_1, a_1, d_1, \delta_1)$ for $(\hat{q}, a', b', \nu_0, \lambda_0) \in \hat{E}$) in stages as follows.

Lemma 2.2. *Let g be the solution of*

$$-g'' + \hat{q}g = \lambda_0 g \qquad (2.11)$$

with initial conditions

$$g(0) = 0 \text{ and } g'(0) = 1. \qquad (2.12)$$

Then $\frac{g'}{g}(1) > \beta$ (resp. $\frac{g'}{g}(1) > a'\lambda_0 + b'$) and $g(x) > 0$ for all $x \in (0, 1]$.

Proof. By the definition of $\hat{E}$, λ_0 is less than the least eigenvalue of (2.11), (2.12) and $\frac{g'}{g}(1) = \beta$ (resp. $\frac{g'}{g}(1) = a'\lambda_0 + b'$). The proof now follows from differential inequality theory applied to the (reversed) differential equation (whose right side is monotone in λ_0) satisfied by $\frac{g'}{g}$. □

Let γ be such that $\beta < \gamma < \frac{g'}{g}(1)$ (resp. $a'\lambda_0 + b' < \gamma < \frac{g'}{g}(1)$) and w be the solution of

$$-w'' + \hat{q}w = \lambda_0 w, \quad w(0) = 1, \quad \frac{w'}{w}(1) = \gamma.$$

Since $\gamma < \frac{g'}{g}(1)$ it follows (as for Lemma 2.2) that $w(x) > 0$ for all $x \in [0,1]$. Thus we can define

$$z_1 = -\frac{w'}{w}$$

$$q_1 = \hat{q} + 2z_1'$$

and

$$a_1 = \tfrac{1}{\gamma - \beta}, \quad b_1 = \tfrac{\lambda_0}{\beta - \gamma} - \gamma \tag{2.13}$$

$$\left(\text{resp.} \quad \begin{array}{ll} a_1 = -\gamma - 1/a', & b_1 = (\lambda_0 - b'\gamma + \gamma^2)/a' \\ c_1 = 1, & d_1 = (b' - \gamma)/a'. \end{array}\right)$$

Note that $a_1 > 0$ (resp. $\delta_1 = a_1 d_1 - b_1 > 0$). This specifies R except for α_1.

It is easily verified that $z_1^2 - z_1' = \hat{q} - \lambda_0$. Let $y = g' + z_1 g$. Then $y(0) = 1$ and it can be verified directly that $-y'' + q_1 y = \lambda_0 y$. We observe from integration by parts and the definitions of a_1, b_1 (resp. a_1, b_1 and d_1) and y, z_1 that

$$
\begin{aligned}
\|y\|_{a_1}^2 &= \int_0^1 y^2 \, dt + a|y(1)|^2 \\
&= \int_0^1 (yg' + yz_1 g) \, dt + \frac{[g'(1) - \gamma g(1)]^2}{\gamma - \beta} \\
&= [(g' + zg)g]_0^1 - \int_0^1 [(g'' + zg')g - yzg] \, dt + \frac{[g'(1) - \gamma g(1)]^2}{\gamma - \beta} \\
&= g(1)^2 \left[\left(\frac{g'}{g}(1) - \gamma\right) + \frac{\left[\frac{g'}{g}(1) - \gamma\right]^2}{\gamma - \beta} \right] \\
&\quad - \int_0^1 [(\hat{q} - \lambda_0 + z')g^2 + zg'g - (g' + zg)zg] \, dt \\
&= \frac{g(1)^2}{\gamma - \beta} \left[\beta - \frac{g'}{g}(1)\right] \left[\gamma - \frac{g'}{g}(1)\right] - \int_0^1 (\hat{q} - \lambda_0 + z' - z^2)g^2 \, dt \\
&= g^2(1) \left[\beta - \frac{g'}{g}(1)\right] \left[\frac{\gamma - \frac{g'}{g}(1)}{\gamma - \beta}\right]
\end{aligned}
$$

$$\left(\text{resp. similarly} \quad \|y\|_{a_1,\delta_1}^2 = g^2(1) \left[(a'\lambda_0 + b') - \frac{g'}{g}(1)\right] \left[\frac{\gamma - \frac{g'}{g}(1)}{\gamma - (a'\lambda_0 + b')}\right]\right).$$

Hence the map taking γ to $\|y\|_{a_1}^2$ (resp. $\|y\|_{a_1,\delta_1}^2$) is continuous and strictly decreasing from $\left(\beta, \frac{g'}{g}(1)\right)$ (resp. $\left(a'\lambda_0 + b', \frac{g'}{g}(1)\right)$) onto $\mathbb{R}^+$. Thus for each positive number ν_0 there exists one and only one $\gamma \in (\beta, \frac{g'}{g}(1))$ (resp. $(a'\lambda_0 + b', \frac{g'}{g}(1))$) such that $\nu_0 = \|y\|_{a_1}^2$ (resp. $\nu_0 = \|y\|_{a_1,\delta_1}^2$). For this γ, let $\alpha_1 = z_1(0)$. This completes the definition of R.

Suppose that $\lambda_1 < \lambda_2 < \cdots$ is the spectrum of $(\hat{q}, \beta; DN)$ (resp. $(\hat{q}, a', b'; DA)$). Observe that the definition of $\hat{E}$ ensures that $\lambda_0 < \lambda_1$. Let $u(x, \lambda)$ be the solution of $-u'' + \hat{q}u = \lambda u, u(0) = 0, u'(0) = 1$. Put $y(x, \lambda) = u'(x, \lambda) + u(x, \lambda)z_1(x)$ and $u_n(x) = u(x, \lambda_n)$. Then

$$y'(x, \lambda) = (\lambda_0 - \lambda)u(x, \lambda) + z_1(x)(u'(x, \lambda) + z_1(x)u(x, \lambda)).$$

For each λ, $y(0, \lambda) = 1$ and $y'(0, \lambda) = \xi_1$ giving $\frac{y'(0, \lambda)}{y(0, \lambda)} = \xi_1$. It is readily verified that $-y'' = (q_1 - \lambda)y$. As in the argument of [3, Theorem 4.1] the terminal condition $\frac{y'(1, \lambda)}{y(1, \lambda)} = a_1\lambda + b_1$ (resp. $\frac{y'(1, \lambda)}{y(1, \lambda)} = \frac{a_1\lambda + b_1}{c_1\lambda + d_1}$) is satisfied if and only if $\lambda_0 = \lambda$ or $\beta = u'(1)/u(1)$ (resp. $\lambda_0 = \lambda$ or $u'(1)/u(1) = a'\lambda + b'$). Accordingly we have shown

Lemma 2.3. *Let the spectrum of $(\hat{q}, \beta; DN)$ (resp. $(\hat{q}, a', b'; DA)$) be $\lambda_1 < \lambda_2 \ldots$ with associated norming constants ξ_j, $j \geq 1$, and let the numbers $\lambda_0 < \lambda_1$ and $\nu_0 > 0$ be given. Then $(q_1, \alpha_1, a_1, b_1; NA)$ (resp. $(q_1, \alpha_1, a_1, d_1, \delta_1; NB)$) where $R(\hat{q}, \beta, \nu_0, \lambda_0) = (q_1, \alpha_1, a_1, b_1)$ (resp. $R(\hat{q}, a', b', \nu_0, \lambda_0) = (q_1, \alpha_1, a_1, d_1, \delta_1)$) has spectrum $\lambda_0 < \lambda_1 < \cdots$ and norming constants ν_0 and $\nu_j = (\lambda_j - \lambda_0)\xi_j$, $j \geq 1$.*

We are now in a position to establish

Lemma 2.4. *The maps R and T are inverses of each other.*

Proof. Using the notation introduced above we have $T(q, \alpha, a, b) = (\hat{q}, \beta, \nu_0, \lambda_0)$ and $R(\hat{q}, \beta, \nu_0, \lambda_0) = (q_1, \alpha_1, a_1, b_1)$ (resp. $T(q, \alpha, a, d, \delta) = (\hat{q}, a', b', \nu_0, \lambda_0)$ and $R(\hat{q}, a', b', \nu_0, \lambda_0) = (q_1, \alpha_1, a_1, \delta_1)$).

Let $v = 1/y_0$; then it is easily verified that $-v'' + \hat{q}v = \lambda_0 v$ and $v(0) = 1$. As $v(x) > 0$ for all $x \in [0, 1]$ it follows that

$$\frac{v'}{v}(1) < \frac{g'}{g}(1).$$

But

$$\frac{v'}{v}(1) = -\frac{y_0'}{y_0}(1) = -(a\lambda_0 + b)$$

(resp. $-(a\lambda_0 + b)/(\lambda_0 + d)$) so $a > 0$ (resp. $\lambda_0 < -d$), gives

$$\beta = -\frac{1}{a} - (a\lambda_0 + b) < \frac{v'}{v}(1)$$

$$\left(\text{resp. } a'\lambda_0 + b' = -\frac{(\lambda_0 + d)^2}{\delta} - \frac{a\lambda_0 + b}{\lambda_0 + d} < \frac{v'}{v}(1)\right).$$

We are thus able to set $\gamma = \frac{v'}{v}(1)$ which gives $a_1 = a$, $b_1 = b$ (resp. $a_1 = a$, $b_1 = b$ and $d_1 = d$), and moreover $v = w$. Thus from the definition of y we have

$$y = \frac{g'v - gv'}{v} \quad \text{and} \quad 1 = y(0) = \frac{[g'v - gv'](0)}{v(0)} = [g'v - gv'](0)$$

giving $g'v - gv' = 1$ and so $y = y_0$. Consequently $z_1 = -\frac{w'}{w} = -\frac{v'}{v} = \frac{y_0'}{y_0} = z$ from which it follows that $q = q_1$, see (2.2), (2.13). This shows that $RT = I|_E$.

The proof that $TR = I|_{\hat{E}}$ is similar. $\qquad\square$

3. Transformation of problems of types DA and DB

We now proceed with the study of isospectral transformations between problems of types DA and NN (resp. DB and NA) in a manner which enables us to give precise expressions for the domains and ranges of the maps, to identify their inverses and to compute the transformed norming constants.

Define F to be the set

$$\{(q,\mu,a,b)|q \in L^1[0,1], b \in \mathbb{R}, a > 0, \mu < \lambda_0(q,a,b;DA)\}$$

(resp. $\{(q,\mu,a,d,\delta)|q \in L^1[0,1], \delta > 0, a,c,\delta \in \mathbb{R}, \mu < \lambda_0(q,a,d,\delta;DB)\}$),

where $\lambda_0(q,a,b;DA)$ (resp. $\lambda_0(q,a,d,\delta;DB)$) is the least eigenvalue of (1.1) with boundary conditions (1.2) with $\alpha = \infty$ and (1.4) (resp. (1.5)). Further define $\hat{F}$ as the set

$$\{(\hat{q},\mu,\gamma,\beta)|\hat{q} \in L^1[0,1], \gamma, \beta \in \mathbb{R}, \mu < \lambda_0(\hat{q},\gamma,\beta;NN)\}$$

(resp. $\{(\hat{q},\mu,\gamma,a',b')|\hat{q} \in L^1[0,1], \gamma, b' \in \mathbb{R}, a' > 0, \mu < \lambda_0(\hat{q},\gamma,a' > 0, b';NA)\}$).

The transformation from a problem of type DA (resp. DB) to one of type NN (resp. NA) in [3] can be viewed as a mapping $S : F \to \hat{F}$ with

$$S(q,\mu,a,b) = (\hat{q},\mu,\gamma,\beta)$$

$$(\text{resp. } S(q,\mu,a,d,\delta) = (\hat{q},\mu,\gamma,a',b'))$$

where

$$\hat{q} = q - 2z_1', \tag{3.1}$$

$$\gamma = -z_1(0), \tag{3.2}$$

$$\beta = -\tfrac{1}{a} - (a\mu + b),$$

$$\left(\text{resp.} \quad \begin{array}{rcl} a' & = & -(\mu + d)/\delta \\ b' & = & -\frac{d(\mu+d)}{\delta} - \frac{a\mu+b}{\mu+d} \end{array}\right).$$

Here $z_1 = w'/w$ and w is the solution of the Cauchy problem

$$-w'' + qw = \mu w, \tag{3.3}$$

$$w(1) = 1, \quad w'(1) = a\mu + b \tag{3.4}$$

$$\left(\text{resp. } w(1) = 1, \quad w'(1) = \frac{a\mu+b}{c\mu+d}\right). \tag{3.5}$$

Note that $w > 0$ on [0,1] since $\mu < \lambda_0$, cf. the proof of Lemma 2.2, so z_1 is defined. Moreover if $(q,a,b;DA)$ (resp. $(q,a,d,\delta;DB)$) has spectrum $\lambda_0 < \lambda_1 < \cdots$, then $(\hat{q},\alpha,\beta;NN)$ (resp. $(\hat{q},\alpha,a',b';NA)$) has spectrum $\lambda_0 < \lambda_1 < \cdots$, i.e. the map S is isospectral. Noting also that $\mu < \lambda_0$ for the transformed problem, and $\mu < -d$ by [1], we see that S maps F into $\hat{F}$.

We define the operator L formally as in (2.1) where $\rho(0) = \alpha$ now means $y(0) = 0$. Again it is known [15] that L is self-adjoint and $(q,\alpha,a,b;DA)$ (resp. $(q,\alpha,a,d,\delta;DB)$) is equivalent to the eigenvalue problem for L.

For $(q,\mu,a,b) \in F$, (resp. $(q,\mu,a,d,\delta) \in F$) denote the spectrum of $(q,a,b;DA)$ (resp. $(q,a,d,\delta;DB)$) by $\lambda_0 < \lambda_1 < \cdots$ and let y_n be the eigenfunction corresponding to λ_n, normalized so that $y'_n(0) = 1$, $n \geq 0$. The norming constants are defined by $\nu_n = \|y_n\|_a^2$ (resp. $\nu_n = \|y_n\|_{a,\delta}^2$).

In addition, the norming constants for $(\hat{q}, \alpha, \beta; NN)$ (resp. $(\hat{q}, \alpha, a', b'; ND)$) where $S(q, \alpha, a, b) = (\hat{q}, \alpha, \lambda_0, \beta)$ (resp. $S(q, \alpha, a, d, \delta) = (\hat{q}, \alpha, \lambda_0, a', b')$) are defined by $\xi_n = (u_n, u_n)_{L^2}$ (resp. $\xi_n = \|u_n\|_{a'}^2$) in which u_n is the eigenfunction corresponding to λ_n normalized by $u_n(0) = 1$.

Lemma 3.1.

$$\xi_n = (\lambda_n - \mu)\nu_n. \tag{3.6}$$

Proof. We obtain from reasoning as in (2.6) (resp. (2.7)) that

$$(\lambda_n - \mu)\nu_n = \int_0^1 \left({y'_n}^2 + (q - \mu)y_n^2 \right) - y_n^2(1)(a\mu + b)$$

$$(\text{resp. } (\lambda_n - \mu)\nu_n = \int_0^1 \left({y'_n}^2 + (q - \mu)y_n^2 \right)$$
$$- [a(a\mu + b)y_n^2 - 2(a\mu + b)y_n y'_n + (\mu + d){y'_n}^2](1)/\delta \,).$$

Now consider $u_n = y'_n - y_n z_1$. We showed in [3] that u_n is an eigenfunction corresponding to λ_n for the NN (resp. NA) problem $(\hat{q}, \beta)$ (resp. $(\hat{q}, a', b')$) and it is readily verified that $u_n(0) = 1$. Proceeding as for (2.8) we find

$$u_n^2 = {y'_n}^2 - (y_n^2 z_1)' + (q - \mu)y_n^2.$$

Hence, reasoning as for (2.9) (resp. (2.10)) we see that $\xi_n = (\lambda_n - \mu)\nu_n$, which establishes the result. $\qquad\square$

Define $W : \hat{F} \to F$ by

$$W(\hat{q}, \mu, \gamma, \beta) = (q_2, \mu, a_2, b_2)$$

$$(\text{resp. } W(\hat{q}, \mu, \gamma, a', b') = (q_2, \mu, a_2, d_2, \delta_2))$$

where

$$a_2 = \frac{-1}{z(1)+\beta}, \quad b_2 = -\left(\beta + a_2\mu + \frac{1}{a_2}\right), \tag{3.7}$$

$$\left(\text{resp. } \begin{array}{ll} a_2 = z(1) - 1/a', & b_2 = (z^2(1) + b'z(1) + \mu)/a' \\ c_2 = 1, & d_2 = (b' + z(1))/a' \end{array} \right) \tag{3.8}$$

$$q_2 = \hat{q} + 2z'. \tag{3.9}$$

Here $z = -v'/v$ and v is the solution of the initial value problem

$$-v'' + \hat{q}v = \mu v, \tag{3.10}$$

$$v'(0) = \gamma, \quad v(0) = 1. \tag{3.11}$$

As $\mu < \lambda_0(\hat{q}, \gamma, \beta; NN)$ (resp. $\mu < \lambda_0(\hat{q}, \gamma, a', b'; NA)$) it follows that $v(x) > 0$ on $[0, 1]$, so indeed z is defined, and $a_2 > 0$ (resp. $a_2 d_2 - b_2 > 0$) follows from a Sturmian argument as for Lemma 2.2. Further

$$z(0) \quad = \quad -\gamma, \tag{3.12}$$

$$z^2 + \mu \quad = \quad z' + \hat{q}. \tag{3.13}$$

We now show that W is an isospectral map, W maps into F, and that W and S are inverses of each other. Let u be the solution of

$$-u'' + \hat{q}u = \lambda u,$$

$$u'(0) = \gamma, \quad u(0) = 1.$$

and let

$$y = u' + zu.$$

It now follows from the argument of [3, Theorem 4.2] that λ is in the spectrum of $(q_2, a_2, b_2; DA)$ (resp. $(q_2, a_2, d_2, \delta_2; DB)$) if and only if it is in the spectrum of $(\hat{q}, \gamma, \beta; NN)$ (resp. $(\hat{q}, \gamma, a', b'; NA)$). Thus W is an isospectral map,

$$\mu < \lambda_0(\hat{q}, \gamma, \beta; NN) = \lambda_0(q_2, a_2, b_2; DA)$$

$$(\text{resp. } \mu < \lambda_0(\hat{q}, \gamma, a', b'; NA) = \lambda_0(q_2, a_2, d_2, \delta_2; DB)),$$

and $W(\hat{q}, \mu, \gamma, \beta) \in F$ (resp. $W(\hat{q}, \mu, \gamma, a', b') \in F$).

With Lemma 3.1 we have thus proved the following result.

Lemma 3.2. *The map $W : \hat{F} \to F$ is isospectral, and if the norming constants for the NN (resp. NA) problem P are ξ_n, then the corresponding norming constants for WP are $\nu_n = \xi_n/(\lambda_n - \mu)$.*

We are now in a position to establish

Lemma 3.3. *The maps W and S are inverse to each other.*

Proof. We present the proof that $WS = I|_F$; the proof that $SW = I|_{\hat{F}}$ is similar.

Adhering to the notation used above we have

$$S(q, \mu, a, b) = (\hat{q}, \mu, \gamma, \beta) \quad (\text{resp. } S(q, \mu, a, d, \delta) = (\hat{q}, \mu, \gamma, a', b'))$$

and

$$W(\hat{q}, \mu, \gamma, \beta) = (q_2, \mu, a_2, b_2) \quad (\text{resp. } W(\hat{q}, \mu, \gamma, a', b') = (q_2, \mu, a_2, d_2, \delta_2)).$$

We must prove that $q = q_2$, $a = a_2$ and $b = b_2$ (resp. $q = q_2$, $a = a_2$, $b = b_2$ and $d = d_2$). The critical step in the proof is showing that $z_1 = z$.

Let $\zeta = 1/w$, where w is as defined in (3.3) and (3.2), then $\zeta'/\zeta = -w'/w = -z_1$ and $\zeta'(0)/\zeta(0) = \gamma$. From (3.4) (resp. 3.5)

$$\zeta'(1)/\zeta(1) = -(a\mu + b), \quad \left(\text{resp. } \zeta'(1)/\zeta(1) = -\frac{a\mu + b}{c\mu + d}\right),$$

and from (3.1) we obtain that $-\zeta'' + \hat{q}\zeta = \mu\zeta$. Hence $\zeta'/\zeta = v'/v$ which, with the initial condition $v(0) = 1 = \zeta(0)$, gives $\zeta(x) = v(x)$ for all $x \in [0,1]$, and, more significantly

$$z = -\frac{\zeta'}{\zeta} = \frac{w'}{w} = z_1.$$

The definitions of q_2 and $\hat{q}$ now yield

$$q_2 = \hat{q} + 2z' = \hat{q} + 2z_1' = q,$$

from which it is straightforward to show $a = a_2$ and $b = b_2$ (resp. $a = a_2$, $b = b_2$ and $d = d_2$). ∎

4. Main theorems

We first show that the maps we have given are the only ones between problems of types (A), (B) and (C) which preserve the spectrum and norming constants as indicated below.

Theorem 4.1.

(i) *There is one and only one map taking problems of type NA (resp. NB), with spectrum and norming constants $\{\lambda_j; \nu_j\}_{j \geq 0}$, to problems of type DN (resp. DA), with spectrum and norming constants $\left\{\lambda_j; \frac{\nu_j}{\lambda_j - \lambda_0}\right\}_{j \geq 1}$.*

(ii) *For each μ, there is one and only map taking problems of type DA (resp. DB), with spectrum and norming constants $\{\lambda_j; \nu_j\}_{j \geq 0}$ where $\lambda_0 > \mu$, to problems of type NN (resp. NA), with spectrum and norming constants $\{\lambda_j; \nu_j(\lambda_j - \mu)\}_{j \geq 0}$.*

(iii) *The above spectra and norming constants determine each such problem uniquely.*

Proof. Existence for (i) and (ii) is ensured by the work of Sections 2 and 3 respectively. Suppose that there are two distinct problems of the form DN (resp. NN) corresponding to one NA (resp. DA) problem. In this case there are two problems of type DN (resp. NN) with the same spectrum and norming constants, which is not possible, by [12, Theorem 2.3.1]. This establishes uniqueness of the mappings from type A to type C problems as well as the final contention for type A problems. We can now repeat the previous argument for mappings from type B to type A problems to establish their uniqueness, and the final contention for type B problems. ∎

Composing the above maps, we obtain direct transformations from problems of type B to type C.

Corollary 4.2. *For each μ, there is one and only map taking problems of type NB (resp. DB), with spectrum and norming constants $\{\lambda_j; \nu_j\}_{j \geq 0}$ where $\lambda_0 > \mu$, to problems of type NN (resp. DN), with spectrum and norming constants $\left\{\lambda_j; \nu_j \frac{\lambda_j - \mu}{\lambda_j - \lambda_0}\right\}_{j \geq 1}$.*

Remark 4.3. *In the transformation from NB to NN problems, the fact that $\lambda_0 < \lambda_1$ allows us to choose $\mu = \lambda_0$, so the transformed spectrum and norming constants coincide with the original ones (but for $j \geq 1$).*

We can now give results paralleling those of Gelfand-Levitan [8], but for problems with eigenparameter dependent boundary conditions.

Theorem 4.4. *The sequences $\lambda_0 < \lambda_1 < \lambda_2 < \cdots$ and $\nu_n > 0, n = 0, 1, 2, \ldots,$ are the eigenvalues and norming constants of a unique NA problem $(q, \alpha, a, b) \in L^1 \times \mathbb{R} \times \mathbb{R}^+ \times \mathbb{R}$ if and only if*

$$\sqrt{\lambda_n} = \pi \left\{ n - \frac{1}{2} + \frac{k}{n} + o\left(\frac{1}{n}\right) \right\}, \tag{4.1}$$

$$\nu_n = \frac{1}{2} + o\left(\frac{1}{n}\right) \tag{4.2}$$

and, for N so large that $\lambda_n \neq 0$ for all $n \geq N$,

$$F(x,t) = \sum_{n=N}^{\infty} \left[\left(1 - \frac{\lambda_0}{\lambda_n}\right) \frac{\sin \sqrt{\lambda_n}x \sin \sqrt{\lambda_n}t}{\nu_n} - 2 \sin \pi x \left(n - \frac{1}{2}\right) \sin \pi t \left(n - \frac{1}{2}\right) \right]$$

has L^1 first derivatives.

Proof. From the Appendix there exists a unique DN problem with spectrum $\lambda_1 < \lambda_2 < \cdots$ and norming constants $\nu_1/(\lambda_1 - \lambda_0), \nu_2/(\lambda_2 - \lambda_0), \ldots$ if and only if (4.1), (4.2) and $F \in W^{1,1}(1,1)^2$ are all satisfied. The sufficiency and necessity of the above conditions follows from $T : E \to \hat{E}$ and $R : \hat{E} \to E$ being surjective.

For uniqueness suppose there to be two NA problems with the same spectra and norming constants. Then as T is one-to-one there are two distinct DN problems with identical spectra and norming constants, which contradicts [12, Theorem 2.3.1]. □

Theorem 4.5. *The sequences $\lambda_0 < \lambda_1 < \lambda_2 < \cdots$ and $\nu_n > 0, n = 0, 1, 2, \ldots,$ are the eigenvalues and norming constants a unique DA problem $(q, a, b) \in L^1 \times \mathbb{R}^+ \times \mathbb{R}$ if and only if*

$$\sqrt{\lambda_n} = \pi \left\{ n + \frac{k}{n} + o\left(\frac{1}{n}\right) \right\},$$

$$\lambda_n \nu_n = \frac{1}{2} + o\left(\frac{1}{n}\right),$$

and, for N so large that $\lambda_n \neq 0$ for all $n \geq N$, there exists $\mu < \lambda_0$ such that

$$F(x,t) = \sum_{n=N}^{\infty} \left[\frac{\cos \sqrt{\lambda_n}x \cos \sqrt{\lambda_n}t}{\nu_n(\lambda_n - \mu)} - 2 \cos \pi x n \cos \pi t n \right]$$

has L^1 first derivatives.

Proof. Let $\mu < \lambda_0$ be as stated in the theorem. From [11, Theorem 1.6.1] there exists an NN problem with spectrum $\lambda_0 < \lambda_1 < \cdots$ and norming constants $\nu_0(\lambda_0 - \mu), \nu_1(\lambda_1 - \mu), \ldots$. By [12, Theorem 2.3.1] and [13, Equ. (4.2)] this problem is unique. The existence of a suitable DA problem follows from $S : F \to \hat{F}$ being surjective. The sufficiency of (4.3), (4.3) and $F \in W^{1,1}(0,1)^2$ follows from [11] and surjectivity of $W : \hat{F} \to F$.

For uniqueness, let $\mu < \lambda_0$ be arbitrary and suppose there to be two DA problems with the same spectra and norming constants. Let $S_\mu(q,a,b) = S(q,\mu,a,b)$. Then, as S is one-to-one, S_μ is one-to-one and there are two distinct NN problems with identical spectra and norming constants, which contradicts [12, Theorem 2.3.1]. $\qquad\square$

The following theorems can be verified in a similar manner by considering compositions of the maps constructed in sections 2 and 3.

Theorem 4.6. *The sequences $\lambda_0 < \lambda_1 < \lambda_2 < \cdots$ and $\nu_n > 0, n = 0, 1, 2, \ldots$, are the eigenvalues and norming constants a unique DB problem $(q,a,d,\delta) \in L^1 \times \mathbb{R}^2 \times \mathbb{R}^+$ if and only if*

$$\sqrt{\lambda_n} = \pi\left\{ n - \frac{1}{2} + \frac{k}{n} + o\left(\frac{1}{n}\right) \right\},$$

$$\lambda_n \nu_n = \frac{1}{2} + o\left(\frac{1}{n}\right)$$

and, for N so large that $\lambda_n \neq 0$ for all $n \geq N$, there exists $\mu < \lambda_0$ such that

$$F(x,t) = \sum_{n=N}^{\infty} \left[\left(1 - \frac{\lambda_0 - \mu}{\lambda_n - \mu}\right) \frac{\sin\sqrt{\lambda_n}x \sin\sqrt{\lambda_n}t}{\nu_n\lambda_n} - 2\sin\pi x \left(n - \frac{1}{2}\right)\sin\pi t \left(n - \frac{1}{2}\right) \right]$$

has L^1 first derivatives.

Theorem 4.7. *The sequences $\lambda_0 < \lambda_1 < \lambda_2 < \cdots$ and $\nu_n > 0, n = 0, 1, 2, \ldots$, are the eigenvalues and norming constants a unique NB problem $(q,\beta,a,d,\delta) \in L^1 \times \mathbb{R}^3 \times \mathbb{R}^+$ if and only if*

$$\sqrt{\lambda_n} = \pi\left\{ n - 1 + \frac{k}{n} + o\left(\frac{1}{n}\right) \right\},$$

$$\nu_n = \frac{1}{2} + o\left(\frac{1}{n}\right),$$

and, for N so large that $\lambda_n \neq 0$ for all $n \geq N$, there exists $\mu < \lambda_0$ such that

$$F(x,t) = \sum_{n=N}^{\infty} \left[\left(1 - \frac{\lambda_0 - \mu}{\lambda_n - \mu}\right) \frac{\cos\sqrt{\lambda_n}x \cos\sqrt{\lambda_n}t}{\nu_n} - 2\cos\pi x n \cos\pi t n \right]$$

has L^1 first derivatives.

5. Appendix

The solution of the inverse spectral problem for non-Dirichlet boundary conditions is well known, [8], [11], but we found no theorem of the refined form presented in [11, Theorem 1.6.1] for the case of Dirichlet boundary conditions at one end and non-Dirichlet boundary conditions at the other. Hence we include the following theorem as an Appendix for the reader's convenience.

Theorem 5.1. *The sequences* $\lambda_0 < \lambda_1 < \lambda_2 < \cdots$ *and* $\nu_n > 0, n = 0, 1, 2, \ldots$, *are the eigenvalues and norming constants of a unique Sturm-Liouville boundary value problem*

$$-y'' + qy \;=\; \lambda y \tag{5.1}$$

$$y(0) = 0, \qquad y'(0) = 1 \tag{5.2}$$

$$y'(1) \;=\; \beta y(1) \tag{5.3}$$

(with finite β) if and only if

$$\sqrt{\lambda_n} \;=\; \pi\left(n + \frac{1}{2} + \frac{k}{n}\right) + o\left(\frac{1}{n}\right) \tag{5.4}$$

$$\nu_n \lambda_n \;=\; \frac{1}{2} + o\left(\frac{1}{n}\right) \tag{5.5}$$

where

$$F(x,t) = \sum_{n=N}^{\infty}\left[\frac{\sin\sqrt{\lambda_n}x \sin\sqrt{\lambda_n}t}{\nu_n\lambda_n} - 2\sin\pi x\left(n + \frac{1}{2}\right)\sin\pi t\left(n + \frac{1}{2}\right)\right] \tag{5.6}$$

has L^1 first derivatives for N sufficiently large that $\lambda_n \neq 0$ for all $n \geq N$.

Proof. Without loss of generality we may assume that $\lambda_n \neq 0$ for all $n = 0, 1, 2, \ldots$.

Given that $\{\lambda_n\}$ and $\{\nu_n\}$ are the spectral characteristics of a boundary value of the form given in (5.1), (5.2) and (5.3), we show that equation (5.6) and the asymptotics (5.4) and (5.5) are satisfied. That (5.4) holds follows from [12], so we are concerned with (5.5) and (5.6).

Let $\varphi(x, \lambda)$ denote the solution of (5.1) with initial conditions (5.2). From [14] there are functions $K(x,t)$ and $H(x,t)$ with L^1 first derivatives such that

$$\varphi(x,\lambda) \;=\; \frac{\sin\sqrt{\lambda}x}{\sqrt{\lambda}} + \int_0^x K(x,t)\frac{\sin\sqrt{\lambda}t}{\sqrt{\lambda}}\,dt \tag{5.7}$$

$$\frac{\sin\sqrt{\lambda}x}{\sqrt{\lambda}} \;=\; \varphi(x,\lambda) + \int_0^x H(x,t)\varphi(x,\lambda)\,dt \tag{5.8}$$

and having

$$K(x,x) = -H(x,x) \;=\; \frac{1}{2}\int_0^x q(t)\,dt \tag{5.9}$$

$$K(x,0) = H(x,0) \;=\; 0 \tag{5.10}$$

$$K(x,t) = H(x,t) \;=\; 0 \text{ for all } t > x. \tag{5.11}$$

From (5.7), and from the first derivatives of $K(x,t)$ being in L^1, for real positive λ we obtain

$$\varphi(x,\lambda) = \frac{\sin\sqrt{\lambda}x}{\sqrt{\lambda}} - \frac{K(x,x)\cos\sqrt{\lambda}x}{\lambda} + o\left(\frac{1}{\lambda}\right). \tag{5.12}$$

Using (5.4) with (5.12) we see that

$$\sqrt{\lambda_n}\varphi(x,\lambda_n) = \sin\pi x\left(n+\frac{1}{2}\right) + \frac{\pi k x}{n}\cos\pi x\left(n+\frac{1}{2}\right) +$$
$$+ \frac{K(x,x)\cos\pi x\left(n+\frac{1}{2}\right)}{\pi\left(n+\frac{1}{2}\right)} + o\left(\frac{1}{n}\right).$$

Squaring and integrating the above we obtain (5.5)

We now verify that $F(x,t)$ has L^1 first partial derivatives. Replacing the $\frac{\sin\sqrt{\lambda_n}x\sin\sqrt{\lambda_n}t}{\lambda_n\nu_n}$ term in (5.6) by means of (5.8) and noting that $\sqrt{2}\sin\pi t\left(n+\frac{1}{2}\right)$ is a complete orthonormal basis for $L^2[0,1]$, one concludes via Parseval's equality, as in [11], that

$$F(x,t) = H(x,t) + \int_0^t H(x,s)H(t,s)\,ds. \tag{5.13}$$

Since $H(x,t)$ is known to have L^1 first partial derivative, the first half of the proof is complete.

We now proceed to the proof of the inverse result. The critical fact here is that [10] ensures that the set $\{\sin\sqrt{\lambda_n}x\}$ is a complete linearly independent family for $L^2[0,1]$. This enables us to conclude, as in [11], that

$$K(x,t) + F(x,t) + \int_0^x K(x,s)F(s,t)\,ds = 0 \tag{5.14}$$

has a unique solution $K(x,t)$ for $0 \le t \le x \le 1$. [We set $K(x,t) = 0$ for $t > x$.] Now that $K(x,t)$ has been defined, we define $\varphi(x,\lambda)$ by (5.7). Proceeding as in [14] (or [11]), we obtain φ as a solution of (5.1) with

$$q(x) = 2\frac{dK(x,x)}{dx} \in L^1[0,1]$$

with the correct boundary conditions. This establishes existence, and uniqueness comes from [12, Theorem 2.3.1] and standard asymptotic estimates. $\square$

Acknowledgements. P. A. Binding and P. J. Browne: Research supported in part by grants from the NSERC of Canada. B. A. Watson: Research conducted while visiting University of Calgary and University of Saskatchewan and supported in part by the Centre for Applicable Analysis and Number Theory.

References

[1] Binding P.A., Browne P.J., Seddighi K., Sturm-Liouville problems with eigenparameter dependent boundary conditions, *Proceedings of the Edinburgh Mathematical Society* **37** (1993), 57–72.

[2] Binding P.A., Browne P.J., Watson B.A., Inverse spectral problems for Sturm-Liouville equations with eigenparameter dependent boundary conditions, *J. London Math. Soc.*, to appear.

[3] Binding P.A., Browne P.J., Watson B.A., Transformations between Sturm-Liouville equations with eigenvalue dependent and independent boundary conditions, submitted.

[4] Fulton C.T., Two-point boundary value problems with eigenvalue parameter contained in the boundary conditions, *Proc. Roy. Soc. Edinburgh* **77A** (1977), 293–308.

[5] Fulton C.T., Singular eigenvalue problems with eigenvalue parameter contained in the boundary conditions, *Proc. Royal Soc. Edinburgh* **87A** (1980), 1–34.

[6] Dijksma A., Langer H., Operator theory and ordinary differential operators, *Fields Institute Monographs* **3** (1996), 75–139.

[7] Dijksma A., Langer H., de Snoo H.S.V., Symmetric Sturm-Liouville operators with eigenvalue depending boundary conditions, *Canadian Math. Soc. Conference Proceedings* **8** (1987), 87–116.

[8] Gelfand I.M., Levitan B.M., On the determination of a differential equation from its spectral function, *Amer. Math. Soc. Translations, Series 2*, **1** (1955), 253–304.

[9] Isaacson E.L., McKean H.P., Trubowitz E., The inverse Sturm-Liouville problem, *Comm. Pure and Applied Math.* **37** (1984), 1–11.

[10] Levinson N., Gap and Density Theorems *Amer. Math. Soc. Colloquium Publications* **26** (1940).

[11] Levitan B.M., Gasymov M.G., Determination of a differential equation by two of it spectra, *Russian Math. Surveys* **19** (1964), 1–64.

[12] Marchenko V.A., Some questions in the theory of one-dimensional linear differential operators of the second order, Part I, *Amer. Math. Soc. Translations Series 2* **101** (1973), 1–104.

[13] McLaughlin J.R., Analytical methods for recovering coefficients in differential equations from spectral data, *SIAM Review* **28** (1986), 53–72.

[14] Thurlow C., A generalisation of the inverse spectral theorem of Levitan and Gasymov, *Proc. Roy. Soc. Edinburgh* **84A** (1979), 185–196.

[15] Walter J., Regular eigenvalue problems with eigenvalue parameter in the boundary conditions, *Math. Z.* **133** (1973), 301–312.

Department of Mathematics and Statistics
University of Calgary
Calgary, Alberta
Canada T2N 1N4

Department of Mathematics and Statistics
University of Saskatchewan
Saskatoon, Saskatchewan
Canada S7N 5E6

Department of Mathematics
University of the Witwatersrand
Private Bag 3, P O WITS 2050
South Africa

1991 Mathematics Subject Classification. Primary 34B24; Secondary 34L20

Received June 30, 2000

Operator Theory:
Advances and Applications, Vol. 130, 153–170
© 2001 Birkhäuser Verlag Basel/Switzerland

Existence and Uniqueness Results for Nonlinear Cooperative Systems

Erich Bohl and Ivo Marek

Dedicated to Peter Lancaster on the occasion of his 70th birthday

Abstract. Several fundamental results of the theory of a class of nonlinear cooperative systems are established such as time-global existence, uniqueness and asymptotic behaviour of the solutions to the underlying dynamical systems. The research is strongly motivated by systems turning up in biology where the results form the basis for pseudo-steady state arguments to obtain expressions for definition of the *speed of chemical networks*. Examples are given.

1. Introduction

In our earlier paper [BM98] we developed a theory for linear problems of the type

$$\frac{d}{dt}w(t) = Tw(t), \ w(0) \text{ given},\tag{1.1}$$

where T is a given infinitesimal generator of a semigroup of operators. We gave several examples mainly from biology and chemistry where the state vectors $w(t)$ of the underlying chemical network follow an evolution (1.1). In these cases the conservation of matter requires the existence of an element f such that the pairing $[w(t), f]$ is constant during all times of the evolution so that we have

$$[w(t), f] = [w(0), f], \ t \geq 0.\tag{1.2}$$

More complicated networks (and we give some examples in Section 3) are described by a state vector $u(t)$ which is formed by finitely many substates in the fashion

$$u(t) = \big(u^1(t), \ldots, u^N(t)\big)\tag{1.3}$$

following an evolution

$$\left.\begin{array}{l} \dfrac{d}{dt}u^j(t) = T^{(j)}(u(t))u^j(t) := B^{(j)}u^j(t) + G^{(j)}(u(t))u^j(t) \\[2mm] [u^j(t), f^j] = [u^j(0), f^j], \ t > 0, \ j = 1, \ldots, N. \end{array}\right\}\tag{1.4}$$

Hence, the subsystems evolve like (1.1) and (1.2) for which we developed the theory in [BM98]. The dependence of the operators $T^{(j)}$ on the data is typically on the complete state (1.3) rather than only on some substates.

If we have $u^j \in X^j$, $j = 1, \ldots, N$, for the states of the subsystems, we can form the product $X = X^1 \times X^2 \times \cdots \times X^N$ and define on X the "block"-diagonal operator

$$B = \mathrm{diag}\left\{B^{(1)}, \ldots, B^{(N)}\right\}, \ G(u) = \mathrm{diag}\left\{G^{(1)}(u), \ldots, G^{(N)}(u)\right\}. \qquad (1.5)$$

Now (1.3) evolves according to

$$\frac{d}{dt}u(t) = Bu(t) + G(u(t))u(t), \ [u(t), f] = [u(0), f], \ t \geq 0, \qquad (1.6)$$

where $f = (f^1, \ldots, f^N) \in X$ and B is the infinitesimal generator of a semigroup of operators of class $\mathcal{C}_0$ [HP, p.321]. Note that structurally (1.6) is very similar to (1.1), (1.2). Only the infinitesimal generator $G = G(u)$ itself depends on the total state (1.3) so that our problem becomes nonlinear.

In this paper we assume the evolution problem in the form (1.6) under very general conditions and prove an existence theorem (concerning mild solutions which in our applications become classical solutions) for all times $t \geq 0$ and settle the question of its long run behaviour: In fact, any solution of (1.6) settles in the long run at a steady state. We give defining equations to determine this steady state. This theorem is important not only in its own right. It also provides the basis for singular perturbation techniques on such systems to obtain analytic expressions which characterize the speed of reaction systems and we refer to [BB], [TGPBB] for the pseudo-steady state process leading to the definition of the speed of the underlying chemical networks. Note that our theory does not need the diagonal form of (1.5). However, we rely heavily on (1.2). (1.6) is only an easy way to fit (1.3), (1.4) under the general pattern (1.7). The original form of the examples is (1.4).

Our motivating examples are particular cases of

$$\frac{d}{dt}u(t) = Bu(t) + G(u(t))u(t) = T(u(t))u(t), \ u(0) = u_0, \qquad (1.7)$$

where B is generally an unbounded linear densely defined operator and $G(u)$ for every $u \in \mathcal{E}$ is a bounded linear map on $\mathcal{E}$, where $\mathcal{E}$ denotes the underlying space to be specified in Section 2 The paper is organized as follows. We begin in Section 2 with some definitions and notation which are basic for the following theory. We continue in Section 3 with the examples just mentioned. Section 4 defines the problem, gathers all our hypotheses and proves the main result on the existence of a time-global solution. Finally, Section 5 settles the asymptotic behaviour.

2. Definitions and notation

As in our earlier paper [BM98] we need the theoretical basis for infinitesimal generators with monotonicity properties. The corresponding basic]definitions are summarized in this section.

Let $\mathcal{E}$ be a Banach space over the field of real numbers. Let $\mathcal{E}'$ denote the dual space of $\mathcal{E}$. Let $\mathcal{F}$, $\mathcal{F}'$ be the corresponding complex extensions of $\mathcal{E}$, $\mathcal{E}'$ respectively and let $\mathbf{B}(\mathcal{E})$ and $\mathbf{B}(\mathcal{F})$ be the spaces of all bounded linear operators mapping $\mathcal{E}$ into $\mathcal{E}$ and $\mathcal{F}$ into $\mathcal{F}$ respectively.

Let $\mathcal{K} \subset \mathcal{E}$ be a closed normal and generating cone, i.e. let

(i) $\mathcal{K} + \mathcal{K} \subset \mathcal{K}$,

(ii) $a\mathcal{K} \subset \mathcal{K}$ for $a \in \mathcal{R}_+$,

(iii) $\mathcal{K} \cap (-\mathcal{K}) = \{0\}$,

(iv) $\overline{\mathcal{K}} = \mathcal{K}$,

where $\overline{\mathcal{K}}$ denotes the norm-closure of $\mathcal{K}$,

(v) $\mathcal{E} = \mathcal{K} - \mathcal{K}$

and

(vi) there exists a $\delta > 0$ such that $||x + y|| \geq \delta ||x||$, whenever x, $y \in \mathcal{K}$. Property (vi) is called *normality of* $\mathcal{K}$.

We let
$$x \leq y \text{ or equivalently } y \geq x \Longleftrightarrow (y - x) \in \mathcal{K}$$
(vii) For every pair $x, y \in \mathcal{K}$ there exist $x \wedge y = \inf\{x, y\}$ and $x \vee y = \sup\{x, y\}$ as elements of $\mathcal{K}$.

A cone $\mathcal{K}$ satisfying condition (vii) is called a *lattice cone* and the partial order on $\mathcal{E}$ a *lattice order*. In the terminology of H.H. Schaefer [S1] $\mathcal{E}$ is called a *Banach lattice*. Our theory is free of hypothesis (vii).

Let
$$\mathcal{K}' = \{x' \in \mathcal{E}' : x'(x) \geq 0 \text{ for all } x \in \mathcal{K}\}$$
and
$$\mathcal{K}^d = \{x \in \mathcal{K} : x'(x) > 0 \text{ for all } 0 \neq x' \in \mathcal{K}'\}.$$

We call $\mathcal{K}'$ the *dual cone* of $\mathcal{K}$ and $\mathcal{K}^d$ the *dual interior* of $\mathcal{K}$, respectively. In the following analysis we assume that the dual interior $\mathcal{K}^d$ is nonemty.

A linear form $\hat{x}' \in \mathcal{K}'$ is called *strictly positive*, if $\hat{x}'(x) > 0$ for all $x \in \mathcal{K}$, $x \neq 0$.

We write $[x, x']$ in place of $x'(x)$, where $x \in \mathcal{E}$ and $x' \in \mathcal{E}'$ respectively. If $\mathcal{E}$ happens to be a Hilbert space then $[x, x']$ denotes the appropriate inner product.

If $T \in \mathbf{B}(\mathcal{E})$ then T' denotes its dual and hence, $T' \in \mathbf{B}(\mathcal{E}')$.

Let $T \in \mathbf{B}(\mathcal{F})$ and let $\sigma(T)$ denote its spectrum. Further, let $T \in \mathbf{B}(\mathcal{E})$. We introduce the operator $\tilde{T}$ by setting $\tilde{T}z = Tx + iTy$, where $z = x + iy$, $x, y \in \mathcal{E}$ and call it *complex extension* of T. By definition, we let $\sigma(T) := \sigma(\tilde{T})$. Similarly, we let $r(T) := r(\tilde{T})$, where $r(\tilde{T}) = \max\{|\mu| : \mu \in \sigma(\tilde{T})\}$ denotes the *spectral radius* of $\tilde{T}$.

In order to simplify notation we will identify T and its complex extension and will thus omit the tilde sign denoting the complex extension.

The set

$$\sigma_\pi(T) = \{\mu \in \sigma(T) : |\mu| = r(T)\}$$

is called *peripheral spectrum of* T. Note that $\sigma_\pi(T)$ is never empty.

If μ is an isolated singularity of $R(\lambda, T) = (\lambda I - T)^{-1}$ we have the following Laurent expansion of $R(\lambda, T)$ around μ [TL], [MZ]

$$R(\lambda, T) = \sum_{k=0}^{\infty} A_k(\mu)(\lambda - \mu)^k + \sum_{k=1}^{\infty} B_k(\mu)(\lambda - \mu)^{-k}, \qquad (2.1)$$

where A_{k-1} and B_k, $k = 1, 2, \ldots$, belong to $\mathbf{B}(\mathcal{F})$. Moreover, it holds [TL]

$$B_1(\mu) = \frac{1}{2\pi i} \int_{\mathcal{C}_0} (\lambda I - T)^{-1} d\lambda, \qquad (2.2)$$

where

$$\mathcal{C}_0 = \{\lambda : |\lambda - \mu| = \rho_0\}$$

and ρ_0 is such that $\{\lambda : |\lambda - \mu| \le \rho_0\} \cap \sigma(T) = \{\mu\}$.

Furthermore,

$$B_{k+1}(\mu) = (T - \mu I)B_k(\mu), \ \ k = 1, 2, \ldots \qquad (2.3)$$

If there is a positive integer $q = q(\mu)$ such that

$$B_q \ne 0, \text{ and } B_k = 0, \text{ for } k > q,$$

then μ is called *a pole of the resolvent operator* and q its *multiplicity*.

We define symbol

$$\text{ind}\,(\mu I - T) = q(\mu)$$

and call it the *index of* T *at* μ. In particular, we call ind(T) the *index of* T instead of index of T at 0.

Let $T \in \mathbf{B}(\mathcal{F})$ and let $\sigma_\pi(T)$ consist of a finite set of poles of the resolvent operator $(\lambda I - T)^{-1}$, then we define

$$S_N(\mu) = \frac{1}{N} \sum_{k=0}^{N} \left(\frac{1}{\mu}T\right)^k k^{-q(\mu)+1},$$

where $\mu \in \sigma_\pi(T)$ and $q(\mu)$ is the multiplicity of μ.

Proposition 2.1. [MZ, Theorem 7.11, p.168] *Let* $T \in \mathbf{B}(\mathcal{E})$ *be such that its peripheral spectrum consists of a set of isolated poles of the resolvent operator* $(\lambda I - T)^{-1}$ *and let* $\mu \in \sigma_\pi(T)$ *have its multiplicity* $q(\mu)$ *maximal among the points in* $\sigma_\pi(T)$. *Then*

$$\lim_{N \to \infty} \|S_N(\mu) - B_q(\mu)\|_{\mathcal{F}} = 0.$$

Corollary 2.2. *If* $T\mathcal{K} \subset \mathcal{K}$ *and* $\mu = r(T)$, *then*

$$B_q(r(T))\,\mathcal{K} \subset \mathcal{K}.$$

Let $R \in \mathbf{B}(\mathcal{E})$ be a $\mathcal{K}$-nonnnegative operator. Every operator T of the form $T = \rho I - R$, with some $\rho \in \mathcal{R}^1$ and $R \in \mathbf{B}(\mathcal{E})$, $R\mathcal{K} \subset \mathcal{K}$, is called an $L(\mathcal{K})$-operator. If $\rho \geq r(R)$, then T is called an $M(\mathcal{K})$-operator.

An operator $T \in \mathbf{B}(\mathcal{E})$ is called cross-$\mathcal{K}$-operator, if relation $[Tx, x'] \geq 0$ holds for every pair $x \in \mathcal{K}$, $x' \in \mathcal{K}'$ such that $[x, x'] = 0$.

3. Motivating examples: Development of a form of enzyme actions

Example 3.1. We begin with the simple network

$$X + E_0 \underset{k_{-0}}{\overset{k_0}{\rightleftharpoons}} E_1 \overset{\kappa}{\rightarrow} E_0 + P \tag{3.1}$$

the classical *Michaelis-Menten Kinetics* which describes the action of an enzyme E_0 in its duty to transform a substrate X into a product P where an intermediate version E_1 of the enzyme is formed. Let $x(t)$, $e_0(t)$, $e_1(t)$, $p(t)$ denote the concentration of the chemical X, E_0, E_1, P at time $t \geq 0$, respectively and let us form the state vectors

$$u^1 := (e_0(t), e_1(t))^T, \quad u^2 := (x(t), e_1(t), p(t))^T. \tag{3.2}$$

Then the dynamics of the network (3.1) is given by

$$\begin{aligned}
\dot{u}^1(t) &= T^{(1)}(u_1^2(t))u^1(t) = B^{(1)}u^1(t) + G^{(1)}(u_1^2(t))u^1(t), \\
\dot{u}^2(t) &= T^{(2)}(u_1^1(t))u^2(t) = B^{(2)}u^2(t) + G^{(2)}(u_1^1(t))u^2(t).
\end{aligned} \tag{3.3}$$

Note the components

$$u_1^1(t) = e_0(t), \quad u_1^2(t) = x(t)$$

which follows from (3.2). All matrices $T^{(j)}$ appearing in (3.3) are given by

$$T^{(1)}(x) = \begin{pmatrix} -k_0 x & k_{-0} + \kappa \\ k_0 x & -k_{-0} - \kappa \end{pmatrix} = \begin{pmatrix} 0 & k_0 + \kappa \\ 0 & -k_{-0} - \kappa \end{pmatrix} + \begin{pmatrix} -k_0 x & 0 \\ k_0 x & 0 \end{pmatrix}$$

$$= B^{(1)} + G^{(1)}(u_1^2)$$

$$T^{(1)}(u_1^2) = B^{(1)} + G^{(1)}(u_1^2)$$

$$T^{(2)}(u_1^1) = \begin{pmatrix} -k_0 e_0 & k_{-0} & 0 \\ k_0 e_0 & -k_{-0} - \kappa & 0 \\ 0 & \kappa & 0 \end{pmatrix} \tag{3.4}$$

$$= \begin{pmatrix} 0 & k_{-0} & 0 \\ 0 & -k_{-0} - \kappa & 0 \\ 0 & \kappa & 0 \end{pmatrix} + \begin{pmatrix} -k_0 e_0 & 0 & 0 \\ k_0 e_0 & 0 & 0 \\ 0 & 0 & 0 \end{pmatrix}$$

$$= B^{(2)} + G^{(2)}(u_1^1)$$

Here we use the reaction constants given in (3.1).

Obviously, the "diagonal-construction" in the Introduction yields the canonical form

$$\dot{u}(t) = T(u(t))u(t) \qquad (3.5)$$

as in (1.7).

Note that in this representation of the dynamical equations the chemical E_1 shows up twice. See the vectors in (3.2). However, the advantage of our represention is that all the matrices in (3.4) satisfy the following basic properties

(i) diagonal elements are ≤ 0

(ii) offdiagonal elements are ≥ 0

(iii) all column sums vanish.

These properties are very important in the generalization and in the proof of our basic results in the last section.

Example 3.2. Our next example is much more complicated and concerns the network (see [TGPBB])

$$X \underset{k}{\overset{k}{\rightleftarrows}} X_p$$

$$X_p + Z \underset{k_{-1}}{\overset{k_1}{\rightleftarrows}} E_1 \qquad (3.6)$$

$$E_1 + R \underset{k_{-2}}{\overset{k_2}{\rightleftarrows}} E_2 \overset{k_4}{\rightarrow} R + Z + X_i, \ R + Z \underset{k_{-3}}{\overset{k_3}{\rightleftarrows}} E_3$$

This network describes the uptake of substrate X through the outer membrane of an E. coli—cell into the periplasm (first line) where the substrate in the periplasm is labeled X_p. Then this chemical is being caught by the binding protein Z which is transformed into the loaded state E_1 (second line). This product finally undergoes the transport step in the last line of (3.6) where with the help of membrane components R the substrat is transported into the interior of the cell and becomes there X_i. To understand the dynamics of this network we form the three state vectors

$$\begin{aligned} u^1(t) &= (z(t), e_1(t), e_2(t), e_3(t)), \\ u^2(t) &= (r(t), e_2(t), e_3(t)) \\ u^3(t) &= (x(t), x_p(t), e_1(t), e_2(t), x_i(t)) \end{aligned} \qquad (3.7)$$

for which the dynamical equations

$$\begin{aligned} \dot{u}^1(t) &= T^{(1)}(r(t), x_p(t))u^1(t) = T^{(1)}(u_1^2(t), u_2^3(t))u^1(t), \\ \dot{u}^2(t) &= T^{(2)}(e_1(t), z(t))u^2(t) = T^{(2)}(u_2^1(t), u_1^1(t))u^2(t) \\ \dot{u}^3(t) &= T^{(3)}(z(t), r(t))u^3(t) = T^{(3)}(u_1^1(t), u_1^2(t))u^3(t) \end{aligned} \qquad (3.8)$$

hold where the matrices $T^{(j)}$ again satisfy the basic conditions (i)–(iii). In the same fashion as in 3.1 we construct the state vector

$$u(t) = (u^1(t), u^2(t), u^3(t)) \in \mathcal{R}^{12}$$

and the appropriate matrices $T^{(j)}$ by setting

$$T^{(1)}(r, x_p) = \begin{pmatrix} -k_1 x_p - k_3 r & k_{-1} & k_4 & k_{-3} \\ k_1 x_p & -k_{-1} - k_2 r & k_{-2} & 0 \\ 0 & k_2 r & -k_4 - k_{-2} & 0 \\ k_3 r & 0 & 0 & -k_{-3} \end{pmatrix} =$$

$$= B^{(1)} + G^{(1)}(r, x_p),$$

where

$$B^{(1)} = \begin{pmatrix} 0 & k_{-1} & k_4 & k_{-3} \\ 0 & -k_{-1} & k_{-2} & 0 \\ 0 & 0 & -k_4 - k_{-2} & 0 \\ 0 & 0 & 0 & -k_{-3} \end{pmatrix}$$

and

$$G^{(1)}(r, x_p) = \begin{pmatrix} -k_1 x_p - k_3 r & 0 & 0 & 0 \\ k_1 x_p & -k_2 r & 0 & 0 \\ 0 & k_2 r & 0 & 0 \\ k_3 r & 0 & 0 & 0 \end{pmatrix}$$

and where

$$r(0) > 0, \quad x_p(0) = 0, \tag{3.9}$$

$$T^{(2)}(e_1, z) = \begin{pmatrix} -k_2 e_1 - k_3 z & k_{-2} + k_4 & k_{-3} \\ k_2 e_1 & -k_{-2} - k_4 & 0 \\ k_3 z & 0 & -k_3 \end{pmatrix} = B^{(2)} + G^{(2)}(e_1, z), \tag{3.10}$$

where

$$B^{(2)} = \begin{pmatrix} 0 & k_{-2} + k_4 & k_{-3} \\ 0 & -k_{-2} - k_4 & 0 \\ 0 & 0 & -k_3 \end{pmatrix}$$

and

$$G^{(2)}(z, e_1) = \begin{pmatrix} -k_2 e_1 - k_3 z & 0 & 0 \\ k_2 e_1 & 0 & 0 \\ k_3 z & 0 & 0 \end{pmatrix}$$

and where

$$z(0) > 0, \quad e_1(0) = 0, \tag{3.11}$$

$$T^{(3)}(z, r) = \begin{pmatrix} -k & k & 0 & 0 & 0 \\ k & -k - k_1 z & k_{-1} & 0 & 0 \\ 0 & k_1 z & -k_2 r - k_{-1} & k_{-2} & 0 \\ 0 & 0 & k_2 r & -k_{-2} - k_4 & 0 \\ 0 & 0 & 0 & k_4 & 0 \end{pmatrix}, \tag{3.12}$$

where

$$
B^{(3)} = \begin{pmatrix}
-k & k & 0 & 0 & 0 \\
k & -k & k_{-1} & 0 & 0 \\
0 & 0 & -k_{-1} & k_{-2} & 0 \\
0 & 0 & 0 & -k_{-2} - k_4 & 0 \\
0 & 0 & 0 & k_4 & 0
\end{pmatrix}
$$

and

$$
G^{(3)}(z,r) = \begin{pmatrix}
0 & 0 & 0 & 0 & 0 \\
0 & -k_1 z & 0 & 0 & 0 \\
0 & k_1 z & -k_2 r & 0 & 0 \\
0 & 0 & k_2 r & 0 & 0 \\
0 & 0 & 0 & 0 & 0
\end{pmatrix}
$$

Again, the "diagonal-construction" in the Introduction yields the canonical form where the negative of the matrices (3.10), (3.12) share the basic properties (i)–(iii).

Remark 3.3. *Note that the matrix-functions $T^{(j)}$ are such that due to (3.9) and (3.11) their exponentials $\exp\{T^{(j)}t\}$, $j = 1, 2, 3$, have the property that $\exp\{T^{(j)}t\}(\mathcal{K}\backslash\{0\}) \subset \mathcal{K}^d$ for $t > 0$, $\mathcal{K}$ being the orthant $\mathcal{R}_+^N = \{x^T = (x_1, ..., x_N) \in \mathcal{R}^N : x_j \geq 0, j = 1, ..., N\}$ and $\mathcal{K}^d$ the interior of $\mathcal{R}_+^N$.*

4. A special evolution problem

In the following text we are going to assume that B is a densely defined linear operator with $\mathcal{D}(B) \subset \mathcal{E}$ as its domain of definition. Assume that B is the generator of a semigroup of operators of class $\mathcal{C}_0$ [HP, p.321] such that each of the operators of the semigroup $T(t; B), t \geq 0$, satisfies

$$
T(t; B)\mathcal{K} \subset \mathcal{K}, \ t \geq 0,
$$

and

$$
\sigma(T(t; B)) \cap \{\lambda : |\lambda| = r(T(t; B))\} = \{1\}.
$$

Definition 4.1. *We say that the semigroup $T(t; B)$ possesses the total concentration preservation property, if there is a strictly positive linear form $\hat{x}' \in \mathcal{K}'$ such that*

$$
B'\hat{x}' = 0, \tag{4.1}
$$

where B' denotes the dual of B, i.e. $B'v' = u'$ which means that $u'(x) = v'(Bx)$ and where $v' \in \mathcal{D}(B')$ means that there is a $u' \in \mathcal{E}'$ such that $u'(x) = v'(Bx)$ for $\forall x \in \mathcal{D}(B)$.

Next, we list the hypotheses under which the problem can be formulated and our theory be developed.

Hypotheses. Besides the hypotheses mentnioned at the beginning of Section 4 let the following hypotheses $B1^0$–$B3^0$ and $G1^0$–$G6^0$ hold.

$B1^0$ Let B be the infinitesimal generator of a semigroup of operators of class $\mathcal{C}_0$ and let the semigroup possess the total concentration property. Let 0 be an isolated pole of the resolvent operator $(\lambda I - B)^{-1}$ and P be the spectral projection onto the null-space with $\dim range(P) < +\infty$. More generally, let P be compact as a map of $\mathcal{E}$ into $\mathcal{E}$. A sufficient condition to tha latter property is compactness of the resolvent operator for some complex λ

$B2^0$ Each operator of the semigroup of operators $T(t; B)$ is such that $T(t; B)u_0 \in \mathcal{K}$ whenever $u_0 \in \mathcal{K}$.

$B3^0$ There are positive real numbers α, γ and τ independent of $t \in [0, +\infty)$, such that

$$\|(I - P)T(t; B)\|_{\mathbf{B}(\mathcal{E})} \le \gamma e^{-\alpha t}, \ t > \tau. \tag{4.2}$$

Let us assume that $G = G(u) \in \mathbf{B}(\mathcal{E})$ for every fixed $u \in \mathcal{E}$ possess the following properties:

$G1^0$ For each $u \in \mathcal{K}$ the operator $-G(u)$ is a singular $M(\mathcal{K})$-operator and therefore, there is an operator $\tilde{G}(u)$ preserving $\mathcal{K}$ such that

$$-G(u) = r(\tilde{G}(u))I - \tilde{G}(u), \ \tilde{G}(u)\mathcal{K} \subset \mathcal{K}.$$

$G2^0$

$$G(u)'\hat{x}' = 0, \ u \in \mathcal{K},$$

where the functional $\hat{x}'$ is identical with that in (4.1).

$G3^0$ The map $G : \mathcal{E} \to \mathbf{B}(\mathcal{E})$ is such that there exist positive reals p and c such that for every $u \in \mathcal{E}$,

$$\|G(u)\|_{\mathbf{B}(\mathcal{E})} \le c\|u\|_{\mathcal{E}}^p.$$

$G4^0$ Let $F = F(t)$ be be a continuous map of $t \in [0, T]$ into $\mathbf{B}(\mathcal{E})$, i.e. $F \in \mathcal{C}([0, T], \mathbf{B}(\mathcal{E}))$ equipped with the norm

$$\|F\|_{\mathcal{C}([0,T], \ \mathbf{B}(\mathcal{E}))} = \max \left\{ \|F(t)\|_{\mathbf{B}(\mathcal{E})} : t \in [0, T] \right\}$$

and similarly,

$$\|F\|_{\mathcal{C}([0,+\infty), \mathbf{B}(\mathcal{E}))} = \sup \left\{ \|F(t)\|_{\mathbf{B}(\mathcal{E})} : t \in [0, +\infty) \right\}.$$

The operator-valued function $G = G(u(.))$ maps $[0, T]$ into $\mathbf{B}(\mathcal{E})$, i.e. $G(u(.)) \in \mathcal{C}([0, T], \mathbf{B}(\mathcal{E}))$ and similarly, $G(u) \in \mathcal{C}([0, +\infty), \mathbf{B}(\mathcal{E}))$ respectively.

$G5^0$ For any positive real $a > 0$ there exists a positive constant L_a such that

$$\|G(u) - G(v)\|_{\mathbf{B}(\mathcal{E})} \le L_a\|u - v\|_{\mathcal{E}}$$

holds for any $u, v \in \mathcal{K}$ and such that $\|u\| \le a$, $\|v\| \le a$.

$G6^0$ For every $u \in \mathcal{K}$ satisfying

$$u(t) = T(t; B)u_{t_0} + \int_{t_0}^t T(t - s; BG(u(s))u(s)ds, \ t \ge t_0,$$

for every element $x' \in \mathcal{K}'$

$$\lim_{T \to \infty} \int_{t_0}^{T} [PG(u(s)), x']ds < +\infty. \tag{4.3}$$

Note that according to hypothesis $B1^0$ the convergence required in (4.3) implies the norm convergence of the integrals $\int_0^T PG(u(s))ds, T > 0$. This is because the weak convergence on the $range(P)$ implies the strong convergence.

Remark 4.2. *In formula (4.3) and similarly in the following text the integrals concern only continuous integrands with respect to the integration variable and are thus understood in the standard sense of Riemann.*

Remark 4.3. *One checks easily that (4.3) is fulfilled in our examples 3.1 and 3.2. Moreover in Example 3.2 the integral (4.3) vanishes for every functional $x' \in \mathcal{R}_+^n$. This is because in this example relation (4.4) holds.*

Remark 4.4. *Let G be such that*

$$[G(u)]' P' = 0, \tag{4.4}$$

where the primes denote the duals of the corresponding maps. Then automatically, $PG(u)$ is the zero-map.

Remark 4.5. *Requirement $G1^0$ expresses in an abstract way the sign conditions (i), (ii) for the examples in Section 3 Furthermore, $G3^0$ generalizes the fact the all matrices in our examples are polynomial in the independent variables (see 3.4). The key condition $G2^0$ is a generalization of (iii) of Section 3 and opens the way for utilizing the Frobenius theory.*

Problem P 4.6 *Under the hypotheses $B1^0 - B3^0$ and $G1^0 - G6^0$ to determine all mild solutions to the following Cauchy problem*

$$\frac{du}{dt}(t) = Bu(t) + G(u(t))u(t), \ u(t_0) = u_{t_0} \in \mathcal{K} \cap \mathcal{D}(B),$$

$$t_0 \in (-\infty, +\infty), \ t > t_0. \tag{4.5}$$

This means to determine all solutions to the problem

$$u(t) = \mathcal{T}(t; B)u_{t_0} + \int_{t_0}^{t} \mathcal{T}(t - s; B)G(u(s))u(s)ds, \ t > t_0, \ u_{t_0} \in \mathcal{K}, \tag{4.6}$$

which we call mild solutions to (4.5).

Remark 4.7. *In the following we describe the situation typical for the examples of Section 3:*

Let the generator B be an $M(\mathcal{K})-$operator and let P be a spectral projection of $\mathcal{E}$ onto the null-space of B such that the subspace $(I - P)\mathcal{E}$ is closed. Let the inverse of the reduction

$$(I - P)B|_{(I-P)\mathcal{E}}$$

be bounded.

Then

$$T(t; B) = \exp\{tB\}.$$

and the situation described by our theory becomes transparent as it is in our finite dimensional examples we started with to motivate our work.

Remark 4.8. *The properties of the dynamical systems investigated in this contribution are natural nonlinear generalizations of those systems studied in* [BM95], [BM94] *and* [BM98]. *Because of very tight relations and analogies between these systems we call the present ones* nonlinear cooperative systems.

Definition 4.9. *Let* $\mathcal{C}([0, \sigma], \mathcal{E})$ *be the space of functions mapping the interval* $[0, \sigma]$ *into* $\mathcal{E}$ *continuous with respect to* $t \in [0, \sigma]$ *with the values in* $\mathcal{E}$. *An element* $u^* = u^*(t) \in \mathcal{C}([0, \sigma], \mathcal{E})$ *will be understood as a solution to (4.5) if (4.5) holds true for all* $t \in [0, \sigma]$, $\sigma \leq +\infty$.

Proposition 4.10. *Assume that* $U = U(t)$ *is a mild solution to (4.5).*
   Then

$$[U(t), \hat{x}'] = [U(0), \hat{x}'], \ U(0) = u_0,$$

holds for all t *in the domain of existence of* $U(t)$.

Proof. We see that

$$[U(t), \hat{x}'] = [T(t; B)U_0, \hat{x}'] + \int_0^t T(t - s; B)G(U(s))U(s), \hat{x}']ds$$

$$= [u_0, \hat{x}'] + \int_0^t [G(U(s))U(s), [T(t - s; B)]'\hat{x}'] \, ds$$

$$= [u_0, \hat{x}'] + \int_0^t [U(s), [G(U(s))]'\hat{x}'] \, ds$$

$$= [u_0, \hat{x}']. \qquad \qquad \square$$

Our theory is developed under the following additional hypothesis.

Hypothesis 4.11 (H(TCP)). *Let* $U \in \mathcal{E}$.
   There exists a positive real κ *independent of* U *such that*

$$\||U\|| \leq \kappa \, [|U|, \hat{x}'] \qquad (4.7)$$

where $\hat{x}'$ *comes from the total concentration preservation property (4.1) and* $|U(t)|$ *is defined by formula*

$$|U| = U_+ + U_- , \qquad (4.8)$$

U_+, U_- *coming from property (v) in the definition of the cone, where*

$$U = U_+ - U_-, \ U_\pm \in \mathcal{K}.$$

Agreement: Hypothesis HTCP is assumed to hold throughout the remainder of the paper.

Remark 4.12. *Formula (4.8) offers us an imitation of the absolute value of an element in $\mathcal{E}$. Such an absolute value is available if $\mathcal{E}$ is e. g. an order complete Banach lattice [S1, p.234].*

Example 4.13. Let $\mathcal{E} = \mathcal{R}^N$, $\mathcal{K} = \mathcal{R}^N_+$, $\|.\|_{\mathcal{R}^N}$ denote any norm on $\mathcal{R}^N$ and $\hat{x}' = (1, \ldots, 1)$. We see that for any element $U \in \mathcal{R}^N$

$$\|U\|_{\mathcal{R}^N} \leq \kappa \sum_{j=0}^{N} |U_j| = \kappa[|U|, \hat{x}']$$

and thus, the hypothesis **H(TCP)** holds in this case.

Proposition 4.14. *There exists a constant c independent of σ, such that the relation*

$$\|U(t)\| \leq c,$$

holds for all $t \in [0, \sigma]$, $0 < \sigma < +\infty$, where $U = U(t) \in \mathcal{K}$ is any mild solution to (4.5) $U = U(t) \in \mathcal{K}$.

Proof. It is easy to see that

$$[|U(t)|, \hat{x}'] = [U(t), \hat{x}'] = [u_0, \hat{x}']. \tag{4.9}$$

To finish the proof we apply hypothesis **H(TPC)** and (4.9) and derive

$$\|U(t)\| \leq \kappa[|U(t)|, \hat{x}'] \leq \kappa[u_0, \hat{x}'] = c.$$

The proof is complete. □

Remark 4.15. *We are going to show that every $\mathcal{K}$-nonnegative solution to problem **P** not only remains bounded in $[0, +\infty)$ but even more: To each such a solution u the limit $u_\infty = \lim_{t \to +\infty} u(t)$ exists (see Section 5).*

Theorem 4.16. *There exists at least one mild solution to (4.5) belonging to the cone $\mathcal{K}$ and every such solution can be continued for all $t \in [0, +\infty)$.*

Proof. The existence of a local solution is guaranteed by the moderate magnitude of the nonlinear term in (4.5) for small t's. Since every $\mathcal{K}$-nonnegative mild solution to (4.5) is uniformly bounded in any time compact $[0, T]$ (Proposition 4.14), it can be continued for all $t > 0$. This completes the proof. □

Lemma 4.17. *Let T be a positive real. Let*

$$(Sw)(t) = \mathcal{T}(t; B)u_0 + \int_0^t \mathcal{T}(t - s; B)G(z(s))w(s)ds, \; z \in \mathcal{C}([0, T], \mathcal{E}).$$

Define S by setting

$$S = \{w \in \mathcal{C}([0, T], \mathcal{E}) : w(t) = (Sz)(t) \text{ for some } z \in \mathcal{C}([0, T], \mathcal{E})\}$$

and

$$S_\nu = \{w \in S : \|w\|_{\mathcal{C}([0,T], \mathcal{E})} \leq \nu\}$$

Let $S^k z_1, S^k z_2$ for some z_1, z_2 both in $\mathcal{U} = \{u \in \mathcal{E} : \|u_0 - u\|_{\mathcal{E}} \leq a\}$ for some $a > 0$, be in S_ν for all $k = 1, 2, \ldots$.

Let
$$M = c\nu^p + L_\nu\nu,$$
where (see G4⁰ and G5⁰)
$$\nu = \|u_0\|_{\mathcal{E}} + a \quad and \quad \omega = \mathrm{Max}\{\gamma, \|P\|_{\mathbf{B}(\mathcal{E})}\}$$
Then
$$\left\|(S^k z_1)(t) - (S^k z_2)(t)\right\|_{\mathcal{E}} \le \frac{(M\omega t)^k}{k!}\|z_1 - z_2\|_{C([0,T],\mathcal{E})}, \quad t \in [0, T].$$

Proof of Lemma 4.17. The proof proceeds by induction on k. Since the induction step is verbally the same as the step $k = 1$, we show just the latter mentioned one.

Let $k = 1$. We see that
$$(S z_1)(t) - (S z_2)(t) = \int_0^t T(t - s; B)\left[G(z_1(s))z_1(s) - G(z_2(s))z_2(s)\right] ds.$$
and hence,
$$\begin{aligned}\|(Sz_1)(t) - (Sz_2)(t)\|_{\mathcal{E}} &= \left\|\int_0^t T(t - s; B)\left[G(z_1(s))z_1(s) - G(z_2(s))z_2(s)\right] ds\right\|_{\mathcal{E}} \\ &\le \int_0^t \omega\left\{c\|z_1(s)\|_{\mathcal{E}}^p + L_\nu\|z_2(s)\|_{\mathcal{E}}\right\}\|z_1(s) - z_2(s)\|_{\mathcal{E}} ds \\ &\le \omega(c\nu^p + L_\nu\nu)\int_0^t \|z_1(s) - z_2(s)\|_{\mathcal{E}} ds \\ &\le \frac{\omega M t}{1!}\|z_1 - z_2\|_{C([0,T],\mathcal{E})}.\end{aligned}$$
The proof is complete. $\qquad\qquad\square$

As a consequence, we derived at

Theorem 4.18. *There exists exactly one solution to (4.6) i.e. exactly one mild solution to Problem (P) for $t \in [0, T], T > 0$.*

Proof. According to Proposition 4.16 it is enough to show uniqueness. Let u_1 and u_2 be two solutions to Problem (4.6). According to the Proposition 4.14 we know that
$$\|u_j\|_{C([0,T],\mathcal{E})} \le \kappa[u_0, \hat{x}'] = \nu$$
and hence, since
$$(Su_j)(t) = u_j(t), \ t > 0, \ j = 1, 2,$$
Lemma 4.17 with suitable $a > 0$ applies for any fixed t in the interval $[0, +\infty)$ of existence for both solution $u_1(t)$ and $u_2(t)$. It follows that
$$u_1(t) - u_2(t) = S^k u_1(t) - S^k u_2(t), \ 0 \le t < +\infty.$$
Moreover, using Lemma 4.17, the following relation holds
$$\begin{aligned}\|u_1(t) - u_2(t)\|_{\mathcal{E}} &= \|S^k u_1(t) - S^k u_2(t)\|_{\mathcal{E}} \\ &\le \frac{(M\omega t)^k}{k!}\|u_1 - u_2\|_{C([0,T],\mathcal{E})}, \ 0 \le t \le T < +\infty,\end{aligned}$$
for any fixed $T > 0$ and $k = 1, 2, \ldots$. Let $t \in [0, T]$ be fixed. For sufficiently large k the right hand side can be made arbitrarily small. Thus, the solutions $u_1(t)$ and $u_2(t)$ must coincide for any $t \in [0, T]$. Since $T \in [0, +\infty)$ is arbitrary our proof is complete. $\qquad\qquad\square$

Theorem 4.19. *Let $u_0 \in \mathcal{K}$.*

Then the unique mild solution u to (4.5) remains in $\mathcal{K}$ for all $t \in [0, T]$.

Proof. To be able to exploit the $\mathcal{K}$-nonnegativity of $\tilde{G}(u(t))$ for $u(t) \in \mathcal{K}$ let us rewrite the original Cauchy problem in the form

$$\dot{u}(t) = Bu(t) - r(\tilde{G}(u(t))u(t)) + \tilde{G}(u(t))u(t), \ u(0) = u_0,$$

and introduce $w(t)$ by setting

$$u(t) = \exp\{-\rho t\}w(t), \tag{4.10}$$

where $0 < \rho \in \mathcal{R}^1$ is chosen sufficiently large. Since we know already that any $\mathcal{K}$−nonnegative mild solution to (4.5) is bounded, we can set

$$\rho = \rho(u_0) = \sup \left\{ r(\tilde{G}(u(t))) : t \in [0, \sigma] \right\}.$$

After making substitution (4.10) we see that the original Cauchy problem takes the form

$$\dot{w}(t) = Bw(t) + \rho w(t) - r(\tilde{G}(\exp\{-\rho t\}w(t)))w(t) + \tilde{G}(exp\{-\rho t\}w(t)))w(t),$$

$$w(0) = u_0.$$

Applying the Piccard-Lindelöf successive approximations with the starting vector $w_0 = u_0$ we let

$$\begin{aligned} w_{k+1}(t) &= \mathcal{T}(t; B)u_0 \\ &+ \int_0^t \mathcal{T}(t - s; B) \left\{ \rho - r(\tilde{G}(\exp\{-\rho s\}w_k(s))) \right\} w_k(s)ds \\ &+ \int_0^t \mathcal{T}(t - s; B)\tilde{G}(exp\{-\rho s\}w_k(s))w_k(s)ds \end{aligned}$$

and conclude that since all the elements of the iteration sequence $\{w_k(t)\}$ are in $\mathcal{K}$, also its limit $w(t)$ belongs to $\mathcal{K}$ for all t's for which the iteration process shown is convergent. This happens to be the case for $0 < t < \delta$, $\delta > 0$. It follows that $w(t)$ being in $\mathcal{K}$ implies that $u(t) \in \mathcal{K}$.

This procedure can be applied starting at any $t_0 > 0$ in place of 0 assuming that $u(t) \in \mathcal{K}$ for $0 < t \leq t_0$. This completes the proof. $\qquad\square$

5. Asymptotic behaviour

Proposition 5.1. *Let P be the Perron eigenprojection of B onto the null-space $\mathcal{N}(B)$. Let $u = u(t)$ be a solution to Problem **P** and let t_0 be a positive real number. Then the system*

$$\{(I - P)u(t)\}_{t \geq t_0}$$

possesses the Cauchy property, i.e. for every $\varepsilon > 0$ there is a real number η such that

$$\|(I - P)[u(t + \Delta) - u(t)]\| < \varepsilon \tag{5.1}$$

holds for $t > \eta$ and $\Delta > 0$.

Proof. Let $0 < s < +\infty$. From

$$
\begin{aligned}
u(t) &= T(t - s; B)u(s) + \int_s^t T(t - \nu; B)G(u(\nu))u(\nu)d\nu \\
&= T(t - s; B)u(s) + \int_s^t T(\nu; B)G(u(t - \nu))u(t - \nu)d\nu
\end{aligned}
$$

we deduce that

$$
\begin{aligned}
u(t + \Delta) - u(t) &= [T(t + \Delta - s; B) - T(t - s; B)]\, u(s) \\
&\quad + \int_s^{t+\Delta} T(\nu; B)G(u(t + \Delta - \nu))u(t + \Delta - \nu)d\nu \\
&\quad - \int_s^t T(\nu; B)G(u(t - \nu))u(t - \nu)d\nu.
\end{aligned}
$$

Let $s = t/2$, where $t/2 > \rho$. Since $\|u(t)\|$ and $\|G(u(t))\|$ are uniformly bounded quantities, we derive that

$$
\begin{aligned}
u(t + \Delta) - u(t) &= [T(t/2 + \Delta; B) - T(t/2; B)]\, u(t/2) \\
&\quad + \int_{t/2}^{t+\Delta} T(\nu; B)G(u(t + \Delta - \nu))u(t + \Delta - \nu)d\nu \\
&\quad - \int_{t/2}^t T(\nu; B)G(u(t - \nu))u(t - \nu)d\nu
\end{aligned}
$$

and thus, with an arbitrary fixed $\delta > 0$,

$$
\|(I - P)\,[u(t + \Delta) - u(t)]\,\| \le c_1 \mathrm{e}^{-\alpha t/2}
$$

with some constant c_1 independent of t. The Cauchy property of the system $\{(I - P)u(t)\}$ then easily follows. $\qquad\square$

As a consequence of Hypothesis $B1^0$ (dim[range(P)] $< +\infty$) and a known formula [BM95]

$$
\lim_{t \to \infty} T(t; B)u_0 = Pu_0,
$$

we obtain

Proposition 5.2. *There exists the limit*

$$
\begin{aligned}
u(\infty) &= Pu_0 \\
&\quad + \lim_{t \to \infty} \int_0^t PG(u(s))u(s)ds \\
&\quad + \lim_{t \to \infty} \int_0^t (I - P)T(t - s; B)G(u(s))u(s)ds \\
&= Pu_0 + u_P + u_{I-P}(\infty).
\end{aligned}
\tag{5.2}
$$

Proof. The limit

$$
u(\infty) = \lim_{t \to +\infty} u(t)
\tag{5.3}
$$

exists according to Proposition 5.1. We already know that

$$
\lim_{t \to +\infty} \int_0^t PG(u(s))u(s)ds
$$

exists as a consequence of Hypothesis $G6^0$. The existence of the limit $u_{I-P}(\infty)$ follows from the fact that

$$
u_{I-P}(\infty) = u(\infty) - u_P(\infty).
$$
$\qquad\square$

6. Concluding remarks

Let us return to our examples in Section 3 In particular, we take the network (3.6) with the dynamical equations (3.8). Our result states that the network stabilizes in the long run ($t \to \infty$) which in natural systems is obtained after a very short period of time. More precisely, the subsystems for $u^1(t)$ and $u^2(t)$ are also stabilizing according to our result and this in reality is observed so that a transient phase cannot be measured. Hence, (3.8) is not the correct description of the evolution. We rather should put

$$
\begin{cases}
\dfrac{d}{ds}u^1(s) = T^{(1)}(u_1^3(s), u_1^1(s))u^1(s) \\[2ex]
\dfrac{d}{ds}u^2(s) = T^{(2)}(u_2^1(s), u_1^1(s))u^2(s),
\end{cases}
\tag{6.1}
$$

where the time scale $s = \mu t$ is much faster than the time scale t i.e. $\mu \gg 1$. So the real fast evolutions happen in the components $u^1 = (z, e_1, e_2, e_3)$, and $u^2 = (r, e_2, e_3)$.

Also $x_p = x_p(s)$ operates on the fast time scale and the remaining components $x(t), x_i(t)$ are represented by the slow time t. Therefore the subsystem for $u^3 = (x, x_p, e_1, e_2, x_i)$ is split into two parts

$$
u^3 = (x(t), x_p(s), e_1(s), e_2(s), x_i(t))
$$

and evolves according to

$$
\begin{pmatrix} \dot{x}(t) \\ x_p'(s) \\ e_1'(s) \\ e_2'(s) \\ \dot{x}_i(t) \end{pmatrix} = D(\mu)T^3(z(s), r(s)) \begin{pmatrix} x(t) \\ x_p(s) \\ e_1(s) \\ e_2(s) \\ x_i(t) \end{pmatrix}
\tag{6.2}
$$

where $D(\mu) = \operatorname{diag}(1, \mu, \mu, \mu, 1)$. Note that (6.1), (6.2) is a singular perturbation problem. To be able to handle its evolution our main theorem provides the necessary ingredients; we must ensure that both subsystems in (6.1) go stationary if $x(t), u_1^1(s)$ in the first system and $u_2^1(s), u_1^1(s)$, in the second system are regarded as parameters and this process must be uniform in these parameters. Our main theorem gives the decisive hints for that. Furthermore we need that the full system (3.8) becomes stationary to keep the trajectories in a compact set of the state space. Again, this is guaranteed by our main result and singular perturbation theory is applicable to settle the question of how the network (3.6) evolves in time. Our result then enables us to define the speed of the network (3.6) (see [BB], [TGPBB]) and to compare the result with measurements available in biology on various transport systems (see [BB], [TGPBB]).

Our theory heavily depends upon the exponential decay of the nontrivial part of the infinitesimal generator B. We check easily that conditions $B1^0 - B3^0$ are satisfied by the matrices denoted by symbol B in Section 3. It is because these matrices have the form $B = C - r(C)I$ with C nonnegative and such that

$$\lim_{k \to \infty} \left(\frac{1}{r(C)} C \right)^k$$

exists and represents a nonzero matrix. In this case the spectrum $\sigma(B)$ satisfies

$$\sigma(B) \subset \{0\} \cup \{\lambda : \Re\lambda < 0\}.$$

At conclusion, let us comment on the concept of a *cooperative system*. Several papers of ours are devoted to problems connected with this concept. However, with the only exception of [BL] we did never use the notion cooperative system. In aggreement with the monograph [BNS] we feel that the title of this contribution is appropriate.

Acknowledgements. The research of the second author was partly supported by the Grant Agency of the Czech Republic under the contract number #201/98/0528 and by grant CEZ J04/98:210000010
The authors are grateful to the referee for his valuable comments.

References

[BNS]   Berman A., Neuman M., Stern R.J., Nonnegative Matrices in Dynamic Systems, J. Wiley Publ. New York 1989..

[BP]   Berman A., Plemmons R., Non-negative Matrices in the Mathematical Sciences, Academic Press 1979..

[B91]   Bohl E., A boundary layer phenomenon for linear systems with a rank deficient matrix, . **ZAMM** (7/8), 1991. 223–231

[B92]   Bohl, E., Constructing amplification via chemical circuits, . In Eisarfeld J., Leonis D. S., Witken M. (Editors), Biomedical modeling simulation. Elsevier Science Publ. B. V.. 1992 331–334

[B90]   Bohl E., *Structural Amplification in Chemical Networks*, In Mosekilde E., Mosekilde L. (Editors), Complexity, Chaos and Biological Evolution, Plenum Press New York 1991. 119–128.

[BB]   Bohl E., Boos W., Quantitative analysis of binding protein-mediated ABC transport system, *J. Theor. Biology* **186** (1997), 65–74.

[BL]   Bohl E., Lancaster P., Perturbation of spectral inverses applied to a boundary layer phenomenon arizing in chemical networks, *Linear Algebra Appl.* **180** (1993), 35–59.

[BM95]   Bohl E., Marek I., A model of amplification, *J. Comput. Appl. Math.* **63** (1995), 27–47.

[BM94]   Bohl E., Marek I., A nonlinear model involving M-operators. An amplification effect measured in the cascade of vision, *J. Comput. Appl. Math.* **60** (1994), 13–28.

[BM98]   Bohl E., Marek I., A stability theorem for a class of linear evolution problems, *Integral Equations Operator Theory* **34** (1999), 251–269.

[C]   Carroll R., *Abstract methods in Partial Differential Equations*, Harper's series
           in Modern Mathematics 1969.

[CB]   Cornish-Bowden A., *Fundamentals of Enzyme Kinetics*, Portland Press, Lon-
           don 1995.

[HP]   Hille E., Phillips R.S., *Functional Analysis and Semigroups*, Amer. Math.
           Socitey Coll. Publ. Vol XXXI, Third printing of Revised Edition Providence,
           Rhode Island 1968.

[KR]   Krein M. G., Rutman M.A., Linear operators leaving invariant a cone in a
           Banach space, *Uspekhi mat. nauk* **III, Nr. 1** (1948), 3–95. (In Russian.) English
           translation in AMS Translations, Vol **26**, 1950..

[MS]   Marek I., Szyld D., Psedoirreducible and pseudoprimitive operators, *Linear
           Algebra Appl.* **154–156** (1990), 779–791.

[MZ]   Marek I., Žitný K., *Analytic Theory of Matrices for Applied Sciences Vol 1.*,
           Teubner Texte zur Mathematik Band 60, Leipzig 1983.

[MWC]   Monod J., Wyman J., Changeux J.P., On the nature of allosteric transitions.
           A plauzible model, *J. Mol. Biol.* **12** (1965), 88–118.

[S1]   Schaefer H.H., *Banach Lattices and Positive Operators*, Springer-Verlag Berlin-
           Heidelberg-New York 1974.

[S2]   Schaefer H.H., *Topological Vector Spaces*, Springer Verlag, New York-Heidel-
           berg-Berlin 1970.

[S]   Sawashima I., On spectral properties of some positive operators, *Rep. Nat.
           Sci. Ochanomizu Univ.* **15** (1964), 53–64.

[SV]   Schneider H., Vidysagar M., Cross-positive matrices, *SIAM J. Numer. Anal.*
           7 (1970), 508–519.

[TL]   Taylor A. E., Lay D.C., *Introduction to Functional Analysis*, Second edition,
           J. Wiley Publ. New York 1980.

[TGPBB]   Tralau C., Greller G., Pajatsch M., Boos W., Bohl E., Mathematical treat-
           ment of transport data of bacterial transport system to estimate limitation in
           diffusion through the outer membrane, *J. theor. Biol.* **207** (2000), 1–14.

Erich Bohl   Ivo Marek
Fachbereich Maythematik   katedra numerické matematiky
Konstanz University   na matematicko – fyzikální fakultě
78434 Konstanz, Germany   University Karlovy
                                            Sokolovská 83
                                            183 00 Praha 8, Czech Republic
                                            and
                                            Faculty of Civil Engineering
                                            Czech University of Technology
                                            Thákurova 7
                                            160 00 Praha 6, Czech Republic

1991 Mathematics Subject Classification. Primary 47H07; Secondary 47J15

Received April 6, 2000

Operator Theory:
Advances and Applications, Vol. 130, 171–184
© 2001 Birkhäuser Verlag Basel/Switzerland

Young's Inequality in Compact Operators

Juliana Erlijman, Douglas R. Farenick and Renying Zeng

Dedicated to Professor Peter Lancaster

Abstract. This paper extends to compact operators an interesting inequality first proved by T. ANDO for matrices: for any compact operators a and b acting on a complex separable Hilbert space, there is a partial isometry u such that $u|ab^*|u^* \leq \frac{1}{p}|a|^p + \frac{1}{q}|b|^q$, for every $p, q \in (1, \infty)$ that satisfy $1/p + 1/q = 1$.

1. Introduction

The first important inequality that can be formed amongst complex numbers is the triangle inequality, and so it is natural to ask if, and how, the triangle inequality might extend to complex matrices, to operators, or to elements in a C*-algebra. These questions have been addressed by R.C. THOMPSON [7] and by C.A. AKEMANN, J. ANDERSON, and G.K. PEDERSEN [1], resulting in interesting, nontrivial formulations of the triangle inequality for matrices and operators.

Another fundamental inequality is the arithmetic-geometric-mean inequality: $|\lambda\mu|^{1/2} \leq \frac{1}{2}(|\lambda| + |\mu|)$, for all complex numbers λ and μ. This important inequality also admits a formulation in complex matrices, as R. BHATIA and F. KITTANEH have demonstrated in [3]. T. ANDO, in an elegant and ingenious paper [2], made a further generalisation by establishing a version of Young's inequality for complex matrices. Ando's result merits further investigation, and so we present in this paper an extension and adaptation of his methods so that they apply to compact operators.

The main result established herein is the following version of Young's inequality, which extends [3] and [2] to compact operators.

Theorem 1.1. *If a and b are compact operators acting on a complex separable Hilbert space, then there is a partial isometry u such that the initial space of u is $(\ker |ab^*|)^{\perp}$ and*

$$u|ab^*|u^* \leq \frac{1}{p}|a|^p + \frac{1}{q}|b|^q, \qquad (1.1)$$

for any $p, q \in (1, \infty)$ that satisfy $\frac{1}{p} + \frac{1}{q} = 1$. Furthermore, if $|ab^*|$ is injective, then the operator u in the inequality above can be taken to be a unitary.

In Corollary 1.2 below, an attractive rephrasing of Theorem 1.1 is made in an important special case. Observe, further, that if H is the 1-dimensional Hilbert space of complex numbers, and if $t = 1/2$, then Corollary 1.2 is precisely the arithmetic-geometric-mean inequality for positive real numbers.

Corollary 1.2. *If a and b are positive compact operators with trivial kernels, then there is a unitary u such that*

$$u|a^t b^{1-t}|u^* \leq ta + (1-t)b, \tag{1.2}$$

for every $t \in [0, 1]$.

In the inequalities (1.1) and (1.2) above, $|x|$ refers to the positive square root of x^*x. If A is a C*-algebra, then we say that $x \in A$ is positive, written $x \geq 0$, if $x^* = x$ and $\sigma(x) \subset [0, \infty)$, and we write $x \leq y$, for hermitian $x, y \in A$, if $y - x$ is positive. For hermitian operators x and y acting on a complex Hilbert space H, the operator inequality $x \leq y$ is equivalent to the statement that the real-number inequalities $\langle x\xi, \xi \rangle \leq \langle y\xi, \xi \rangle$ hold for all $\xi \in H$.

The operator inequality (1.1) has two necessary twists to it: the first appears in the use of the adjoint of b, the second is by way of conjugation by a partial isometry u. These are forced upon us by the noncommutativity of operator algebra in the first case, and by the infinite dimensionality of the Hilbert space in the second case—but they can be dropped if a and b are commuting normal operators (see Lemma 2.2).

Theorem 1.1 differs from the matricial Young inequality [2] in that the operator u that arises here is typically a partial isometry, whereas in the matrix case u can always be taken to be a unitary. This is explained by the fact that, for operators acting on infinite-dimensional spaces, the dimension of the kernel of a positive operator is not in general determined by the dimension of the operator's range. If $|ab^*|$ has trivial kernel, then this partial isometry u is in fact an isometry; however, as we shall show in Section 5, the operator u can, in this case, be taken to be unitary.

The remainder of the paper is devoted to the proof of Theorem 1.1. We follow Ando's model, but make modifications to it in many places to account for spaces of infinite dimension. Ando's methods are of two sorts. The first is through the use of operator inequalities, which we develop further for non-compact operators in the section that follows. The second is by means of spectral inequalities, which is made possible here by the tractable spectral theory of compact hermitian operators. The two methods are then coupled by way of an approximation of compact operators by non-compact operators.

It is assumed henceforth that H is an infinite-dimensional, separable Hilbert space.

2. A compression inequality

Young's inequality for positive real numbers is easily proved by exploiting the elementary fact that the logarithm is a concave function. Not surprisingly, then, operator-concave functions have a role in the proof of Theorem 1.1. What is needed for the proof is encapsulated in the well-known result below.

Lemma 2.1. *Suppose that* $r \in (0,1]$, $s \in [1,2]$, *and that* x, y, *and* z *are hermitian operators such that* $0 \le x \le y$ *and* $||z|| \le 1$. *Then*

(a) $x^r \le y^r$,

(b) $z^* x^r z \le (z^* x z)^r$, *and*

(c) $(z^* x z)^s \le z^* x^s z$.

Proof. If $r \in (0,1]$, then the function $t \mapsto t^r$, defined on $[0, \infty)$, is operator-monotone and operator-concave (by [6: 1.3.8, 1.3.11]); hence, $0 \le x \le y$ implies that $x^r \le y^r$. Let $f(t) = -t^r$, which is operator-convex on $[0, \infty)$ and satisfies $f(0) = 0$. By [5; 2.1], $f(z^* x z) \le z^* f(x) z$, and so $z^* x^r z \le (z^* x z)^r$, proving (b). Next, let $s \in (1,2]$, $g(t) = t^s$, and $h(t) = \frac{1}{t} g(t) = t^{s-1} = t^r$, where $r = s-1 \in (0,1]$. As h is operator-monotone on $(0, \infty)$ [6; 1.3.11], g is necessarily operator-convex on $[0, \infty)$, implying that $(z^* x z)^s = g(z^* x z) \le z^* g(x) z = z^* x^s z$. The case $s = 1$ is trivial. $\square$

The next lemma shows that Young's inequality extends easily to commuting operators.

Lemma 2.2. *If* A *is a unital commutative* C^*-*algebra, and if* $p, q \in (1, \infty)$ *satisfy* $\frac{1}{p} + \frac{1}{q} = 1$, *then* $ab \le \frac{1}{p} a^p + \frac{1}{q} b^q$ *for all positive* $a, b \in A$.

Proof. Let Ω be the maximal ideal space of A. By the Gelfand theory, there is a $*$-isomorphism $\phi : A \to C(\Omega)$, where $C(\Omega)$ is the C^*-algebra of all continuous complex-valued functions on Ω. Given positive $a, b \in A$, let $f = \phi(a)$ and $y = \phi(b)$. Then for every $w \in \Omega$, we have Young's inequality in real numbers:

$$0 \le f(w)g(w) \le \frac{1}{p} f(w)^p + \frac{1}{q} g(w)^q.$$

Thus, as elements of $C(\Omega)$, $0 \le fg \le \frac{1}{p} f^p + \frac{1}{q} g^q$, which, via ϕ, implies that $0 \le ab \le \frac{1}{p} a^p + \frac{1}{q} b^q$. $\square$

An argument similar to that used to prove Lemma 2.2 shows that in any unital commutative C^*-algebra A, $|ab| \le \frac{1}{p} |a|^p + \frac{1}{q} |b|^q$ for all $a, b \in A$.

A key technique in Ando's paper [2] appears here as the following proposition. Its formulation below is more technical than Ando's original because of a significant new complicating factor: compact operators on infinite-dimensional spaces do not have inverses. Furthermore, the result below is (in infinite dimensions) slightly weaker than Ando's result for matrices, for what is lost is finer information about the geometric multiplicities of individual eigenvalues.

Proposition 2.3. *Suppose that a and b are positive operators acting on a separable Hilbert space H, and assume that b is invertible. Let e denote the spectral resolution of the identity for the positive operator $(ba^2b)^{1/2}$. For each $t \in [0, \infty)$, set $e_t = e([t, \infty))$ and let h_t denote the projection onto the (closed) subspace $H_t = b^{-1}e_t(H)$. Then, given $p \in (1, 2]$ and $q = \dfrac{p}{(p-1)}$,*

$$th_t \leq h_t \left(\frac{1}{p} a^p + \frac{1}{q} b^q \right) h_t, \text{ for every } t \in [0, \infty).$$

Proof. Fix $t \in [0, \infty)$, and choose any $\eta \in H$. Then $h_t \eta \in H_t$ and so $h_t \eta = b^{-1}e_t\xi$ for some $\xi \in H$. Therefore, $bh_t\eta = e_t\xi = e_t^2\xi = e_tbh_t\eta$, which implies that $e_tbh_t = bh_t$. Passing to adjoints yields $h_tb = h_tbe_t$. Similarly, $b^{-1}e_t = h_tb^{-1}$ and $e_tb^{-1} = b^{-1}h_t$. Thus,

$$(h_tb^2h_t)(b^{-1}e_tb^{-1}) = h_t \text{ and } (b^{-1}e_tb^{-1})(h_tb^2h_t) = h_t.$$

From this we make the following conclusions:

(i) that $b^{-1}e_tb^{-1}$ and $h_tb^2h_t$ map H_t bijectively onto H_t;

(ii) that, as operators on H_t, $b^{-1}e_tb^{-1}$ and $h_tb^2h_t$ are inverse to each other; and

(iii) $b^{-1}e_tb^{-1}$ and $h_tb^2h_t$ vanish on $H_t^\perp$.

Using e, express $(ba^2b)^{1/2}$ as $\displaystyle\int_0^\infty \lambda\, de$; then, by the spectral theorem,

$$ba^2b = \int_0^\infty \lambda^2 de \geq \int_t^\infty \lambda^2 de \geq t^2 \int_t^\infty de = t^2 e_t. \tag{2.1}$$

Therefore, $t^2(b^{-1}e_tb^{-1}) \leq a^2$. Now set $r = \dfrac{p}{2}$; because $r \in (0, 1]$ by the hypothesis on p, Lemma 2.1 implies that

$$t^p(b^{-1}e_tb^{-1})^{p/2} = [t^2(b^{-1}e_tb^{-1})]^r \leq (a^2)^r = a^p.$$

From

$$h_t(b^{-1}e_tb^{-1})h_t = (h_tb^{-1}e_t)b^{-1}h_t = (b^{-1}e_t)b^{-1}h_t = b^{-1}(e_tb^{-1}h_t) = b^{-1}e_tb^{-1},$$

the inequality above yields

$$t^p(b^{-1}e_tb^{-1})^{p/2} = t^2h_t(b^{-1}e_tb^{-1})^{p/2}h_t \leq h_ta^ph_t.$$

Let x and y be operators on H_t; we write $x \leq_t y$ to denote the inequalities $\langle x\eta, \eta \rangle \leq \langle y\eta, \eta \rangle$ for every $\eta \in H_t$. Because as an operator on H_t the inverse of $b^{-1}e_tb^{-1}$ is $h_tb^2h_t$ [by (ii)], the inequality above becomes

$$t^p(h_tb^2h_t)^{-p/2} \leq_t h_ta^ph_t. \tag{2.2}$$

The remainder of this proof splits into two cases.

Case # 1: $q \in [2, 4]$. Set $s = \frac{q}{2}$. Because $s \in [1, 2]$ and $||h_t|| = 1$, Lemma 2.1(c) implies that $(h_t b^2 h_t)^{q/2} = (h_t b^2 h_t)^s \leq h_t(b^2)^s h_t = h_t b^q h_t$. This inequality, coupled with (2.2), results in the following inequality for operators on H_t:

$$\frac{1}{p} \left[t(h_t b^2 h_t)^{-1/2} \right]^p + \frac{1}{q} \left[(h_t b^2 h_t)^{1/2} \right]^q \leq_t h_t \left(\frac{1}{p} a^p + \frac{1}{q} b^q \right) h_t. \qquad (2.3)$$

As the positive operators $(h_t b^2 h_t)^{-1/2}$ and $(h_t b^2 h_t)^{1/2}$ commute when acting on H_t, Lemma 2.2 gives the first of the two inequalities that follow:

$$\begin{aligned} t(h_t b^2 h_t)^{-1/2}(h_t b^2 h_t)^{1/2} &\leq_t \frac{1}{p} \left[t(h_t b^2 h_t)^{-1/2} \right]^p + \frac{1}{q} \left[(h_t b^2 h_t)^{1/2} \right]^q \\ &\leq_t h_t \left(\frac{1}{p} a^p + \frac{1}{q} b^q \right) h_t. \end{aligned}$$

The second inequality above follows from (2.3). On both sides of the inequality above, the operators in question vanish on $H_t^{\perp}$, implying that the inequality extends from H_t to H:

$$t h_t \leq h_t \left(\frac{1}{p} a^p + \frac{1}{q} b^q \right) h_t,$$

as desired.

Case # 2: $q \in (4, \infty)$. Set $u = \frac{q}{2}$ so that $\frac{2}{u} \in (0, 1)$ and $\frac{q}{u} = 2$. By Lemma 2.1(a), using $r = \frac{q}{u}$, we have

$$(h_t b^u h_t)^2 = (h_t b^u h_t)^{q/u} \leq h_t(b^u)^{q/u} h_t = h_t b^q h_t. \qquad (2.4)$$

Using $r = \frac{2}{u}$, Lemma 2.1(b) yields

$$h_t b^2 h_t = h_t(b^u)^r h_t \leq (h_t b^u h_t)^r = (h_t b^u h_t)^{2/u}.$$

Apply Lemma 2.1(a), using $r = \frac{p}{2}$, to the inequality above to obtain

$$(h_t b^2 h_t)^{p/2} \leq (h_t b^u h_t)^{(2/u)(p/2)} = (h_t b^u h_t)^{p/u}.$$

Because $h_t b^2 h_t$ is invertible as an operator on H_t, the inequality above implies that $h_t b^u h_t$ is also invertible as an operator on H_t. Thus,

$$(h_t b^u h_t)^{-p/u} \leq_t (h_t b^2 h_t)^{-p/2}. \qquad (2.5)$$

Now $(h_t b^u h_t)^{-1/u}$ and $(h_t b^u h_t)^{1/u}$ are positive operators commuting on H_t, and so Lemma 3 can be invoked to obtain the first inequality below:

$$\begin{aligned} t(h_t b^u h_t)^{-1/u}(h_t b^u h_t)^{1/u} &\leq_t \frac{1}{p} \left[t(h_t b^u h_t)^{-1/u} \right]^p + \frac{1}{q} \left[(h_t b^u h_t)^{1/u} \right]^q \\ &\leq_t \frac{1}{p} \left[t^p (h_t b^u h_t)^{-p/2} \right] + \frac{1}{q} \left[(h_t b^u h_t)^2 \right] \\ &\leq_t h_t \left(\frac{1}{p} a^p + \frac{1}{q} b^q \right) h_t. \end{aligned}$$

The middle inequality above follows from (2.5) and the fact that $\frac{q}{u} = 2$; the third inequality above follows from (2.2) and (2.4). The left-hand side of the inequality above is the operator th_t on H_t, and therefore the entire inequality extends to H (because both sides vanish on $H_t^\perp$) to get

$$th_t \leq h_t \left(\frac{1}{p} a^p + \frac{1}{q} b^q \right) h_t,$$

as desired.   □

3. Young's inequality in nonzero eigenvalues

In this section we use Proposition 2.3 to establish an inequality between the nonzero eigenvalues of $(ba^2b)^{1/2}$ and those of $\frac{1}{p}a^p + \frac{1}{q}b^q$, where a and b are positive compact operators. Because compact operators are not invertible, we first need to consider perturbations of b by $\varepsilon 1$ (for $\varepsilon > 0$) in order to invoke Proposition 2.3.

Lemma 3.1. *If A is a unital C^*-algebra, if $b \in A$ is positive, and if $q \in (1, \infty)$, then there is a function $\gamma : (0, \infty) \to [0, \infty)$ such that*

(a) $\lim_{\varepsilon \to 0^+} \gamma(\varepsilon) = 0$, *and*

(b) *for each $\varepsilon > 0$ there is a positive $y_\varepsilon \in A$ with $(b + \varepsilon 1)^q = b^q + y_\varepsilon$ and $\|y_\varepsilon\| \leq \gamma(\varepsilon)$.*

Proof. Let $\beta > 0$ be such that $\sigma(b) \subseteq [0, \beta]$, and for each $\varepsilon > 0$ let g_ε be the function $g_\varepsilon(t) = t + \varepsilon$. Then $g_\varepsilon(t) \to t$ uniformly on $[0, \beta]$ as $\varepsilon \to 0^+$; hence, $(t + \varepsilon)^q \to t^q$ uniformly on $[0, \beta]$. Set $f_\varepsilon(t) = (t + \varepsilon)^q - t^q$, so that $f_\varepsilon \to 0$ uniformly on $[0, \beta]$ as $\varepsilon \to 0^+$. Define $\gamma(\varepsilon) = \max_{t \in [0, \beta]} f_\varepsilon(t)$ and, using functional calculus, let $y_\varepsilon = f_\varepsilon(b)$. Then y_ε is positive, $\|y_\varepsilon\| \leq \gamma(\varepsilon)$, and $\gamma(\varepsilon) \to 0$. As $y_\varepsilon = (b + \varepsilon 1)^q - b^q$, we have $(b + \varepsilon 1)^q = b^q + y_\varepsilon$.   □

Let x be any positive compact operator on separable Hilbert space. Enumerate the nonzero eigenvalues of x so that $\lambda_1(x) \geq \lambda_2(x) \geq \cdots > 0$, with eigenvalues repeated according to their geometric multiplicity. Let e be any projection.

The following results are well known. Assume that k is fixed.

(i) The Min-Max Principle:

$$\lambda_k(x) = \min_{\substack{L \subseteq H \\ \dim L = k-1}} \max_{\substack{\xi \in L^\perp \\ \|\xi\| = 1}} \langle x\xi, \xi \rangle.$$

(ii) The Poincaré Inequality: $\lambda_k(exe) \leq \lambda_k(x)$.

Lemma 3.2. *If a and b are positive compact operators, and if $0 < \varepsilon < 1$ and $b_\varepsilon = b + \varepsilon 1$, then, for any k,*

$$\lim_{\varepsilon \to 0^+} \lambda_k((b_\varepsilon a^2 b_\varepsilon)^{1/2}) = \lambda_k((ba^2b)^{1/2}).$$

Proof. Fix k. For every $\varepsilon > 0$, $b_\varepsilon a^2 b_\varepsilon = ba^2b + \varepsilon(ba^2 + a^2b + \varepsilon a^2) = ba^2b + z_\varepsilon$, where z_ε is compact and $\|z_\varepsilon\| \le \varepsilon M$, given $M = \|ba^2\| + \|a^2b\| + \|a^2\|$. Thus

$$\|b_\varepsilon a^2 b_\varepsilon - ba^2 b\| \le \varepsilon M$$

which implies that $\left|\langle b_\varepsilon a^2 b_\varepsilon \xi, \xi \rangle - \langle ba^2 b\xi, \xi\rangle\right| \le \varepsilon M$, for every unit vector $\xi \in H$. Therefore, by the min-max principle, $\left|\lambda_k(b_\varepsilon a^2 b_\varepsilon) - \lambda_k(ba^2 b)\right| \le \varepsilon M$. Thus, $\lim_{\varepsilon \to 0^+} \lambda_k(b_\varepsilon a^2 b_\varepsilon) = \lambda_k(ba^2 b)$. On passing to the positive square roots of the positive compact operators $b_\varepsilon a^2 b_\varepsilon$ and $ba^2 b$, the geometric multiplicities of the nonzero eigenvalues are unchanged. Hence, $\lim_{\varepsilon \to 0^+} \lambda_k((b_\varepsilon a^2 b_\varepsilon)^{1/2}) = \lambda_k((ba^2 b)^{1/2})$. $\quad\square$

We now arrive at the crucial spectral inequality between nonzero eigenvalues that leads to Theorem 1.1.

Proposition 3.3. *If a and b are positive compact operators, and if $p \in (1,2]$ and $q = \dfrac{p}{(p-1)}$, then, for every k,*

$$\lambda_k((ba^2 b)^{1/2}) \le \lambda_k\left(\frac{1}{p}a^p + \frac{1}{q}b^q\right).$$

Proof. Let $\varepsilon > 0$ and consider $b_\varepsilon = b + \varepsilon 1$. By Lemma 3.1, there is a positive operator y_ε such that $(b_\varepsilon)^q = b^q + y_\varepsilon$ and $\|y_\varepsilon\| \to 0$ as $\varepsilon \to 0^+$. Because the operator $(b_\varepsilon a^2 b_\varepsilon)^{1/2}$ is positive and compact, there is an orthonormal sequence $\{\phi_j^{(\varepsilon)}\}_j$ in H consisting of eigenvectors:

$$(b_\varepsilon a^2 b_\varepsilon)^{1/2}\phi_j^{(\varepsilon)} = \lambda_j((b_\varepsilon a^2 b_\varepsilon)^{1/2})\phi_j^{(\varepsilon)}.$$

Fix k. Let e_ε denote the spectral resolution of the identity for $(b_\varepsilon a^2 b_\varepsilon)^{1/2}$, and set $t_\varepsilon = \lambda_k((b_\varepsilon a^2 b_\varepsilon)^{1/2})$. Let e'_ε be the projection onto the span of $\{\phi_1^{(\varepsilon)}, \cdots, \phi_k^{(\varepsilon)}\}$ and let p_ε denote the projection $p_\varepsilon = c_\varepsilon([t_\varepsilon, \infty))$. Then

$$e'_\varepsilon(H) \subseteq \sum_{\mu \ge t_\varepsilon} {}^{\oplus} \ker((b_\varepsilon a^2 b_\varepsilon)^{1/2} - \mu 1) = e_\varepsilon([t_\varepsilon, \infty))(H) = p_\varepsilon(H).$$

Therefore, $b_\varepsilon^{-1} e'_\varepsilon(H) \subseteq b_\varepsilon^{-1} p_\varepsilon(H)$. Let h'_{t_ε} and h_{t_ε} be the projections onto $b_\varepsilon^{-1} e'_\varepsilon(H)$ and $b_\varepsilon^{-1} p_\varepsilon(H)$ respectively. Then,

$$t_\varepsilon h'_{t_\varepsilon} \le t_\varepsilon h_{t_\varepsilon} \le h_{t_\varepsilon}\left(\frac{1}{p}a^p + \frac{1}{q}b_\varepsilon^q\right)h_{t_\varepsilon} \le h_{t_\varepsilon}\left(\frac{1}{p}a^p + \frac{1}{q}b^q\right)h_{t_\varepsilon} + \frac{1}{q}h_{t_\varepsilon}y_\varepsilon h_{t_\varepsilon}.$$

The second inequality above is from Proposition 2.3.

Recall (from linear algebra) that if $M, N \subseteq H$ are subspaces such that $\dim M < \dim N < \infty$, then $M^\perp \bigcap N \ne \{0\}$. Therefore, because the projection h'_{t_ε} has rank k, $L^\perp \bigcap(h'_{t_\varepsilon}(H)) \ne \{0\}$ for every $(k-1)$-dimensional subspace $L \subset H$. Thus,

$$\min_{\dim L = k-1} \max_{\substack{\xi \in L^\perp \\ \|\xi\|=1}} \langle h'_{t_\varepsilon}\xi, \xi \rangle = 1.$$

What this implies further is:

$$
\begin{aligned}
\lambda_k((b_\varepsilon a^2 b_\varepsilon)^{1/2}) &= t_\varepsilon = \min_{\dim L = k-1} \max_{\substack{\xi \in L^\perp \\ ||\xi||=1}} \langle t_\varepsilon h'_{t_\varepsilon} \xi, \xi \rangle \\
&\leq \min_{\dim L = k-1} \max_{\substack{\xi \in L^\perp \\ ||\xi||=1}} \left\langle \left(\frac{1}{p} a^p + \frac{1}{q} b_\varepsilon^q \right) \xi, \xi \right\rangle \\
&\leq \left(\min_{\dim L = k-1} \max_{\substack{\xi \in L^\perp \\ ||\xi||=1}} \left\langle \left(\frac{1}{p} a^p + \frac{1}{q} b^q \right) \xi, \xi \right\rangle \right) + \frac{1}{q} ||y_\varepsilon|| \\
&= \lambda_k \left(\frac{1}{p} a^p + \frac{1}{q} b^q \right) + \frac{1}{q} ||y_\varepsilon|| .
\end{aligned}
$$

Hence, by the inequalities above and Lemmas 3.1 and 3.2,

$$
\begin{aligned}
\lambda_k((ba^2 b)^{1/2}) &= \lim_{\varepsilon \to 0^+} \lambda_k((b_\varepsilon a^2 b_\varepsilon)^{1/2}) \\
&\leq \lambda_k \left(\frac{1}{p} a^p + \frac{1}{q} b^q \right) + \lim_{\varepsilon \to 0^+} \frac{1}{q} ||y_\varepsilon|| \\
&= \lambda_k \left(\frac{1}{p} a^p + \frac{1}{q} b^q \right),
\end{aligned}
$$

which is precisely the sought after inequality. □

4. Transfer of point spectra

To this stage we have worked only with positive operators. The present section facilitates the transition from positive operators of the type $(ba^2 b)^{1/2}$ to positive operators of the form $|xy^*|$.

Proposition 4.1. *For any operators a and b acting on a Hilbert space,*

$$
\dim(\ker(|ab^*| - \lambda 1)) = \dim \left(\ker \left(\left| |a||b| \right| - \lambda 1 \right) \right), \quad \text{for every nonzero } \lambda.
$$

Proof. Let $b = w|b|$ denote the polar decomposition of b. Observe that $\left| |a||b| \right| = (|b||a|^2|b|)^{1/2}$. Thus, because the closures of the ranges of a positive operator and its square root coincide, the closures of the ranges of $|b||a|^2|b|$ and $\left| |a||b| \right|$ are equal. Moreover, as $w^* w \left| |a||b| \right| = \left| |a||b| \right|$, we have that

$$
f(w(|b||a|^2|b|)w^*) = wf(|b||a|^2|b|)w^*, \tag{4.1}
$$

for all polynomials f. Choose $\delta > 0$ so that $\sigma(|b||a|^2|b|) \subseteq [0, \delta]$. By the Weierstrass approximation theorem, there is a sequence of polynomials f_n such that $f_n(t) \to$

$\sqrt{t}$ uniformly on $[0, \delta]$. Thus, from (4.1) and functional calculus,

$$(w(|b||a|^2|b|)w^*)^{1/2} = w(|b||a|^2|b|)^{1/2}w^* = w \left| \, |a||b| \, \right| w^*.$$

Let $a = v|a|$ be the polar decomposition of a. Then the left-hand term in the equalities above expands as follows:

$$\begin{aligned} (w(|b||a|^2|b|)w^*)^{1/2} &= (w|b||a|v^*v|a||b|w^*)^{1/2} \\ &= (ba^*ab^*)^{1/2} \\ &= |ab^*|. \end{aligned}$$

Thus, $|ab^*| = w \left| \, |a||b| \, \right| w^*$.

Next, let λ be a nonzero eigenvalue of $\left| \, |a||b| \, \right|$ and let $\{\phi_j\}_j$ be a sequence of orthonormal basis vectors for $\ker \left(\left| \, |a||b| \, \right| - \lambda 1 \right)$. Fix j. Then $\left| \, |a||b| \, \right| \phi_j = \lambda \phi_j$ implies that $\lambda^2 \phi_j = \left| \, |a||b| \, \right|^2 \phi_j \in (|b||a|^2|b|)(H) \subseteq |b|(H) \subseteq [b^*(H)]^-$. Because ϕ_j belongs to the closure of the range of $b^*(H)$ and because w^*w is the projection onto this closed range, $\phi_j = w^*w\phi_j$. Thus, $\|w\phi_j\| = 1 \neq 0$. We aim to show that $w\phi_j$ is an eigenvector of $|ab^*|$ and that λ is its corresponding eigenvalue. This is a straightforward verification:

$$|ab^*|w\phi_j = w \left| \, |a||b| \, \right| w^*(w\phi_j) = w \left| \, |a||b| \, \right| \phi_j = w(\lambda\phi_j) = \lambda(w\phi_j).$$

The above relation is true for every j. If a finite sum $\sum_j \alpha_j(w\phi_j) = 0$, then

$$0 = w^* \left(\sum_j \alpha_j(w\phi_j) \right) = \sum_j \alpha_j(w^*w\phi_j) = \sum_j \alpha_j \phi_j,$$

and so $\alpha_j = 0$ for all j. This proves that

$$\dim \left(\ker \left| \, |a| \cdot |b| \, \right| - \lambda 1 \right) \leq \dim(\ker(|ab^*| - \lambda 1)).$$

Conversely, let λ be a nonzero eigenvalue of $|ab^*|$, and let $\{\psi_i\}_i$ be an orthonormal basis of $\ker(|ab^*| - \lambda 1)$. Fix i. Then $|ab^*|\psi_i = \lambda\psi_i$ implies that

$$\lambda^2 \psi_i = (ba^*ab^*)\psi_i \in b(H).$$

Thus, $ww^*\psi_i = \psi_i$, because ww^* is the projection onto the closure of $b(H)$; in particular, $\|w^*\psi_i\| = 1 \neq 0$. Now $\lambda\psi_i = |ab^*|\psi_i = w \left| \, |a||b| \, \right| w^*\psi_i$ implies that

$$\lambda(w^*\psi_i) = w^*w \left| \, |a||b| \, \right| w^*\psi_i = \left| \, |a||b| \, \right| w^*\psi_i,$$

and hence $w^*\psi_i \in \ker\left(\Big|\,|a||b|\,\Big| - \lambda 1\right)$. As we argued earlier, the vectors $\{w^*\psi_i\}_i$ are linearly independent and so

$$\dim(\ker(|ab^*| - \lambda 1)) \leq \dim\left(\ker\left(\Big|\,|a||b|\,\Big| - \lambda 1\right)\right).$$

These arguments also show that a nonzero complex number λ is an eigenvalue of $|ab^*|$ if and only if λ is an eigenvalue of $\Big|\,|a||b|\,\Big|$, which completes the proof. $\square$

It is also true that the nonzero essential spectra of $|ab^*|$ and $\Big|\,|a||b|\,\Big|$ coincide, but proving so is considerably more involved; in any case, this fact is not needed here.

Lemma 4.2. *For any two operators x and y acting on a Banach space,*

$$\dim(\ker(xy^2x - \lambda 1)) = \dim(\ker(yx^2y - \lambda 1)), \text{ for all nonzero } \lambda.$$

Proof. If ξ is a nonzero vector, $\lambda \neq 0$, and $xy^2x\xi = \lambda\xi$, then $yx\xi \neq 0$. Thus $yx^2y(yx\xi) = yx(xy^2x)\xi = \lambda(yx\xi)$, implying that λ is an eigenvalue of yx^2y. Assume that $\xi_1, \cdots, \xi_k$ are linearly independent eigenvectors of xy^2x corresponding to the eigenvalue λ, and assume that $\alpha_1, \cdots, \alpha_k$ are scalars such that

$$0 = \sum_j \alpha_j(yx\xi_j).$$

Then

$$0 = xy\left(\sum_j \alpha_j yx\xi_j\right) = \sum_j \alpha_j(xy^2x\xi_j) = \lambda\sum_j \alpha_j\xi_j.$$

Because $\xi_1, \cdots, \xi_k$ are linearly independent and $\lambda \neq 0$, we must have $\alpha_1 = \cdots = \alpha_k = 0$. Thus $yx\xi_1, \cdots, yx\xi_k$ are linearly independent. This proves that

$$\dim(\ker(xy^2x - \lambda 1)) \leq \dim(\ker(yx^2y - \lambda 1)) \quad \forall \lambda \neq 0.$$

The reverse inequality is proved similarly. $\square$

For the proposition that follows, recall that $\lambda_k(x)$ refers only to the *nonzero* eigenvalues of a positive compact operator x.

Proposition 4.3. *If a and b are compact operators, then $\lambda_k(|ab^*|) = \lambda_k(|ba^*|)$.*

Proof.

$$
\begin{aligned}
\lambda_k(|ab^*|)^2 &= \lambda_k\left(\left|\,|a||b|\,\right|\right)^2 && \text{[by Proposition 4.1]}\\
&= \lambda_k\left(\left|\,|a||b|\,\right|^2\right) && \text{[by the Spectral Mapping Theorem]}\\
&= \lambda_k(|b||a|^2|b|)\\
&= \lambda_k(|a||b|^2|a|) && \text{[by Lemma 4.2]}\\
&= \lambda_k\left(\left|\,|b||a|\,\right|\right)^2\\
&= \lambda_k(|ba^*|)^2 && \text{[by Proposition 4.1].}
\end{aligned}
$$

Thus, $\lambda_k(|ab^*|) = \lambda_k(|ba^*|)$. $\qquad\qquad\qquad\qquad\qquad\qquad\qquad\qquad\qquad\square$

5. Young's inequality in compact operators

This section brings together the results that have been thus far established to prove the main theorem (Theorem 1.1) concerning Young's inequality.

Theorem 5.1. *If a and b are compact, then there is a partial isometry u such that the initial space of u is $(\ker |ab^*|)^{\perp}$ and*

$$
u|ab^*|u^* \le \frac{1}{p}|a|^p + \frac{1}{q}|b|^q, \tag{5.1}
$$

for any $p, q \in (1, \infty)$ that satisfy $\frac{1}{p} + \frac{1}{q} = 1$.

Proof. Assume that a and b are arbitrary compact operators acting on H, and fix any pair $p, q \in (1, \infty)$ for which $\frac{1}{p} + \frac{1}{q} = 1$. Enumerating the nonzero eigenvalues of positive compact operators in the usual descending order, we have by Propositions 4.1 and 4.3 that

$$
\lambda_k(|ab^*|) = \lambda_k(|ba^*|) = \lambda_k\left(\left|\,|a||b|\,\right|\right) = \lambda_k\left(\left|\,|b||a|\,\right|\right), \quad \forall k.
$$

First assume that $p \in (1, 2]$; then Proposition 3.3 states that $\lambda_k\left(\left|\,|a||b|\,\right|\right) \le \lambda_k\left(\frac{1}{p}|a|^p + \frac{1}{q}|b|^q\right)$. Second, if $p > 2$, then $q < 2$; so, Proposition 3.3 states that $\lambda_k\left(\left|\,|b||a|\,\right|\right) \le \lambda_k\left(\frac{1}{q}|b|^q + \frac{1}{p}|a|^p\right)$. In either case, then end result, for each k, is:

$$
\lambda_k(|ab^*|) \le \lambda_k\left(\frac{1}{p}|a|^p + \frac{1}{q}|b|^q\right).
$$

This means that $\frac{1}{p}|a|^p + \frac{1}{q}|b|^q$ has at least as many nonzero eigenvalues as $|ab^*|$ has. Therefore, there are orthonormal sequences $\{\phi_k\}_k$ and $\{\psi_k\}_k$ in H such that, for every $\xi \in H$,

$$|ab^*|\xi = \sum_k \lambda_k(|ab^*|)\langle \xi, \phi_k \rangle \phi_k,$$

$$\left(\frac{1}{p}|a|^p + \frac{1}{q}|b|^q\right)\xi = \sum_k \lambda_k\left(\frac{1}{p}|a|^p + \frac{1}{q}|b|^q\right)\langle \xi, \psi_k \rangle \psi_k.$$

Let $u : H \to H$ be the operator for which $u\phi_k = \psi_k$ for every k, and for which $u\xi = 0$ for every $\xi \in \{\phi_k : k = 1, 2, \cdots\}^\perp$. Then u is a partial isometry and, for every $\xi \in H$,

$$\langle u|ab^*|u^*\xi, \xi \rangle = \sum_k \lambda_k(|ab^*|)\,|\langle \xi, u\phi_k \rangle|^2$$

$$= \sum_k \lambda_k(|ab^*|)\,|\langle \xi, \psi_k \rangle|^2$$

$$\leq \sum_k \lambda_k\left(\frac{1}{p}|a|^p + \frac{1}{q}|b|^q\right)|\langle \xi, \psi_k \rangle|^2$$

$$= \left\langle \left(\frac{1}{p}|a|^p + \frac{1}{q}|b|^q\right)\xi, \xi \right\rangle.$$

Hence, $u|ab^*|u^* \leq \frac{1}{p}|a|^p + \frac{1}{q}|b|^q$. $\square$

To establish Young's inequality via unitary conjugation in the case where $|ab^*|$ has trivial kernel, the following variant of Lemma 2.2 of [4] is necessary.

Lemma 5.2. *Suppose that z is a contraction and a is a positive compact operator. If $\ker a = \ker z = \{0\}$, then there is a unitary operator u such that $z^*az \leq u^*au$.*

Proof. We proceed as in Lemma 2.2 of [4]. Because z is a contraction, the operator matrix w below is a unitary operator on $H \oplus H$:

$$w = \begin{bmatrix} z & (1 - zz^*)^{1/2} \\ (1 - z^*z)^{1/2} & -z^* \end{bmatrix}.$$

The operator z^*az appears as the $(1,1)$-entry of the (compact) operator matrix

$$w^*(a \oplus 0)w = \begin{bmatrix} z^*az & z^*a(1 - zz^*)^{1/2} \\ (1 - z^*z)^{1/2}az & (1 - zz^*)^{1/2}a(1 - zz^*)^{1/2} \end{bmatrix}.$$

Because $\ker a = \ker z = \{0\}$ and a is positive, the compact positive operator z^*az has trivial kernel, and so its eigenvalues may be enumerated in decreasing order $\lambda_1(z^*az) \geq \lambda_2(z^*az) \geq \cdots > 0$, with eigenvalues repeated in this list in accordance with their geometric multiplicity. The nonzero eigenvalues of $w^*(a \oplus 0)w$ are precisely the eigenvalues of a (as $\ker a = \{0\}$) and, moreover, the multiplicity of any nonzero eigenvalue of $w^*(a \oplus 0)w$ is the same as the multiplicity of that eigenvalue of a. Thus, in listing the nonzero eigenvalues of $w^*(a \oplus 0)w$ and the eigenvalues of a in descending order, we have that $\lambda_k(w^*(a \oplus 0)w) = \lambda_k(a)$, for all positive integers k. Now invoke the Poincaré inequality to obtain

$$\lambda_k(a) = \lambda_k(w^*(a \oplus 0)w) \geq \lambda_k(z^*az)$$

for all k. Because a and z^*az are injective, there exist orthonormal bases $\{\phi_k\}_k$ and $\{\psi_k\}_k$ of H comprising eigenvectors of a and z^*az respectively. That is, for all k, $a\phi_k = \lambda_k(a)\phi_k$ and $z^*az\psi_k = \lambda_k(z^*az)\psi_k$. Let u be the unique unitary opeator on H for which $u\psi_k = \phi_k$, for every k. Then the inequalities $\lambda_k(z^*az) \leq \lambda_k(a)$ imply, by standard arguments, that $z^*az \leq u^*au$.   □

Corollary 5.3. *If a and b are compact operators and if $|ab^*|$ is injective, then there is a unitary u_0 such that, for any $p, q \in (1, \infty)$ that satisfy $\frac{1}{p} + \frac{1}{q} = 1$,*

$$u_0|ab^*|u_0^* \leq \frac{1}{p}|a|^p + \frac{1}{q}|b|^q.$$

Proof. Because $|ab^*|$ has trivial kernel, the partial isometry u that occurs in Young's inequality (5.1) is an isometry. Thus, $u^*u = 1$ and $\ker u = \{0\}$. From

$$u^*(u|ab^*|u^*)u \leq u^*(\frac{1}{p}|a|^p + \frac{1}{q}|b|^q)u$$

we obtain

$$|ab^*| \leq u^*(\frac{1}{p}|a|^p + \frac{1}{q}|b|^q)u.$$

This inequality implies that $u^*(\frac{1}{p}|a|^p + \frac{1}{q}|b|^q)u$ is an injective compact operator. Therefore, by Lemma 5.2, there is a unitary operator u_0 such that

$$u^*(\frac{1}{p}|a|^p + \frac{1}{q}|b|^q)u \leq u_0^*(\frac{1}{p}|a|^p + \frac{1}{q}|b|^q)u_0.$$

Hence,

$$u_0|ab^*|u_0^* \leq \frac{1}{p}|a|^p + \frac{1}{q}|b|^q,$$

thereby completing the proof.   □

Acknowledgements. The work of the first two authors is supported in part by the Natural Sciences and Engineering Research Council of Canada.

References

[1] Akemann, C.A., Anderson, J., and Pedersen, G.K., Triangle inequalities in operator algebras, *Linear and Multilinear Algebra* **11** (1982), 167–178.

[2] Ando, T., Matrix Young inequalities, *Oper. Theory Adv. Appl.* **75** (1995), 33–38.

[3] Bhatia, R., and Kittaneh, F., On the singular values of a product of operators, *SIAM J. Matrix Anal. Appl.* **11** (1990), 272–277.

[4] Farenick, D.R, and Psarrakos, P.J., A triangle inequality in Hilbert modules over matrix algebras, (preprint), 2000.

[5] Hansen, F., and Pedersen, G.K., Jensen's inequality for operators and Löwner's theorem, *Math. Ann.* **258** (1982), 229–241.

[6] Pedersen, G.K., *C*-algebras and their Automorphism Groups*, Academic Press, New York 1979.

[7] Thompson, R.C., Convex and concave functions of singular values of matrix sums, *Pacific J. Math.* **66** (1976), 285–290.

Department of Mathematics and Statistics
University of Regina
Regina, Saskatchewan   S4S 0A2
Canada

1991 Mathematics Subject Classification. Primary 47A63

Received July 17, 2000

Operator Theory:
Advances and Applications, Vol. 130, 185–195

Partial Indices of Small Perturbations of a Degenerate Continuous Matrix Function

Israel Feldman, Naum Krupnik and Alexander Markus

Dedicated to Peter Lancaster with friendship and admiration

Abstract. If $A(t)$ is a continuous on the unit circle $n \times n$ matrix function such that $\det A(t_0) = 0$ for some t_0, $|t_0| = 1$, and $s = (s_1, s_2, \ldots, s_n)$ is a given vector in $\mathbf{Z}^n$, then in any neighborhood of $A(t)$ there exists a rational matrix function with partial indices $(s_1, s_2, \ldots, s_n)$.

1. Introduction

Factorization of matrix functions and corresponding factorization indices (or partial indices) have various applications in analysis. We mention here singular integral equations, Wiener Hopf equations, Toeplitz operators, finite section method for convolution equations (see, e.g., [1, 2, 4, 7]).

One of the most important results in this theory is the description of partial indices for small perturbations of a non-degenerate continuous matrix function (see Theorem 5.1 below and the paragraph following it). This result was obtained by I. Gohberg and M. Krein [3] more than 40 years ago.

The purpose of this paper is to describe the partial indices of non-degenerate matrix functions close enough to a given continuous matrix function which degenerates at least at one point of the contour. There are no any visible restrictions for these indices, and therefore it is naturally to expect that they can be arbitrary. We prove that this hypothesis is true. Somewhat surprising that its proof which we have found turned out to be rather complicated.

We are grateful to Professor V. A. Marchenko for drawing our attention to the considered problem.

2. Scalar case

2.1. For the sake of simplicity we consider only the unit circle $\mathbf{T} = \{t : |t| = 1\}$ although the results remain true for rather wide class of closed contours.

Let $C(\mathbf{T})$ be the algebra of all continuous functions on $\mathbf{T}$ with the norm

$$\|f\| = \max_{t \in \mathbf{T}} |f(t)|. \tag{2.1}$$

We use all topological notions (neighborhood, convergence etc.) with respect either to this norm or to matrix norm (3.1).

Let $C_+(\mathbf{T})$ $(C_-(\mathbf{T}))$ be the subalgebra of $C(\mathbf{T})$ which consists of functions admiting analytic extension to the disk $|t| < 1$ (to the domain $|t| > 1$, including ∞). Denote by $GC(\mathbf{T})$, $GC_+(\mathbf{T})$, $GC_-(\mathbf{T})$ the sets of invertible elements in corresponding algebras.

The *index* of the function $f(t) \in GC(\mathbf{T})$ is defined to be the integer

$$\mathrm{ind} f(t) = \frac{1}{2\pi} \left[\arg f(e^{i\theta}) \right]_{\theta=0}^{2\pi}.$$

A function $f(t) \in GC(\mathbf{T})$ is said to admit a *factorization* (with respect to $\mathbf{T}$) if

$$f(t) = f_-(t) t^\kappa f_+(t)$$

where $f_\pm \in GC_\pm(\mathbf{T})$ and $\kappa \in \mathbf{Z}$. It is easy to see that $\kappa = \mathrm{ind}\ f(t)$.

It is well known that any non-degenerate sufficiently smooth (e.g., Hölder continuous) function admits a factorization.

Denote by $\mathcal{R}$ the set of all rational functions which have no poles on $\mathbf{T}$, and let

$$\mathcal{R}_\pm = \mathcal{R} \cap C_\pm(\mathbf{T}), \quad G\mathcal{R} = \mathcal{R} \cap GC(\mathbf{T}), \quad G\mathcal{R}_\pm = \mathcal{R}_\pm \cap GC_\pm(\mathbf{T}).$$

2.2. We begin with an elementary statement.

Lemma 2.1. *The following equalities hold:*

$$\lim_{\epsilon \to 0^+} \left\| \left[\left(\frac{t-1-\epsilon}{t-1+\epsilon} \right)^k - 1 \right] (t-1) \right\| = 0 \quad (k \in \mathbf{Z}), \tag{2.2}$$

$$\lim_{\epsilon \to 0^+} \left\| \frac{(t-1-\epsilon)^{k+1}}{(t-1+\epsilon)^k} - (t-1) \right\| = 0 \quad (k \in \mathbf{Z}), \tag{2.3}$$

$$\lim_{\epsilon \to 0^+} \left\| \left[\left(\frac{t-1}{t-1-\epsilon} \right)^k - 1 \right] (t-1) \right\| = 0 \quad (k \in \mathbf{N}). \tag{2.4}$$

Proof. Denote

$$\alpha_\epsilon(t) = \frac{t-1-\epsilon}{t-1+\epsilon} \quad (-1 < \epsilon < 1).$$

The relations

$$|t-1+\epsilon| \geq ||t| - (1-\epsilon)| = |\epsilon| \quad (|t| = 1), \quad \alpha_\epsilon(t) = 1 - 2\epsilon(t-1+\epsilon)^{-1}$$

imply

$$|\alpha_\epsilon(t)| \leq 1 + 2|\epsilon| \frac{1}{|\epsilon|} = 3. \tag{2.5}$$

Analogously

$$\left|\frac{t-1}{t-1-\epsilon}\right| = |1+\epsilon(t-1-\epsilon)^{-1}| \le 1 + |\epsilon|\frac{1}{|\epsilon|} = 2. \tag{2.6}$$

Now we prove (2.2), at first for $k > 0$ (for $k = 0$ it is trivially):

$$(\alpha_\epsilon^k(t) - 1)(t-1) = (t-1)(\alpha_\epsilon(t) - 1)\sum_{j=0}^{k-1}\alpha_\epsilon^j(t) = -2\epsilon\frac{t-1}{t-1+\epsilon}\sum_{j=0}^{k-1}\alpha_\epsilon^j(t). \tag{2.7}$$

This equality and (2.5), (2.6) imply (2.2).

If $k < 0$, then $\alpha_\epsilon^k(t) = \alpha_{-\epsilon}^{|k|}(t)$, and we again obtain (2.2) from (2.7).

Pass now to the proof of (2.3). We have

$$\alpha_\epsilon^k(t)(t-1-\epsilon) - (t-1) = -\epsilon\alpha_\epsilon^k(t) + (\alpha_\epsilon^k(t) - 1)(t-1).$$

The second term in the right hand side tends to 0 as $\epsilon \to 0$ by the first part of the proof. The first term tends to zero by (2.5), and we obtain (2.3).

Finally consider (2.4). Like in (2.7) we have for $k > 0$

$$\left[\left(\frac{t-1}{t-1-\epsilon}\right)^k - 1\right](t-1) = (t-1)\left(\frac{t-1}{t-1-\epsilon} - 1\right)\sum_{j=0}^{k-1}\left(\frac{t-1}{t-1-\epsilon}\right)^j =$$

$$= \epsilon\sum_{j=0}^{k-1}\left(\frac{t-1}{t-1-\epsilon}\right)^{j+1}$$

and we obtain (2.4) from (2.6). □

Lemma 2.1 immediately implies the following statements.

Lemma 2.2. Let $k \in \mathbf{Z}$. In any neighborhood of the function $t - 1$ there exists a function of the form $g(t)(t-1)$ where $g(t) \in GR$ and ind $g(t) = k$.

Lemma 2.3. Let $k \in \mathbf{Z}$. In any neighborhood of the function $t - 1$ there exists a function $f(t) \in GR$ such that ind $f(t) = k$.

Lemma 2.4. Let $k \in \mathbf{N}$. In any neighborhood of the function $t - 1$ there exists a function of the form $h(t)(t-1)^k$ where $h(t) \in GR$ and ind $h(t) = 0$.

2.3. Here we consider the scalar case of the problem.

Theorem 2.1. Let $f(t) \in C(\mathbf{T})$ and $f(t_0) = 0$ for some $t_0 \in \mathbf{T}$. For arbitrary $k \in \mathbf{Z}$ in any neighborhood of the function $f(t)$ there exists a function $h(t) \in GR$ such that ind $h(t) = k$.

Proof. Suppose for convenience that $t_0 = 1$, and let $\epsilon > 0$ be arbitrary. Choose a rational function $r(t)$ such that $\|f - r\| < \epsilon/6$ and denote $r_0(t) = r(t) - r(1)$. We may assume that $r_0(t) \not\equiv 0$ (even if $f(t) \equiv 0$). Obviously

$$\|f - r_0\| < \epsilon/3 \tag{2.8}$$

and $r_0(1) = 0$. Represent $r_0(t) = r_1(t)(t-1)$ and choose a function $r_2(t) \in G\mathcal{R}$ such that

$$\|r_2 - r_1\| < \epsilon/6. \tag{2.9}$$

(For example, we can set $r_2(t) = r_1(\rho t)$ where the real number ρ is sufficiently close to 1.) Denote ind $r_2(t) = s$. By Lemma 2.3 there exists a function $r_3(t) \in G\mathcal{R}$ such that

$$\|r_3(t) - (t-1)\| < \epsilon/3\|r_2\| \tag{2.10}$$

and ind $r_3(t) = k - s$. Set $h(t) = r_2(t)r_3(t)$. Then $h(t) \in G\mathcal{R}$, ind $h(t) = k$ and by (2.8)–(2.10)

$$\|h - f\| < \|r_2 r_3 - r_0\| + \epsilon/3 = \|r_2 r_3 - (t-1)r_1\| + \epsilon/3 = \|r_2 r_3 - (t-1)r_2 + (t-1)(r_2 - r_1)\| + \epsilon/3 \le \|r_2\| \, \|r_3 - (t-1)\| + \|t-1\| \, \|r_2 - r_1\| + \epsilon/3 < \epsilon. \quad \square$$

In what follows we will use the following statement.

Lemma 2.5. *Let $f(t) \in C(\mathbf{T})$ and $f(1) = 0$. Then for arbitrary $k \in \mathbf{Z}$ in any neighborhood of the function $f(t)$ there exists a function of the form $h(t)(t-1)$ where $h(t) \in G\mathcal{R}$ and rm ind $h(t) = k$.*

Proof. We have shown in the proof of Theorem 2.1 that $f(t)$ can be approximated by a function of the form $r_2(t)(t-1)$ where $r_2 \in G\mathcal{R}$. Denote ind r_2 by s. By Lemma 2.2 in any neighborhood of $t-1$ there exists a function of he form $g(t)(t-1)$ where $g(t) \in G\mathcal{R}$ and ind $g(t) = k - s$. Then the function $h(t) = r_2(t)g(t)$ has all the required properties. $\quad \square$

3. Matrices of second order

3.1. Let $C^{n \times n}(\mathbf{T})$ $\left(C_{\pm}^{n \times n}(\mathbf{T})\right)$ be the vector space of all $n \times n$ matrix functions with entries from $C(\mathbf{T})$ $(C_{\pm}(\mathbf{T}))$. In the space $C^{n \times n}(\mathbf{T})$ we use the norm

$$\|A\| = \max_{1 \le j,k \le n} \|a_{jk}\| \tag{3.1}$$

where $A(t) = [a_{jk}(t)]_{j,k=1}^n$ and $\|a_{jk}\|$ is defined by (2.1).

Denote by $GC^{n \times n}(\mathbf{T})$ $\left(GC_{\pm}^{n \times n}(\mathbf{T})\right)$ the set of all matrix functions $A(t)$ from $C^{n \times n}(\mathbf{T})$ $\left(C_{\pm}^{n \times n}(\mathbf{T})\right)$ such that det $A(t) \in GC(\mathbf{T})$ $(GC_{\pm}(\mathbf{T}))$.

Let $A(t) \in GC^{n \times n}(\mathbf{T})$. The *right factorization* of $A(t)$ (with respect to $\mathbf{T}$) is its representation in the form

$$A(t) = A_-(t)D(t)A_+(t)$$

where

$$A_{\pm}(t) \in GC_{\pm}^{n \times n}(\mathbf{T}), \quad D(t) = \mathrm{diag}[t^{r_1}, t^{r_2}, \ldots, t^{r_n}]$$

and $r_1, r_2, \ldots, r_n$ are integers which are called the *right indices* (or right partial indices) of $A(t)$.

The definition of left factorization and left indices is similar. We will not mention them in what follows. Note that all results of this paper remain true for the left indices as well.

It is well known that any sufficiently smooth (e.g., Hölder continuous) matrix function admits a right factorization. It is possible to define the right indices for any matrix function $A(t) \in C^{n \times n}(\mathbf{T})$ (and even for more wider class of matrix functions) but we do not need this definition. As a rule, we will use the notion of right indices only for rational matrix functions.

Denote by $\mathcal{R}^{n \times m}$ $\left(\mathcal{R}_{\pm}^{n \times m}\right)$ the set of all $n \times m$ matrix functions with elements from $\mathcal{R}$ $(\mathcal{R}_{\pm})$. In the case $m = n$ we denote

$$GR^{n \times n} = \mathcal{R}^{n \times n} \cap GC^{n \times n}(\mathbf{T}), \quad GR_{\pm}^{n \times n} = \mathcal{R}_{\pm}^{n \times n} \cap GC_{\pm}^{n \times n}(\mathbf{T}).$$

3.2. We consider here a model example of 2×2 matrix function degenerate on $\mathbf{T}$:

$$Q(t) = \begin{bmatrix} 1 & 0 \\ 0 & t-1 \end{bmatrix}. \tag{3.2}$$

Lemma 3.1. *For arbitrary natural numbers k and l in any neighborhood of the matrix function $Q(t)$ there exists a matrix function $H(t) \in GR^{2 \times 2}$ such that the pair of the right indices of $H(t)$ is $(l, -k)$.*

Proof. By Lemma 2.4 the function $t - 1$ can be approximated by a function of the form $g(t)(t-1)^{3l+1}$ where $g(t) \in GR$ and ind $g(t) = 0$. Therefore it is enough to approximate the matrix function

$$W(t) = \begin{bmatrix} 1 & 0 \\ 0 & (t-1)^{3l+1} \end{bmatrix}. \tag{3.2}$$

by a matrix function $H(t) \in GR^{2 \times 2}$ with right indices $(l, -k)$.

As in Section 2, denote

$$\alpha_\epsilon(t) = \frac{t-1-\epsilon}{t-1+\epsilon} \quad (0 < \epsilon < 1).$$

Let

$$B_\epsilon(t) = \begin{bmatrix} 1 & \delta(\epsilon)\alpha_\epsilon^{k+3l+1}(t) \\ \delta(\epsilon)\alpha_\epsilon^{-3l-1}(t) & \alpha_\epsilon^k(t)(t-1-\epsilon)^{3l+1} \end{bmatrix}$$

where the positive function $\delta(\epsilon)$, which tends to 0 if $\epsilon \to 0$, will be defined later. It follows from (2.3) and (2.5) that $\|B_\epsilon - W\| \to 0$ if $\epsilon \to 0$. The matrix function $B_\epsilon(t)$ can be factorized as follows:

$$B_\epsilon(t) = \begin{bmatrix} 1 & \frac{\delta(\epsilon)}{(t-1+\epsilon)^{3l+1}} \\ 0 & 1 \end{bmatrix} \begin{bmatrix} g_\epsilon(t) & 0 \\ 0 & (t-1+\epsilon)^{-k} \end{bmatrix} \begin{bmatrix} 1 & 0 \\ \frac{\delta(\epsilon)(t-1+\epsilon)^k}{\alpha_\epsilon^{3l+1}(t)} & (t-1-\epsilon)^{k+3l+1} \end{bmatrix}$$

where $g_\epsilon(t) = 1 - \delta^2(\epsilon)(t-1-\epsilon)^{-3l-1}$. Since ind $[(t-1+\epsilon)^{-k}] = -k$ it is enough to define the function $\delta(\epsilon)$ in such a way that ind $g_\epsilon(t) = l$. Obviously

$$\mathrm{ind}\, g_\epsilon(t) = \mathrm{ind}\left[(t-1-\epsilon)^{3l+1} - \delta^2(\epsilon)\right]. \tag{3.3}$$

Denote $r(\epsilon) = [\delta(\epsilon)]^{2/(3l+1)}$. Polynomial

$$p_\epsilon(t) = (t-1-\epsilon)^{3l+1} - \delta^2(\epsilon) = (t-1-\epsilon)^{3l+1} - r^{3l+1}(\epsilon)$$

has roots

$$t_j = 1 + \epsilon + r(\epsilon)\exp\frac{2\pi i j}{3l+1} \quad (j = 0, 1, \ldots, 3l)$$

which are located on the circle with the center $1+\epsilon$ and radius $r(\epsilon)$. We will choose the radius $r(\epsilon)$ in such a way that exactly l roots t_j will lie inside of the unit disk, and there will not be any root on the unit circle. By (3.3) in this case the required equality ind $g_\epsilon(t) = l$ will hold.

It is clear that the mentioned properties of the radius $r(\epsilon)$ will be satisfied if the angle θ between the two radii of the circle $|z - 1 - \epsilon| = r(\epsilon)$ directed to the points of its intersection with the unit circle equals $2\pi/3$. Of course, we assume that these circles intersect, i.e. $r(\epsilon) > \epsilon$. It is easy to check that

$$\cos\frac{\theta}{2} = \frac{2\epsilon + \epsilon^2 + r^2(\epsilon)}{2(1+\epsilon)r(\epsilon)}.$$

Hence, for

$$r(\epsilon) = \frac{1}{2}(1 + \epsilon - \sqrt{1 - 6\epsilon - 3\epsilon^2})$$

we obtain $\cos\theta/2 = 1/2$ and $\theta = 2\pi/3$. Since $r(\epsilon) \to 0$ if $\epsilon \to 0$ the lemma is proved. □

Lemma 3.2. *Let* $s \in \mathbf{Z}$ *and*

$$Q_s(t) = \begin{bmatrix} t^s & 0 \\ 0 & t - 1 \end{bmatrix}.$$

If $k, l \in \mathbf{Z}$ *and* $k + s > 0$, $l - s > 0$ *then in any neighborhood of* $Q_s(t)$ *there exists a matrix function* $H_s(t) \in G\mathcal{R}^{2\times 2}$ *such that the pair of right indices of* $H_s(t)$ *is* $(l, -k)$.

Proof. By Lemma 2.2 we can approximate the function $t - 1$ by a function of the form $(t - 1)a_-a_+t^s$ where $a_\pm \in G\mathcal{R}_\pm$. Then the matrix function

$$F_s(t) = \begin{bmatrix} t^s & 0 \\ 0 & (t-1)a_-a_+t^s \end{bmatrix}$$

approximates $Q_s(t)$. Obviously

$$F_s(t) = t^s \begin{bmatrix} 1 & 0 \\ 0 & a_- \end{bmatrix} \begin{bmatrix} 1 & 0 \\ 0 & t-1 \end{bmatrix} \begin{bmatrix} 1 & 0 \\ 0 & a_+ \end{bmatrix}.$$

By Lemma 3.1 in any neighborhood of the matrix function (3.2) there exists a matrix function $H(t) \in G\mathcal{R}^{2\times 2}$ with right indices $(l - s, -s - k)$. Then the matrix function

$$H_s(t) = t^s \begin{bmatrix} 1 & 0 \\ 0 & a_- \end{bmatrix} H(t) \begin{bmatrix} 1 & 0 \\ 0 & a_+ \end{bmatrix} =$$

$$= t^s \begin{bmatrix} 1 & 0 \\ 0 & a_- \end{bmatrix} H_-(t) \begin{bmatrix} t^{l-s} & 0 \\ 0 & t^{-s-k} \end{bmatrix} H_+(t) \begin{bmatrix} 1 & 0 \\ 0 & a_+ \end{bmatrix} =$$

$$= \begin{bmatrix} 1 & 0 \\ 0 & a_- \end{bmatrix} H_-(t) \begin{bmatrix} t^l & 0 \\ 0 & t^{-k} \end{bmatrix} H_+(t) \begin{bmatrix} 1 & 0 \\ 0 & a_+ \end{bmatrix}$$

has all the required properties. □

4. Main lemma

Lemma 4.1. *Let* $A(t) \in C^{n \times n}(\mathbf{T})$ *and* $\det A(1) = 0$. *In any neighborhood of the matrix function* $A(t)$ *there exists a matrix function* $B(t) \in \mathcal{R}^{n \times n}$ *which admits representation*

$$B(t) = B_-(t)D(t)B_+(t)$$

where $B_\pm(t) \in G\mathcal{R}_\pm^{n \times n}$,

$$D(t) = \text{diag}[t^{m_1}, t^{m_2}, \dots, t^{m_{n-1}}, t - 1]$$

and $m_j \in \mathbf{Z}$ $(j = 1, 2, \dots, n-1)$.

Proof. We use induction on n. If $n = 1$ the assertion follows from Lemma 2.5 (for $k = 0$). For arbitrary positive number ϵ choose a matrix function $E(t) \in \mathcal{R}^{n \times n}$ such that $\|A - E\| < \epsilon/2$. Denote $S(t) = E(t) - E(1) + A(1)$. Then $\|A - S\| < \epsilon$ and $S(1) = A(1)$. There exists a polynomial matrix $P(t)$ such that $\det P(t) \equiv 1$ and the matrix function $R(t) = S(t)P(t)$ is lower triangular (see, e.g., [6], Section 4.7). If $R(t) = [r_{jk}(t)]_{j,k=1}^n$, then at least one of the numbers $r_{jj}(1) = 0$. At first consider the case when at least one of the numbers $r_{22}(1), \dots, r_{nn}(1)$ equals 0. Then in the representation

$$\begin{bmatrix} r_{11} & 0 \\ M & U \end{bmatrix}$$

the $(n-1) \times (n-1)$ matrix function $U(t)$ satisfies the condition $\det U(1) = 0$. By inductive hypothesis we can approximate $U(t)$ by a matrix function of the form $U_-(t)D_0(t)U_+(t)$, where $U_\pm(t) \in G\mathcal{R}_\pm^{(n-1) \times (n-1)}$ and

$$D_0(t) = \text{diag}[t^{s_1}, \dots, t^{s_{n-2}}, t - 1] \quad (s_j \in \mathbf{Z}).$$

We can also assume that $r_{11}(t) \neq 0$ ($|t| = 1$). Otherwise we replace r_{11} with a close rational function having this property . Hence we approximate $R(t)$ by a matrix function of the form $F_-(t)L(t)F_+(t)$ where

$$F_\pm = \begin{bmatrix} 1 & 0 \\ 0 & U_\pm \end{bmatrix}, \quad L = \begin{bmatrix} r_{11} & 0 \\ U_-^{-1}M & D_0 \end{bmatrix}.$$

Rewrite $L(t)$ in another form:

$$L = \begin{bmatrix} S & 0 \\ Y & t-1 \end{bmatrix}$$

where $S(t) \in G\mathcal{R}^{(n-1) \times (n-1)}$. Let $S(t) = S_-(t)D_1(t)S_+(t)$ be the right factorization of matrix function $S(t)$ and let p be its minimal right index. By Corollary 1 we can approximate the function $t - 1$ by a function of the form $a_-a_+t^p(t - 1)$ where $a_\pm \in G\mathcal{R}_\pm$. Hence $R(t)$ is approximated by a matrix function of the form

$$H(t) = t^p F_- \begin{bmatrix} S_- & 0 \\ 0 & a_- \end{bmatrix} \begin{bmatrix} D_2 & 0 \\ W & t-1 \end{bmatrix} \begin{bmatrix} S_+ & 0 \\ 0 & a_+ \end{bmatrix} F_+$$

where $D_2 = \mathrm{diag}[t^{q_1},\ldots,t^{q_{n-1}}]$ and all integers q_k are non-negative. We can represent the vector function $W(t)$ in the form $W = W_- + W_+$ where $W_\pm \in \mathcal{R}_\pm^{1\times(n-1)}$ and $W_+(1) = 0$. Then

$$\begin{bmatrix} D_2 & 0 \\ W & t-1 \end{bmatrix} = \begin{bmatrix} I & 0 \\ Z_- & 1 \end{bmatrix} \begin{bmatrix} D_2 & 0 \\ 0 & t-1 \end{bmatrix} \begin{bmatrix} I & 0 \\ Z_+ & 1 \end{bmatrix}$$

where $Z_- = W_- D_2^{-1} \in \mathcal{R}_-^{1\times(n-1)}$ and $Z_+ = W_+(t-1)^{-1} \in \mathcal{R}_+^{1\times(n-1)}$. Hence

$$H = \widetilde{B}_- \begin{bmatrix} t^p D_2 & 0 \\ 0 & t^p(t-1) \end{bmatrix} \widetilde{B}_+,$$

where $\widetilde{B}_\pm \in GR_\pm^{n\times n}$. By Lemma 2.2 we can approximate function $t^p(t-1)$ by a function of the form $f_-(t)f_+(t)(t-1)$ where $f_\pm \in GR_\pm$. Then we obtain the required approximation $B(t)$ of $A(t)$ where $B_- = \widetilde{B}_-\,\mathrm{diag}\,[1,\ldots,1,f_-]$, $B_+ = \mathrm{diag}\,[1,\ldots,1,f_+]\widetilde{B}_+ P^{-1}$ and

$$D = \begin{bmatrix} t^p D_2 & 0 \\ 0 & t-1 \end{bmatrix}.$$

Now consider the second case, when $r_{jj}(1) \neq 0$ for $j = 2,\ldots,n$ and hence $r_{11}(1) = 0$. Then

$$R = \begin{bmatrix} Q & 0 \\ K & u \end{bmatrix},$$

where $Q \in \mathcal{R}^{(n-1)\times(n-1)}$ and $\det Q(1) = 0$. By inductive hypothesis we can approximate $Q(t)$ by a matrix function of the form $Q_-(t)D'(t)Q_+(t)$ where $Q_\pm(t) \in GR_\pm^{(n-1)\times(n-1)}$ and

$$D'(t) = \mathrm{diag}[t^{d_1},\ldots,t^{d_{n-2}},t-1] \quad (d_k \in \mathbf{Z}).$$

Then

$$\begin{bmatrix} Q_- D' Q_+ & 0 \\ K & u \end{bmatrix} = \begin{bmatrix} Q_- & 0 \\ 0 & 1 \end{bmatrix} V \begin{bmatrix} Q_+ & 0 \\ 0 & 1 \end{bmatrix}$$

where

$$V = \begin{bmatrix} D' & 0 \\ KQ_+^{-1} & u \end{bmatrix}.$$

If $n > 2$ then the matrix function $V(t)$ satisfies the condition of the first case, because in $V(t)$ the element $v_{n-1,n-1}(t) = t-1$ and $n-1 \geq 2$. Thus it remains to prove our assertion for the matrix function of second order

$$V = \begin{bmatrix} t-1 & 0 \\ x & u \end{bmatrix}$$

where $u(t) \neq 0$. Since

$$\begin{bmatrix} 0 & 1 \\ 1 & 0 \end{bmatrix} V \begin{bmatrix} 0 & 1 \\ 1 & 0 \end{bmatrix} = \begin{bmatrix} u & x \\ 0 & t-1 \end{bmatrix} \tag{4.1}$$

it is enough to consider the last matrix. Let ind $u = \kappa$. By Lemma 2.2 we can approximate matrix (4.1) by matrices of the form

$$W = \begin{bmatrix} u & x \\ 0 & (t-1)g \end{bmatrix}.$$

where ind $g = \kappa$. Let $u = u_- t^\kappa u_+$, $g = g_- t^\kappa g_+$ be the factorizations of u and g. Then

$$W = \begin{bmatrix} u_- & 0 \\ 0 & g_- \end{bmatrix} \begin{bmatrix} t^\kappa & yt^\kappa \\ 0 & (t-1)t^\kappa \end{bmatrix} \begin{bmatrix} u_+ & 0 \\ 0 & g_+ \end{bmatrix} =$$

$$\begin{bmatrix} u_- & 0 \\ 0 & g_- \end{bmatrix} \begin{bmatrix} 1 & \frac{y_-}{t-1} \\ 0 & 1 \end{bmatrix} \begin{bmatrix} t^\kappa & 0 \\ 0 & (t-1)t^\kappa \end{bmatrix} \begin{bmatrix} 1 & y_+ \\ 0 & 1 \end{bmatrix} \begin{bmatrix} u_+ & 0 \\ 0 & g_+ \end{bmatrix},$$

where $y = u_-^{-1} g_+^{-1} t^{-\kappa} x$ is represented in the form $y = y_- + y_+$ with $y_\pm \in \mathcal{R}_\pm$ and $y_-(1) = 0$. It remains to approximate the function $(t-1)t^\kappa$ by a function of the form $h(t)(t-1)$ with ind $h = 0$. This is possible by Lemma 2.2.   □

5. Main result

5.1. For a vector $a = (a_j)_1^n \in \mathbf{R}^n$ we denote by $a^* = (a_j^*)_1^n$ its decreasing rearrangement: $a_1^* \geq a_2^* \geq \cdots a_n^*$. The following definition belongs to Hardy, Littlewood and Polya [5].

Let $a, b \in \mathbf{R}^n$. We say that the vector a is *majorized* by the vector b ($a \prec b$) if

$$\sum_{j=1}^n a_j = \sum_{j=1}^n b_j \text{ and } \sum_{j=1}^k a_j^* \leq \sum_{j=1}^k b_j^* \ (k = 1, 2, \ldots, n-1).$$

This important notion has many aplications in various areas of mathematics (see [8]).

In the proof of the main theorem we will use the following result by I. Gohberg and M. Krein [3].

Theorem 5.1. *Let $A(t) \in G\mathcal{R}^{n \times n}$ and let r be the collection of its right indices. If $s \in \mathbf{Z}^n$ and $s \prec r$ then for any $\epsilon > 0$ there exists a matrix function $A_\epsilon(t) \in G\mathcal{R}^{n \times n}$ such that $\|A - A_\epsilon\| < \epsilon$ and the collection of the right indices of $A_\epsilon(t)$ is s.*

Note that in [3] the inverse result was also obtained, but we will not use it. Another remark is that in [3] both $A(t)$ and $A_\epsilon(t)$ belong to some wider class of matrix functions. It is easy to see by inspecting the proof from [3] that in the case $A(t) \in \mathcal{R}^{n \times n}$ the matrix function $A_\epsilon(t)$ can be also chosen from $\mathcal{R}^{n \times n}$.

5.2. Now we can obtain the main result.

Theorem 5.2. *Let $A(t) \in C^{n \times n}(\mathbf{T})$ and $\det A(t_0) = 0$ for some $t_0 \in \mathbf{T}$. For arbitrary collection of n integers $s = (s_1, \ldots, s_n)$ in any neighborhood of $A(t)$ there exists a matrix function $S(t) \in G\mathcal{R}^{n \times n}$ such that the collection of right indices of $S(t)$ is s.*

Proof. Without loss of generality we may assume that $t_0 = 1$. By Lemma 4.1 in any neighborhood of $A(t)$ there exists a matrix function $B(t) \in \mathcal{R}^{n \times n}$ which admits representation $B(t) = B_-(t)D(t)B_+(t)$ where $B_\pm(t) \in G\mathcal{R}_\pm^{n \times n}$ and

$$D(t) = \text{diag}[t^{m_1}, \ldots, t^{m_{n-1}}, t - 1] \quad (m_k \in \mathbf{Z}).$$

By Lemma 3.2 we can approximate the 2×2 matrix function

$$\begin{bmatrix} t^{m_{n-1}} & 0 \\ 0 & t - 1 \end{bmatrix}$$

by a matrix function of the form

$$M_-(t) \begin{bmatrix} t^l & 0 \\ 0 & t^{-k} \end{bmatrix} M_+(t)$$

where $M_\pm(t) \in G\mathcal{R}_\pm^{2 \times 2}$ and l, k are arbitrary integers satisfying the relations

$$l - m_{n-1} > 0, \quad k + m_{n-1} > 0. \tag{5.1}$$

Then the matrix function

$$Q_{lk}(t) =$$
$$= B_-(t) \begin{bmatrix} I_{n-2} & 0 \\ 0 & M_-(t) \end{bmatrix} \text{diag}[t^{m_1}, \ldots, t^{m_{n-2}}, t^l, t^{-k}] \begin{bmatrix} I_{n-2} & 0 \\ 0 & M_+(t) \end{bmatrix} B_+(t)$$

approximates $A(t)$, and its collection of right indices equals $(m_1, \ldots, m_{n-2}, l, -k)$.

Demonstrate that for given vector $s \in \mathbf{Z}^n$ it is possible to choose numbers l and k in such a way that

$$s \prec (m_1, \ldots, m_{n-2}, l, -k). \tag{5.2}$$

In fact it is enough to choose arbitrary sufficient large number l and to put

$$k = l + \sum_{j=1}^{n-2} m_j - \sum_{j=1}^n s_j. \tag{5.3}$$

For example, if

$$l > \sum_{j=1}^n |s_j| + \sum_{j=1}^{n-1} |m_j| + \max_{1 \le j \le n-2} |m_j| \tag{5.4}$$

and k is defined by (5.3) then it is easy to check that the conditions (5.1) and (5.2) are satisfied. (The last term in (5.4) guarantees inequalities $-k < m_j, 1 \le j \le n-2$, which we use in verification of (5.2).)

By theorem of Gohberg and Krein (see Theorem 5.1 above) in any neighborhood of the matrix function $Q_{lk}(t)$ (for l and k satisfying (5.1) and (5.2)) there exists a matrix function $S(t) \in G\mathcal{R}^{n \times n}$ such that its collection of right indices is s. $\qquad \square$

Acknowledgements. This research was supported by the Israel Science Foundation founded by the Israel Academy of Sciences and Humanities.

References

[1] Clancey, K., Gohberg, I., *Factorization of matrix functions and singular integral operators*, Birkhäuser, Basel 1981.

[2] Gohberg, I. C., Feldman, I.A., *Convolution equations and projection methods for their solutions*, Amer. Math. Soc., Providence, R.I. 1974.

[3] Gohberg, I. C., Krein, M. G., Systems of integral equations on a half line with kernels depending on the difference of arguments, *Amer. Math. Soc. Transl. (2)* **14** (1960), 217-287.

[4] Hagen, R., Roch, S., Silbermann, B., *Spectral theory of approximation methods for convolution equations*, Birkhäuser, Basel 1995.

[5] Hardy, G. H., Littlewood, J. E., Polya, G., *Inequalities*, Cambridge University Press, Cambridge 1934.

[6] Lancaster, P., *Theory of Matrices*, Academic Press, New York and London 1969.

[7] Litvinchuk, G. S., Spitkovskii, I. M., *Factorization of measurable matrix functions*, Birkhäuser, Basel 1995.

[8] Marshall, A. W., Olkin, I., *Inequalities: Theory of majorization and its applications*, Academic Press, New York 1979.

I. Feldman and N. Krupnik
Dept. of Math. and Comp. Sci.
Bar-Ilan University
Ramat-Gan, 52900
Israel

A. Markus
Dept. of Mathematics
Ben-Gurion University of the Negev
Beer-Sheva, 84105
Israel

1991 Mathematics Subject Classification. 47A68

Received October 2, 2000

Operator Theory:
Advances and Applications, Vol. 130, 197–207
© 2001 Birkhäuser Verlag Basel/Switzerland

Finite Section Method for Difference Equations

I. Gohberg, M.A. Kaashoek and F. van Schagen

*Dedicated to Peter Lancaster, an outstanding colleague and
a good friend, on the occasion of his seventieth birthday*

Abstract. A finite section method is developed for linear difference equations over an infinite time interval. A necessary and sufficient condition is given in order that the solutions of such equations may be obtained as limits of solutions of corresponding equations over a finite time interval. Both the time-variant and the time-invariant case are considered. For the time-invariant case the condition reduces to the requirement that two subspaces defined in terms of the equations should be complementary. The results obtained extend those derived earlier for linear ordinary differential equations.

1. Introduction

In this paper solutions of linear difference equations over an infinite time interval are obtained as limits of solutions of corresponding equations over a finite time interval. Consider the equation

$$
\begin{cases}
x_{n+1} - A_n x_n = f_n, & n = 0, 1, 2, \ldots, \\
P_0 x_0 = 0,
\end{cases}
\tag{1.1}
$$

and a corresponding equation over a finite time interval

$$
\begin{cases}
x_{n+1} - A_n x_n = f_n, & n = 0, 1, 2, \ldots, N-1, \\
P_0 x_0 + Q_N x_N = 0.
\end{cases}
\tag{1.2}
$$

Our aim is to obtain the solutions of (1.1) as limits of solutions of (1.2) for $N \to \infty$. As one may expect this is not always possible, and in this paper we give a condition, which is necessary and sufficient for this approximation to work. The results obtained are discrete analogues of those derived earlier for linear ordinary differential equations [7], [8].

To state our main theorem, let us first list our hypotheses on the coefficients and boundary value matrices of (1.1) and (1.2). We shall refer to these hypotheses as our *standing conditions*. The coefficients $A_0, A_1, A_2, \ldots$ form a bounded sequence of invertible $r \times r$ matrices, and the initial value matrix P_0 is an exponential dichotomy for the homogeneous equation

$$x_{n+1} - A_n x_n = 0, \quad n = 0, 1, 2, \ldots. \tag{1.3}$$

The latter means (see [3] or [6], page 686) that P_0 is a projection and there exists constants $M \geq 0$ and $0 \leq a < 1$ such that

$$\|U_n P_0 U_m^{-1}\| \leq Ma^{n-m} \quad (n \geq m \geq 0), \tag{1.4}$$

$$\|U_n (I - P_0) U_m^{-1}\| \leq Ma^{m-n} \quad (m \geq n \geq 0). \tag{1.5}$$

where $U_0 = I$ (the $r \times r$ identity matrix) and $U_n = A_{n-1} A_{n-2} \cdots A_1 A_0$ for $n \geq 1$. The fact that P_0 is an exponential dichotomy implies (see Proposition 2.2 below and the paragraph preceding it) that for each $f = (f_0, f_1, f_2, \ldots)$ in ℓ_r^2 the equation (1.1) has a unique solution $x = (x_0, x_1, x_2, \ldots)$ in ℓ_r^2. For each N the boundary value matrix Q_N in (1.2) is assumed to be such that rank $P_0 +$ rank $Q_N = r$. We also assume that for N sufficiently large equation (1.2) has a unique solution for each right hand side $f_0, f_1, \ldots, f_{N-1}$. The latter happens if and only if the boundary conditions in (1.2) are well-posed for N sufficiently large, i.e., we assume that for some N_0

$$\det(P_0 + Q_N U_N) \neq 0 \quad (N \geq N_0). \tag{1.6}$$

Now, let $f = (f_0, f_1, f_2, \ldots) \in \ell_r^2$ be given. Our aim is to approximate the unique solution

$$x = (x_0, x_1, \ldots) \quad \text{of (1.1)}$$

in ℓ_r^2 by the solution

$$x^{(N)} = (x_0^{(N)}, x_1^{(N)}, \ldots, x_N^{(N)}) \quad \text{of (1.2)}.$$

More precisely, we look for conditions guaranteeing that $x^{(N)}$ converges in the ℓ^2 norm to x, i.e.,

$$\sum_{\nu=0}^{N} \|x_\nu^{(N)} - x_\nu\|^2 \to 0 \quad (N \to \infty).$$

If this happens for each right hand side $f = (f_0, f_1, f_2, \ldots) \in \ell_r^2$ we shall say that for (1.1) the *finite section method relative to the boundary value matrices* $\{Q_N\}$ *converges in* ℓ^2 . Our main theorem is the following result.

Theorem 1.1. *Assume that our standing conditions on equations (1.1) and (1.2) are fulfilled. In particular, P_0 is an exponential dichotomy for (1.3), and for the sequence of boundary value matrices $\{Q_N\}_{N=0}^{\infty}$ condition (1.6) holds. Then for equation (1.1) the finite section method relative to the boundary value matrices $\{Q_N\}$ converges in ℓ^2 if and only if*

$$\sup_{N \geq N_0} \|(P_0 U_N^{-1} + Q_N)^{-1} Q_N U_N P_0 U_N^{-1}\| < \infty. \tag{1.7}$$

From the above theorem it follows that, given the exponential dichotomy P_0, one can always choose the boundary value matrices $\{Q_N\}$ in (1.2) such that the corresponding finite section method converges. In fact, if $Q_N = (I - P_0)U_N^{-1}$ for each N, then (1.7) holds.

Theorem 1.1 is specified further for the time invariant case, when the coefficients and the boundary value matrices do not depend on N, i.e., $A_N = A$ and $Q_N = Q$ for all N. In this case, we show that condition (1.7) is equivalent to the requirement

$$\mathcal{M} \oplus \operatorname{Ker} Q = \mathbb{C}^r, \tag{1.8}$$

where $\mathcal{M}$ is the spectral subspace consisting of all eigenvectors and generalized eigenvectors of A corresponding to eigenvalues inside the open unit disk.

The paper consists of three sections, the first being the present introduction. In the second section we prove Theorem 1.1. The third section deals with the time invariant case.

2. Proof of Theorem 1.1

The proof of Theorem 1.1 will be based on several intermediate results. We begin with a lemma (cf., [5], Proposition I.1.1) that will be repeatedly used in the sequel.

Lemma 2.1. *Let U be an invertible $r \times r$ matrix. If P is a projection and Q is an operator on $\mathbb{C}^r$ such that* $\operatorname{rank} P + \operatorname{rank} Q = r$ *and $P + QU$ is invertible, then $(PU^{-1} + Q)^{-1}Q$ is the projection of $\mathbb{C}^r$ along $\operatorname{Ker} Q$ onto $\operatorname{Ker} PU^{-1}$.*

Proof. If $Qx = 0$ and $PU^{-1}x = 0$, then $(P + QU)U^{-1}x = 0$ and thus $x = 0$. Since also $\operatorname{rank} P + \operatorname{rank} Q = r$, this implies that $\operatorname{Ker} Q \oplus \operatorname{Ker} PU^{-1} = \mathbb{C}^r$. Now for $x \in \mathbb{C}^r$ we can write $x = y + z$ with $y \in \operatorname{Ker} Q$ and $z \in \operatorname{Ker} PU^{-1}$. Then $(PU^{-1} + Q)^{-1}Qx = (PU^{-1} + Q)^{-1}Qz = z$ because $(PU^{-1} + Q)z = Qz$. Thus $(PU^{-1} + Q)^{-1}Q$ is the projection along $\operatorname{Ker} Q$ onto $\operatorname{Ker} PU^{-1}$. $\square$

Notice that in the above lemma the fact that P is a projection is not essential.

We shall need some auxiliary spaces and operators. By W we shall denote the subspace of ℓ_r^2 consisting of all vectors $x = (x_0, x_1, \ldots)$ in ℓ_r^2 such that $x_0 \in \operatorname{Ker} P_0$. Obviously, W is closed in ℓ_r^2, and W endowed with the ℓ_r^2-norm is a Hilbert space in its own right. We define T to be the operator from W into ℓ_r^2 given by

$$T \begin{pmatrix} x_0 \\ x_1 \\ x_2 \\ \vdots \end{pmatrix} = \begin{pmatrix} x_1 - A_0 x_0 \\ x_2 - A_1 x_1 \\ x_3 - A_2 x_2 \\ \vdots \end{pmatrix}. \tag{2.1}$$

Since the sequence $A_0, A_1, A_2, \ldots$ is bounded in norm, T is a bounded operator.

The operator T defined by (2.1) is associated to (1.1) in a natural way. For instance, the statement that for each $(f_0, f_1, f_2, \ldots)$ in ℓ_r^2 the equation (1.1) has a unique solution $(x_0, x_1, x_2, \ldots)$ in ℓ_r^2 is equivalent to the requirement that T is invertible. The next proposition, which is a special case of Theorem 1.1 in [1], shows that our standing conditions imply the invertibility of T.

Proposition 2.2. *Assume that $A_0, A_1, A_2, \ldots$ is a bounded sequence of invertible matrices, and let P_0 be an exponential dichotomy for (1.3). Then the operator T in (2.1) is boundedly invertible.*

Proof. We shall derive the proposition as a corollary of Theorem 1.1 in [1]. As before, let $U_0 = I$ and $U_n = A_{n-1}A_{n-2}\cdots A_1 A_0$ for $n \geq 1$. Put $P_n = U_n P_0 U_n^{-1}$ for $n = 0, 1, 2, \ldots$. Then by (1.4) and (1.5) the operators $P_0, P_1, P_2, \ldots$ form a bounded sequence of projections of constant rank. Since $A_n P_n = P_{n+1} A_n$, we have

$$A_n \mathrm{Im}\, P_n = \mathrm{Im}\, P_{n+1}, \quad A_n \mathrm{Ker}\, P_n = \mathrm{Ker}\, P_{n+1}.$$

Together with (1.4) and (1.5), this shows that $\{P_n\}$ is a dichotomy of (1.3) in the sense of [1].

To derive our proposition as a corollary of Theorem 1.1 in [1] it remains to prove that $\{P_n\}$ is a normal dichotomy of (1.3). For this purpose it is sufficient to show that

$$\sup_{n=0,1,2,\ldots} \|A_n^{-1}|_{\mathrm{Ker}\, P_{n+1}}\| < \infty. \tag{2.2}$$

Take $y \in \mathrm{Ker}\, P_{n+1}$. Then

$$A_n^{-1} y = A_n^{-1}(I - P_{n+1})y = A_n^{-1} U_{n+1}(I - P_0) U_{n+1}^{-1} y = U_n(I - P_0) U_{n+1}^{-1} y.$$

So $\|A_n^{-1} y\| \leq Ma\|y\|$ by (1.5), and hence (2.2) is proved. $\qquad\square$

Next we introduce the operator associated with the equation (1.2). First notice that $P_0 x_0 + Q_N x_N = 0$ if and only if $P_0 x_0 = 0$ and $Q_N x_N = 0$. To see this we use that $P_0 x_0 + Q_N x_N = 0$ if and only if

$$(P_0 U_N^{-1} + Q_N)^{-1} P_0 U_N^{-1} U_N x_0 + (P_0 U_N^{-1} + Q_N)^{-1} Q_N x_N = 0,$$

and thus it follows from Lemma 2.1 that $U_N x_0 \in \mathrm{Ker}\, P_0 U_N^{-1}$ and $x_N \in \mathrm{Ker}\, Q_N$. Now put

$$V_N = \mathrm{Ker}\, P_0 \oplus (\mathbb{C}^r)^{N-1} \oplus \mathrm{Ker}\, Q_N \subset (\mathbb{C}^r)^{N+1}.$$

Define $T_N : V_N \to (\mathbb{C}^r)^N$ by setting $(T_N x)_k = x_{k+1} - A_k x_k$ for $k = 0, \ldots, N-1$, where $x = (x_0, x_1, \ldots, x_{N-1}, x_N) \in V_N$. Then $T_N x = f$ for $x \in V_N$ if and only if x is the solution of (1.2). Clearly the space V_N depends on the particular choice of the boundary value matrix Q_N. If we choose $Q_N = (I - P_0) U_N^{-1}$, then we write $V_{S,N}$ in place of V_N and $T_{S,N}$ in place of T_N, where the index S stands for the word special.

Finally, let $L_N : l_r^2 \to l_r^2$ be the natural projection onto the first N components. Whenever convenient we will identify the range of L_N with $(\mathbb{C}^r)^N$.

Proposition 2.3. *Under our standing conditions the operator T_N is invertible. Furthermore, for each $f \in \ell_r^2$ the vector $T_{S,N}^{-1} L_N f$ is equal to $L_{N+1} T^{-1} L_N f$. In particular, if for each N we choose $Q_N = (I - P_0) U_N^{-1}$, then for (1.1) the finite section method relative to the boundary value matrices $\{Q_N\}$ converges in ℓ^2.*

Proof. Since $\operatorname{Ker} P_0 U_N^{-1} \oplus \operatorname{Ker} Q_N = \mathbb{C}^r$, we have $\dim V_N = r$. So to prove that T_N is invertible, it suffices to show that $\operatorname{Ker} T_N$ is trivial. Assume $x = (x_0, x_1, x_2, \ldots, x_N) \in V_N$ and $T_N x = 0$. Then $x_j = U_j x_0$ for $j = 0, 1, \ldots, N$. Hence $(P_0 + Q_N U_N)x_0 = P_0 x_0 + Q_N x_N = 0$, and thus $x_0 = 0$ by (1.6). This shows that T_N is invertible.

Fix $f = (f_0, f_1, f_2, \ldots)$ in ℓ_r^2, and set $x = T^{-1}f$. Then $x = (x_0, x_1, x_2, \ldots)$ is the unique solution of (1.1). Using this connection, one verifies by direct substitution that

$$(T^{-1}f)_k = \sum_{\ell=0}^{k-1} U_k P_0 U_{\ell+1}^{-1} f_\ell - \sum_{\ell=k}^{\infty} U_k (I - P_0) U_{\ell+1}^{-1} f_\ell, \quad k = 0, 1, \ldots.$$

Next, consider (1.2) with $Q_N = (I - P_0)U_N^{-1}$. Put $f^{(N)} = L_N f = (f_0, \ldots, f_{N-1})$, and set $x^{(N)} = T_{S,N}^{-1} f^{(N)}$. Then $x^{(N)} = (x_0^{(N)}, \ldots, x_N^{(N)})$ is the unique solution of (1.2), and hence by direct verification one shows that

$$(T_{S,N}^{-1} L_N f)_k = \sum_{\ell=0}^{k-1} U_k P_0 U_{\ell+1}^{-1} f_\ell - \sum_{\ell=k}^{N-1} U_k (I - P_0) U_{\ell+1}^{-1} f_\ell, \quad k = 0, 1, \ldots, N.$$

$$(2.3)$$

The second statement of the proposition is immediate from these formulas.

To prove the final statement, first note that by the result of the previous paragraph

$$L_{N+1} T^{-1} f - T_{S,N}^{-1} L_N f = L_{N+1} T^{-1} (I - L_N) f.$$

Since $\|L_N\| = 1$, the operator T^{-1} is bounded, and $L_N f \to f$ if $N \to \infty$, it follows that $L_{N+1} T^{-1}(I - L_N)f \to 0$ if $N \to \infty$. Hence

$$\lim_{N \to \infty} \|L_{N+1} T^{-1} f - T_{S,N}^{-1} L_N f\| = 0,$$

which completes the proof. $\square$

The next lemma is well-known (see, for instance, [4], Theorem II.2.1, or [2], Proposition 3.2); for the sake of completeness we include its proof.

Lemma 2.4. *Let $\mathcal{X}$ and $\mathcal{X}'$ be Banach spaces, and let $P_1, P_2, \ldots$ be a sequence of projections on $\mathcal{X}$ and $P_1', P_2', \ldots$ on $\mathcal{X}'$. Assume that both sequences converge pointwise to the respective identity operators. Let A be an invertible operator from $\mathcal{X}$ onto $\mathcal{X}'$, and for each N let B_N be an invertible operator from $\operatorname{Im} P_N$ onto $\operatorname{Im} P_N'$ such that*

$$\lim_{N \to \infty} B_N P_N x = A x, \qquad x \in \mathcal{X}, \tag{2.4}$$

Then, in order that

$$\lim_{N \to \infty} B_N^{-1} P_N' x' = A^{-1} x', \qquad x' \in \mathcal{X}', \tag{2.5}$$

it is necessary and sufficient that $\sup_N \|B_N^{-1}\| < \infty$.

Proof. The necessity part is a straightforward application of the principle of uniform boundedness. To prove the sufficiency, assume that $\sup_N \|B_N^{-1}\| < \infty$. Fix $x' \in \mathcal{X}'$. Put $x = A^{-1}x'$ and $z_N = B_N^{-1}P_N'x'$. We have to show that $z_N \to x$ if $N \to \infty$. Notice that

$$B_N z_N = P_N'x' = P_N'Ax.$$

Since $(I - P_N')Ax \to 0$ if $N \to \infty$, we can use (2.4) to show that

$$
\begin{aligned}
B_N(z_N - P_N x) &= P_N'Ax - B_N P_N x \\
&= Ax - B_N P_N x - (I - P_N')Ax \\
&\to 0 \quad (N \to \infty).
\end{aligned}
$$

From $\sup_N \|B_N^{-1}\| < \infty$ we now conclude that $z_N - P_N x \to 0$ if $N \to \infty$. But $P_N x \to x$ if $N \to \infty$, and therefore $z_N \to x$ if $N \to \infty$, which completes the proof. $\qquad\square$

Proof of Theorem 1.1. In what follows we identify a vector $(x_0, \ldots, x_N)$ in $(\mathbb{C}^r)^{N+1}$ with the vector $(x_0, \ldots, x_N, 0, 0, \ldots)$ in ℓ_r^2. This allows us to view V_N as a subspace of W. Let L_N' be the projection of W onto V_N given by

$$L_N'(x_0, x_1, \ldots) = (x_0, \ldots, x_{N-1}, (I - Q_N')x_N, 0, \ldots),$$

where Q_N' is the orthogonal projection with $\operatorname{Ker} Q_N' = \operatorname{Ker} Q_N$. Then the sequence of projections $\{L_N'\}_{N=1}^\infty$ converges pointwise to the identity on W. Also $\{L_N\}_{N=1}^\infty$ converges pointwise to the identity on ℓ_r^2 and $T_N : \operatorname{Im} L_N' \to \operatorname{Im} L_N$. Furthermore,

$$
(Tx)_k - (T_N L_N'x)_k = \begin{cases} 0 & k = 0, \ldots, N-2, \\ Q_N'x_N & k = N-1, \\ x_{k+1} - A_k x_k & k = N, N+1, \ldots, \end{cases}
$$

which gives that $\lim_{N\to\infty} T_N L_N'x = Tx$. Therefore, we are in the position to apply Lemma 2.4, and conclude that $\lim_{N\to\infty} T_N^{-1}L_N f = T^{-1}f$ for each $f \in \ell_r^2$ if and only if the sequence $\|T_N^{-1}\|$ is bounded. In the special case when $Q_N = (I-P_0)U_N^{-1}$, we already know that $\lim_{N\to\infty} T_{S,N}^{-1}L_N f = T^{-1}f$. Therefore, it follows that $\|T_{S,N}^{-1}\|$ is a bounded sequence. We conclude that the finite section method relative to the boundary value matrices $\{Q_N\}$ converges if and only if

$$\sup_{N \geq N_0} \|T_{S,N}^{-1} - T_N^{-1}\| < \infty, \tag{2.6}$$

where we compute the difference $T_{S,N}^{-1} - T_N^{-1}$ by considering both V_N and $V_{S,N}$ as subspaces of $(\mathbb{C}^r)^{N+1}$. We will finish the proof by showing that (1.7) is equivalent to (2.6).

For $x = (x_0, \ldots, x_N) \in (\mathbb{C}^r)^{N+1}$ put $(M_N x)_0 = P_0 x_0 + Q_N x_N$, and $(M_N x)_{k+1} = x_{k+1} - A_k x_k$ for $k = 0, 1, \ldots, N-1$. In the case when $Q_N = (I-P_0)U_N^{-1}$, we write $M_{S,N}$ in stead of M_N. Notice that for $x \in V_N$ we have that $M_N x = (0, T_N x)$, and similarly $M_{S,N}x = (0, T_{S,N}x)$ for $x \in V_{S,N}$. From the block matrix representation of $M_N : (\mathbb{C}^r)^{N+1} \to (\mathbb{C}^r)^{N+1}$ one obtains that M_N is invertible if and only if

$P_0 + Q_N U_N$ is invertible. Thus for $f \in (\mathbb{C}^r)^N$ one gets $T_N^{-1} f = M_N^{-1}(0, f)$. So for each $f \in (\mathbb{C}^r)^N$ it follows that

$$(T_{S,N}^{-1} - T_N^{-1})f = (M_{S,N}^{-1} - M_N^{-1})(0, f) = M_N^{-1}(M_N - M_{S,N})M_{S,N}^{-1}(0, f).$$

Observe that for any $x \in (\mathbb{C}^r)^{N+1}$ one has $((M_N - M_{S,N})x)_k = 0$ if $k \geq 1$ and $((M_N - M_{S,N})x)_0 = (Q_N - (I - P_0)U_N^{-1})x_N$. Hence, using (2.3), we obtain

$$((M_N - M_{S,N})M_{S,N}^{-1}(0, f))_0$$

$$= ((Q_N - (I - P_0)U_N^{-1}) \left(U_N P_0 U_1^{-1} \quad \cdots \quad U_N P_0 U_N^{-1} \right) f,$$

and the other coordinates of $((M_N - M_{S,N})(M_{S,N}^{-1}(0, f))$ are equal to 0. From the block matrix representation of M_N one obtains that $M_N^{-1}(M_N - M_{S,N})M_{S,N}^{-1}(0, f)$ is equal to

$$\begin{pmatrix} I \\ U_1 \\ \vdots \\ U_N \end{pmatrix} (P_0 + Q_N U_N)^{-1}(Q_N - (I - P_0)U_N^{-1}) \left(U_N P_0 U_1^{-1} \quad \cdots \quad U_N P_0 U_N^{-1} \right) f.$$

Notice that $(I - P_0)U_N^{-1}U_N P_0 = 0$ and $U_N P_0 U_N^{-1} U_N P_0 = P_0$. Furthermore, from Lemma 2.1 it follows that $(P_0 + Q_N U_N)^{-1} Q_N U_N$ is a projection onto $\text{Ker } P_0$, and hence

$$(I - P_0)(P_0 + Q_N U_N)^{-1} Q_N U_N = (P_0 + Q_N U_N)^{-1} Q_N U_N.$$

Thus we find that $T_N^{-1} f - T_{S,N}^{-1} f$ is equal to

$$\begin{pmatrix} I \\ U_1 \\ \vdots \\ U_N \end{pmatrix} (I - P_0)U_N^{-1}(P_0 U_N^{-1} + Q_N)^{-1} Q_N U_N P_0 U_N^{-1} \left(U_N P_0 U_1^{-1} \cdots U_N P_0 U_N^{-1} \right) f.$$

Since P_0 is an exponential dichotomy, in order for $T_{S,N}^{-1} - T_N^{-1}$ to be a norm bounded sequence it is sufficient that $(P_0 U_N^{-1} + Q_N)^{-1} Q_N U_N P_0 U_N^{-1}$ is a norm bounded sequence.

Conversely, if the sequence $T_{S,N}^{-1} - T_N^{-1}$ is bounded in norm, we use that $T_{S,N}^{-1} f - T_N^{-1} f$ is also equal to

$$\begin{pmatrix} U_N^{-1} \\ U_1 U_N^{-1} \\ \vdots \\ I \end{pmatrix} (P_0 U_N^{-1} + Q_N)^{-1} Q_N \left(U_N P_0 U_1^{-1} \quad \cdots \quad U_N P_0 U_N^{-1} \right) f.$$

In particular it follows that the right lower corner of the matrix that we apply to f must be bounded in N. Hence, $(P_0 U_N^{-1} + Q_N)^{-1} Q_N U_N P_0 U_N^{-1}$ is norm bounded. $\square$

3. Time invariant case

In this section we consider the time-invariant case, i.e., the case when the coefficients and boundary value matrices in (1.1) and (1.2) do not depend on n. So we consider the equation

$$\begin{cases} x_{n+1} - Ax_n = f_n, & n = 0, 1, 2, \ldots, \\ P_0 x_0 = 0, \end{cases} \tag{3.1}$$

and a corresponding equation over a finite time interval

$$\begin{cases} x_{n+1} - Ax_n = f_n, & n = 0, 1, 2, \ldots, N-1, \\ P_0 x_0 + Q x_N = 0. \end{cases} \tag{3.2}$$

We assume that A is an invertible $r \times r$ matrix and P_0 is an exponential dichotomy for the associated homogeneous equation

$$x_{n+1} - Ax_n = 0, \quad n = 0, 1, 2, \ldots. \tag{3.3}$$

In this time-invariant setting the latter just means that A has no eigenvalue on the unit circle, and that P_0 is a projection of $\mathbb{C}^r$ such that $\operatorname{Im} P_0 = \operatorname{Im} P_A$, where P_A is the spectral projection of A corresponding to the eigenvalues in the unit disk. We say that for (3.1) *the finite section method relative to the boundary value matrix Q converges in ℓ^2* if for each $f = (f_0, f_1, f_2, \ldots) \in \ell_r^2$ the unique solution $x = (x_0, x_1, x_2, \ldots)$ of (3.1) in ℓ_r^2 is obtained as the limit in ℓ^2 of the solution $x^{(N)} = (x_0^{(N)}, x_1^{(N)}, \ldots, x_N^{(N)})$ of (3.2), i.e.,

$$\sum_{\nu=0}^{N} \|x_\nu^{(N)} - x_\nu\|^2 \to 0 \quad (N \to \infty).$$

We shall prove the following theorem.

Theorem 3.1. *Let P_0 be an exponential dichotomy for (3.3), and assume that Q is a projection of $\mathbb{C}^r$ with the property that $\operatorname{rank} P_0 + \operatorname{rank} Q = r$, and that $P_0 + QA^N$ is invertible for $N \geq N_0$. Then for (3.1) the finite section method relative to the boundary value matrix Q converges in ℓ^2 if and only if*

$$\operatorname{Ker} P_A \oplus \operatorname{Ker} Q = \mathbb{C}^r.$$

Theorem 3.1 is an immediate corollary of Theorem 1.1 and the next proposition.

Proposition 3.2. *Let P_0 be an exponential dichotomy for (3.3), and assume that Q is a projection of $\mathbb{C}^r$ with the property that $\operatorname{rank} P_0 + \operatorname{rank} Q = r$, and that $P_0 + QA^N$ is invertible for $N \geq N_0$. Then the following are equivalent:*

(a) $\sup_{N \geq N_0} \|(P_0 A^{-N} + Q)^{-1} Q\| < \infty$;

(b) $\sup_{N \geq N_0} \|(P_0 A^{-N} + Q)^{-1} Q A^N P_0 A^{-N}\| < \infty$;

(c) $\operatorname{Ker} P_A \oplus \operatorname{Ker} Q = \mathbb{C}^r$.

Here P_A is the spectral projection of A corresponding to the eigenvalues in the unit disk.

For the proof of Proposition 3.2 it will be convenient to use the following lemma.

Lemma 3.3. *Let $\mathbb{C}^r = \mathcal{X}_1 \oplus \mathcal{X}_2$ be a direct sum decomposition, and for $i = 1, 2$ let $\mathcal{M}_i$ be a subspace of $\mathbb{C}^r$ with $\dim \mathcal{M}_i = \dim \mathcal{X}_i$. Then $\mathcal{M}_1$ and $\mathcal{M}_2$ can be represented in the following way*

$$\mathcal{M}_1 = \mathrm{Ker} \begin{pmatrix} S_1 & S_2 \end{pmatrix}, \qquad \mathcal{M}_2 = \mathrm{Im} \begin{pmatrix} R_1 \\ R_2 \end{pmatrix},$$

where

(a) *the mapping $\begin{pmatrix} S_1 & S_2 \end{pmatrix}$ from $\mathcal{X}_1 \oplus \mathcal{X}_2$ into $\mathcal{X}_2$ is surjective,*

(b) *the mapping $\begin{pmatrix} R_1 \\ R_2 \end{pmatrix}$ from $\mathcal{X}_2$ into $\mathcal{X}_1 \oplus \mathcal{X}_2$ is injective.*

Furthermore, $\mathbb{C}^r = \mathcal{M}_1 \oplus \mathcal{M}_2$ if and only if $S_1 R_1 + S_2 R_2$ is invertible, and in this case the projection Γ of $\mathbb{C}^r$ along $\mathcal{M}_1$ onto $\mathcal{M}_2$ is given by

$$\Gamma = \begin{pmatrix} R_1 \\ R_2 \end{pmatrix} (S_1 R_1 + S_2 R_2)^{-1} \begin{pmatrix} S_1 & S_2 \end{pmatrix}.$$

Proof of Proposition 3.2. We begin with some preliminaries. Decompose the space $\mathbb{C}^r$ as $\mathbb{C}^r = \mathrm{Im}\, P_A \oplus \mathrm{Ker}\, P_A$. With respect to this decomposition we write P_A, P_0 and A as operator matrices

$$P_A = \begin{pmatrix} I & 0 \\ 0 & 0 \end{pmatrix}, \quad P_0 = \begin{pmatrix} I & R \\ 0 & 0 \end{pmatrix}, \quad A = \begin{pmatrix} A_1 & 0 \\ 0 & A_2 \end{pmatrix}. \tag{3.4}$$

Notice that A_1 has all its eigenvalues in the open unit disk and A_2 has all its eigenvalues outside the closure of the unit disk. From the operator matrix representations in (3.4) it follows that

$$\mathrm{Ker}\, P_0 A^{-N} = \mathrm{Im} \begin{pmatrix} -A_1^N R A_2^{-N} \\ I \end{pmatrix}. \tag{3.5}$$

In particular, $\dim \mathrm{Ker}\, P_0 A^{-N} = \dim \mathrm{Ker}\, P_A$.

From Lemma 2.1 we know that $(P_0 A^{-N} + Q)^{-1} Q$ is the projection along $\mathrm{Ker}\, Q$ onto $\mathrm{Ker}\, P_0 A^{-N}$. Hence,

$$\dim \mathrm{Ker}\, Q = r - \dim \mathrm{Ker}\, P_0 A^{-N} = \dim \mathrm{Im}\, P_A.$$

According to the first part of Lemma 3.3, this allows us to represent $\mathrm{Ker}\, Q$ as $\mathrm{Ker}\, Q = \mathrm{Ker} \begin{pmatrix} S & T \end{pmatrix}$, where $\begin{pmatrix} S & T \end{pmatrix}$ is a mapping from $\mathrm{Im}\, P_A \oplus \mathrm{Ker}\, P_A$ into $\mathrm{Ker}\, P_A$ and this mapping is surjective. By combining this with (3.5), we can use the second part of Lemma 3.3 to show that

$$(P_0 A^{-N} + Q)^{-1} Q = \begin{pmatrix} -A_1^N R A_2^{-N} \\ I \end{pmatrix} (T - S A_1^N R A_2^{-N})^{-1} \begin{pmatrix} S & T \end{pmatrix}. \tag{3.6}$$

We are now ready to prove the the equivalence of (a), (b) and (c).

(c) $\Rightarrow$ (a). Since

$$\mathrm{Ker}\, Q = \mathrm{Ker}\, \begin{pmatrix} S & T \end{pmatrix}, \qquad \mathrm{Ker}\, P_A = \mathrm{Im}\, \begin{pmatrix} 0 \\ I \end{pmatrix},$$

Lemma 3.3 shows that (c) is equivalent to the invertibility of T. Assume (c) holds. Since

$$\lim_{N \to \infty} A_1^N R A_2^{-N} = 0, \tag{3.7}$$

the invertibility of T and (3.6) yield

$$\lim_{N \to \infty} (P_0 A^{-N} + Q)^{-1} Q = \begin{pmatrix} 0 \\ I \end{pmatrix} T^{-1} \begin{pmatrix} S & T \end{pmatrix},$$

which proves (a). The implication (a) $\Rightarrow$ (b) is trivial.

(b) $\Rightarrow$ (c). Assume (b) holds. From (3.6) we see that

$$(P_0 A^{-N} + Q)^{-1} Q A^N P_0 A^{-N}$$

$$= \begin{pmatrix} -A_1^N R A_2^{-N} \\ I \end{pmatrix} (T - S A_1^N R A_2^{-N})^{-1} \begin{pmatrix} S & S A_1^N R A_2^{-N} \end{pmatrix}.$$

Property (b) gives that in particular the left lower component of this block matrix, i.e. $(T - S A_1^N R A_2^{-N})^{-1} S$, is uniformly bounded with respect to N for N sufficiently large. It follows that

$$\sup_{N \ge N_0} \| (T - S A_1^N R A_2^{-N})^{-1} \begin{pmatrix} S & T \end{pmatrix} \begin{pmatrix} I & -A_1^N R A_2^{-N} \\ 0 & I \end{pmatrix} \| < \infty. \tag{3.8}$$

Now recall that $\begin{pmatrix} S & T \end{pmatrix}$ is surjective. So (3.8) implies that $(T - S A_1^N R A_2^{-N})^{-1}$ is uniformly bounded in N for N sufficiently large. By (3.7) the latter can only happen when T is invertible. Indeed, assume $Tx = 0$. Then, using (3.7) and the fact that $(T - S A_1^N R A_2^{-N})^{-1}$ is bounded, we have

$$x = (T - S A_1^N R A_2^{-N})^{-1} S A_1^N R A_2^{-N} x \to 0 \quad (N \to \infty).$$

Hence $x = 0$ and T is invertible. But the invertibility of T is equivalent with (c), and therefore (c) is proved. $\qquad \square$

References

[1] Ben-Artzi, A., Gohberg, I., Kaashoek, M.A., Invertibility and dichotomy of singular difference equations, in: *Topics in Operator Theory. Ernst D. Hellinger Memorial Volume* (Eds. L. de Branges, I. Gohberg and J. Rovnyak), Birkhäuser Verlag, Basel, **OT 48** (1990), pp. 157–184.

[2] Böttcher, A., Infinite Matrices and Projection Methods, in: *Lectures on Operator Theory and Its Applications* (Ed. P. Lancaster), Amer. Math. Soc., Providence (RI), (1996), pp. 1–72.

[3] Coffman, C.V., Schäffer, J.J., Dichotomies for linear difference equations, *Math. Ann.* **172** (1967), 139–166.

[4] Gohberg, I., Feldman, I.A., *Convolution Equations and Projection Methods for Their Solution*, Amer. Math. Soc. Transl. of Math. Monographs 41, Providence (RI), (Russian Original: Nauka, Moscow, 1971), 1974.

[5] Gohberg, I., Kaashoek, M.A., Time varying systems with boundary conditions and integral operators, I. The transfer operator and its properties, *Integral Equations and Operator Theory* **7** (1984), 325–391.

[6] Gohberg, I., Kaashoek, M.A., Schagen, F. van, Non-compact integral operators with semi-separable kernels and their discrete analogues: inversion and Fredholm properties, *Integral Equations and Operator Theory* **7** (1984), 642–703.

[7] _____ , Finite section method for linear ordinary differential equations, *Journal of Differential Equations* **163** (2000), 312–334.

[8] _____ , Finite section method for linear ordinary differential equations on the full line, *submitted for publication* (2000), –.

School of Mathematical Sciences,
Tel Aviv University
Ramat Aviv 69978
Israel
gohberg@math.tau.ac.il

Division of Mathematics and Computer Science,
Vrije Universiteit
De Boelelaan 1081a
1081 HV Amsterdam
The Netherlands
kaash@cs.vu.nl, freek@cs.vu.nl

1991 Mathematics Subject Classification.
Primary 39A10, 15A06; Secondary 39A12, 45L05, 47A50

Received October 2, 2000

Operator Theory:
Advances and Applications, Vol. 130, 209–221
© 2001 Birkhäuser Verlag Basel/Switzerland

Iterative Solution of a Matrix Riccati Equation Arising in Stochastic Control

Chun-Hua Guo

Dedicated to Peter Lancaster on the occasion of his 70th birthday

Abstract. We consider iterative methods for finding the maximal Hermitian solution of a matrix Riccati equation arising in stochastic control. Newton's method is very expensive when the size of the problem is large. A much less expensive iteration is introduced and shown to have several convergence properties similar to those of Newton's method. In ordinary situations, the convergence of the new iteration is linear while the convergence of Newton's method is quadratic. In extreme cases, the convergence of the new iteration may be sublinear while the convergence of Newton's method may be linear. We also show how the performance of Newton's method can be improved when its convergence is not quadratic.

1. Introduction

Let $\mathcal{H}$ be the linear space of all $n \times n$ Hermitian matrices over the field $\mathbb{R}$. For any $X, Y \in \mathcal{H}$, we write $X \geq Y$ (or $Y \leq X$) if $X - Y$ is positive semidefinite. For any $A \in \mathbb{C}^{n \times n}$, the spectrum of A will be denoted by $\sigma(A)$. The transpose and the conjugate transpose of A will be denoted by A^T and A^*, respectively. We denote by $\mathbb{C}_<$ (resp. $\mathbb{C}_\leq$) the set of complex numbers with negative (resp. nonpositive) real parts. A matrix A is said to be *stable* if $\sigma(A) \subset \mathbb{C}_<$. For any matrices $A, B, C \in \mathbb{C}^{n \times n}$, the pair (A, B) is *stabilizable* if $A - BK$ is stable for some $K \in \mathbb{C}^{n \times n}$. The pair (C, A) is *detectable* if (A^*, C^*) is stabilizable.

In this paper, we are concerned with the numerical solution of the matrix Riccati equation

$$\mathcal{R}(X) = A^*X + XA + \Pi(X) + C - XDX = 0, \qquad (1.1)$$

where $A, C, D \in \mathbb{C}^{n \times n}$, $C^* = C$, $D^* = D$, $D \geq 0$, and Π is a *positive* linear operator from $\mathcal{H}$ into itself, i.e., $\Pi(X) \geq 0$ whenever $X \geq 0$. The Riccati function $\mathcal{R}$ is thus a mapping from $\mathcal{H}$ into itself.

Matrix Riccati equations of this type were first studied by W. M. WONHAM [11]. The following result is a slight different presentation of a result in [11]. It establishes the existence of a positive semidefinite solution to (1.1) under some additional conditions.

Theorem 1.1. *If $C \geq 0$, (A, D) is stabilizable, (C, A) is detectable, and*

$$\inf_{K \in \mathcal{H}} \left\| \int_0^\infty e^{t(A-DK)^*} \Pi(I) e^{t(A-DK)} dt \right\| < 1, \qquad (1.2)$$

where $\| \cdot \|$ is the spectral norm, then (1.1) has at least one solution $\hat{X} \geq 0$ such that $A - D\hat{X}$ is stable.

The above result was proved in [11] by using an iterative procedure, which is in fact the Newton iteration. For the Riccati function $\mathcal{R}$, the first Fréchet derivative of $\mathcal{R}$ at a matrix $X \in \mathcal{H}$ is a linear map $\mathcal{R}'_X : \mathcal{H} \to \mathcal{H}$ given by

$$\mathcal{R}'_X(H) = (A - DX)^* H + H(A - DX) + \Pi(H). \qquad (1.3)$$

Also the second derivative at X, $\mathcal{R}''_X : \mathcal{H} \times \mathcal{H} \to \mathcal{H}$, is given by

$$\mathcal{R}''_X(H_1, H_2) = -H_1 D H_2 - H_2 D H_1. \qquad (1.4)$$

The Newton method for the solution of (1.1) is

$$X_{i+1} = X_i - (\mathcal{R}'_{X_i})^{-1} \mathcal{R}(X_i), \quad i = 0, 1, \ldots, \qquad (1.5)$$

given that the maps $\mathcal{R}'_{X_i}$ are all invertible. In view of (1.3), the iteration (1.5) is equivalent to

$$(A - DX_i)^* X_{i+1} + X_{i+1}(A - DX_i) + \Pi(X_{i+1}) = -X_i D X_i - C, \qquad (1.6)$$
$$i = 0, 1, \ldots.$$

Newton's method has been studied recently by T. DAMM and D. HINRICHSEN [2] for a rational matrix equation which includes (1.1) as a special case. We first give a definition from [2].

Definition 1.2. *A matrix $X \in \mathcal{H}$ is called* stabilizing *for $\mathcal{R}$ if $\sigma(\mathcal{R}'_X) \subset \mathbb{C}_<$ and* almost stabilizing *if $\sigma(\mathcal{R}'_X) \subset \mathbb{C}_{\leq}$.*

When $\Pi = 0$, it is readily seen that $\sigma(\mathcal{R}'_X) \subset \mathbb{C}_<$ (resp. $\mathbb{C}_{\leq}$) if and only if $\sigma(A - DX) \subset \mathbb{C}_<$ (resp. $\mathbb{C}_{\leq}$).

A solution X_+ of (1.1) is called maximal if $X_+ \geq X$ for any solution X. The maximal solution is the most desirable solution in applications. When $\Pi = 0$, the maximal solution may be found by subspace methods (see [10], for example). However, those methods are not applicable when $\Pi \neq 0$.

The following result, given in [2], is a generalization of Theorem 9.1.1 of [9]. It shows that the maximal solution of (1.1) can be found by Newton's method under mild conditions.

Theorem 1.3. *Assume that there exist a solution $\hat{X}$ to $\mathcal{R}(X) \geq 0$ and a stabilizing matrix X_0. Then the Newton sequence is well defined and, moreover, the following are true:*

1. $X_k \geq X_{k+1}, \quad X_k \geq \hat{X}, \quad \mathcal{R}(X_k) \leq 0, \quad k \geq 1.$
2. $\sigma(\mathcal{R}'_{X_k}) \subset \mathbb{C}_<, \quad k \geq 0.$
3. $\lim_{k \to \infty} X_k = X_+$ *is the maximal solution of* (1.1).
4. $\sigma(\mathcal{R}'_{X_+}) \subset \mathbb{C}_\leq.$

Remark 1.4. If (1.2) is true, then

$$\left\| \int_0^\infty e^{t(A-DX_0)^*} \Pi(I) e^{t(A-DX_0)} dt \right\| < 1$$

for some $X_0 \in \mathcal{H}$. It is noted in [2] that this X_0 is necessarily stabilizing for $\mathcal{R}$. The assumption that $C \geq 0$ and (C, A) is detectable is not needed for the above theorem. As a result, the maximal solution is not necessarily positive semidefinite. It can be seen that Theorem 1.3 is also a generalization of Theorem 3.3 of [4].

Note that the solution of the linear equation (1.6) is required in each step of the Newton iteration. The presence of the linear operator Π on the left hand side will make solving this equation very expensive when n is large. For example, if Π is given by $\Pi(H) = B^* H B$, then we will need to solve a linear matrix equation of the form

$$A^* X + X A + B^* X B = C$$

in each step of the Newton iteration. This equation is equivalent to

$$(I \otimes A^* + A^T \otimes I + B^T \otimes B^*) \text{vec} \, X = \text{vec} \, C,$$

where $\otimes$ is the Kronecker product and the vec operator stacks the columns of a matrix into a long vector (see [9], for example). A direct solution of this equation would require $O(n^6)$ operations. On the other hand, a matrix equation of the form

$$A^* X + X A = C$$

can be solved by the Bartels-Stewart algorithm [1] in $O(n^3)$ operations when it has a unique solution.

This observation leads us to consider the iteration

$$(A - DX_i)^* X_{i+1} + X_{i+1}(A - DX_i) = -\Pi(X_i) - X_i DX_i - C, \qquad (1.7)$$
$$i = 0, 1, \dots.$$

Iteration (1.7) is obtained by replacing $\Pi(X_{i+1})$ with $\Pi(X_i)$ in iteration (1.6). A new convergence analysis will be needed for iteration (1.7).

2. Convergence of the iteration (1.7)

In this section, we will show that iteration (1.7) has several convergence properties similar to those of the Newton iteration. First, we note that iteration (1.7) can be rewritten as

$$(A - DX_i)^* (X_{i+1} - X_i) + (X_{i+1} - X_i)(A - DX_i) = -\mathcal{R}(X_i), \qquad (2.1)$$
$$i = 0, 1, \dots.$$

212 Chun-Hua Guo

We will also need the following well known result (see [9], for example).

Lemma 2.1. *Let* $A, C \in \mathbb{C}^{n \times n}$ *with* A *stable and* C *Hermitian. Then the Lyapunov equation* $A^*X + XA = C$ *has a unique solution* X *(necessarily Hermitian). If* $C \leq 0$, *then* $X \geq 0$.

Theorem 2.2. *Assume that there exist a solution* $\hat{X}$ *to* $\mathcal{R}(X) \geq 0$ *and a Hermitian matrix* X_0 *such that* $X_0 \geq \hat{X}$, $\mathcal{R}(X_0) \leq 0$, *and* $A - DX_0$ *is stable. Then the iteration* (1.7) *defines a sequence* $\{X_k\}$ *such that*

1. $X_k \geq X_{k+1}$, $X_k \geq \hat{X}$, $\mathcal{R}(X_k) \leq 0$, $k \geq 0$.
2. $\sigma(A - DX_k) \subset \mathbb{C}_<$, $k \geq 0$.
3. $\lim_{k \to \infty} X_k = \tilde{X}$ *is a solution of* (1.1) *and* $\tilde{X} \geq \hat{X}$.
4. $\sigma(A - D\tilde{X}) \subset \mathbb{C}_\leq$.

Proof. We prove by induction that for each $i \geq 0$, X_{i+1} is uniquely determined and
$$X_i \geq X_{i+1}, \quad X_i \geq \hat{X}, \quad \mathcal{R}(X_i) \leq 0, \quad \sigma(A - DX_i) \subset \mathbb{C}_<. \tag{2.2}$$
For $i = 0$, we already have $X_0 \geq \hat{X}$, $\mathcal{R}(X_0) \leq 0$, and $\sigma(A - DX_0) \subset \mathbb{C}_<$. By (2.1) with $i = 0$ and Lemma 2.1, X_1 is uniquely determined and $X_0 \geq X_1$. We now assume that X_{k+1} is uniquely determined and (2.2) is true for $i = k$ ($k \geq 0$). By (1.7) with $i = k$,
$$\begin{aligned} & (A - DX_k)^*(X_{k+1} - \hat{X}) + (X_{k+1} - \hat{X})(A - DX_k) \\ = & -\Pi(X_k) - X_k DX_k - C - A^*\hat{X} - \hat{X}A + X_k D\hat{X} + \hat{X}DX_k \\ \leq & -\Pi(X_k) - X_k DX_k + \Pi(\hat{X}) - \hat{X}D\hat{X} + X_k D\hat{X} + \hat{X}DX_k \\ = & -\Pi(X_k - \hat{X}) - (X_k - \hat{X})D(X_k - \hat{X}) \leq 0. \end{aligned}$$
Therefore, $X_{k+1} \geq \hat{X}$ by Lemma 2.1. To show that $A - DX_{k+1}$ is stable, we will use an argument in [5]. Note first that, by writing $A - DX_{k+1} = A - DX_k + D(X_k - X_{k+1})$,
$$\begin{aligned} & (A - DX_{k+1})^*(X_{k+1} - \hat{X}) + (X_{k+1} - \hat{X})(A - DX_{k+1}) \\ \leq & -\Pi(X_k - \hat{X}) - (X_k - \hat{X})D(X_k - \hat{X}) \\ & +(X_k - X_{k+1})D(X_{k+1} - \hat{X}) + (X_{k+1} - \hat{X})D(X_k - X_{k+1}) \\ = & -\Pi(X_k - \hat{X}) - (X_{k+1} - \hat{X})D(X_{k+1} - \hat{X}) \qquad (2.3) \\ & -(X_k - X_{k+1})D(X_k - X_{k+1}) \\ \leq & -(X_k - X_{k+1})D(X_k - X_{k+1}). \end{aligned}$$
If $A - DX_{k+1}$ is not stable, we let λ be an eigenvalue of $A - DX_{k+1}$ with $\text{Re}(\lambda) \geq 0$ and $(A - DX_{k+1})x = \lambda x$ for some $x \neq 0$. Now, by (2.3),
$$2\text{Re}(\lambda)x^*(X_{k+1} - \hat{X})x \leq -x^*(X_k - X_{k+1})D(X_k - X_{k+1})x.$$
Therefore, $x^*(X_k - X_{k+1})D(X_k - X_{k+1})x = 0$ and thus $D(X_k - X_{k+1})x = 0$. Now, $(A - DX_k)x = (A - DX_{k+1})x = \lambda x$, which is contradictory to the stability

of $A - DX_k$. We have thus proved that $A - DX_{k+1}$ is stable. So, X_{k+2} is uniquely determined and

$$
\begin{aligned}
& (A - DX_{k+1})^*(X_{k+1} - X_{k+2}) + (X_{k+1} - X_{k+2})(A - DX_{k+1}) \\
= {} & (A - DX_k + D(X_k - X_{k+1}))^* X_{k+1} + X_{k+1}(A - DX_k + D(X_k - X_{k+1})) \\
& + \Pi(X_{k+1}) + X_{k+1} DX_{k+1} + C \\
= {} & -\Pi(X_k - X_{k+1}) - X_k DX_k + X_{k+1} DX_{k+1} \\
& + (X_k - X_{k+1})DX_{k+1} + X_{k+1}D(X_k - X_{k+1}) \\
= {} & -\Pi(X_k - X_{k+1}) - (X_k - X_{k+1})D(X_k - X_{k+1}) \leq 0.
\end{aligned}
$$

Therefore, $X_{k+1} \geq X_{k+2}$. Since

$$
(A - DX_{k+1})^*(X_{k+1} - X_{k+2}) + (X_{k+1} - X_{k+2})(A - DX_{k+1}) = \mathcal{R}(X_{k+1}),
$$

we also get $\mathcal{R}(X_{k+1}) \leq 0$. The induction process is now complete. Thus, the sequence $\{X_k\}$ is well defined, monotonically decreasing, and bounded below by $\hat{X}$. Let $\lim_{k \to \infty} X_k = \tilde{X}$. We have $\tilde{X} \geq \hat{X}$. By taking limits in (1.7), we see that $\tilde{X}$ is a solution of (1.1). Since $\sigma(A - DX_k) \subset \mathbb{C}_<$ for each k, $\sigma(A - D\tilde{X}) \subset \mathbb{C}_\leq$. □

Remark 2.3. If, in addition, X_0 is an upper bound for the solution set of (1.1) (i.e., $X_0 \geq X$ for all solutions of (1.1)), then $\tilde{X}$ is the maximal solution of (1.1).

To further study the convergence behaviour of iteration (1.7), we need some results from [2].

We first note that $\mathcal{H}$ is a Hilbert space with the Frobenius inner product $\langle X, Y \rangle = \text{trace}(XY)$. For a linear operator $\mathcal{L}$ on $\mathcal{H}$, let $\rho(\mathcal{L}) = \max\{|\lambda| : \lambda \in \sigma(\mathcal{L})\}$ denote the spectral radius, and $\beta(\mathcal{L}) = \max\{\text{Re}(\lambda) : \lambda \in \sigma(\mathcal{L})\}$ the spectral abscissa. The identity map is denoted by I. As for matrices, $\mathcal{L}$ is called *stable* if $\sigma(\mathcal{L}) \subset \mathbb{C}_<$.

Definition 2.4. *A linear operator $\mathcal{L}$ on $\mathcal{H}$ is called* positive *if $\mathcal{L}(H) \geq 0$ whenever $H \geq 0$. $\mathcal{L}$ is called* inverse positive *if $\mathcal{L}^{-1}$ exists and is positive. $\mathcal{L}$ is called* resolvent positive *if the operator $\alpha I - \mathcal{L}$ is inverse positive for all sufficiently large $\alpha > 0$.*

Theorem 2.5. (cf. [2]) *Let $\mathcal{L} : \mathcal{H} \to \mathcal{H}$ be resolvent positive and $\Pi : \mathcal{H} \to \mathcal{H}$ be positive. Then $\mathcal{L} + \Pi$ is also resolvent positive. Moreover, the following are equivalent.*

1. *$\mathcal{L} + \Pi$ is stable.*
2. *$-(\mathcal{L} + \Pi)$ is inverse positive.*
3. *$\mathcal{L}$ is stable and $\rho(\mathcal{L}^{-1}\Pi) < 1$.*

Theorem 2.6. (cf. [2]) *If $\mathcal{L} : \mathcal{H} \to \mathcal{H}$ is resolvent positive, then $\beta(\mathcal{L}) \in \sigma(\mathcal{L})$ and there exists a nonzero matrix $V \geq 0$ such that $\mathcal{L}(V) = \beta(\mathcal{L})V$.*

As noted in [2], if $\mathcal{L}$ is resolvent positive, then the adjoint operator $\mathcal{L}^*$ is also resolvent positive and $\beta(\mathcal{L}^*) = \beta(\mathcal{L})$.

Lemma 2.7. *For any $A \in \mathbb{C}^{n \times n}$, the linear operator $\mathcal{L} : \mathcal{H} \to \mathcal{H}$ defined by*

$$\mathcal{L}(H) = A^*H + HA$$

is resolvent positive. The adjoint operator of $\mathcal{L}$ is given by

$$\mathcal{L}^*(H) = AH + HA^*.$$

Proof. The first part of the lemma is proved in [2]. For any $U, V \in \mathcal{H}$, $\langle \mathcal{L}U, V \rangle = \text{trace}(\mathcal{L}UV) = \text{trace}(A^*UV) + \text{trace}(UAV) = \text{trace}(UVA^*) + \text{trace}(UAV) = \langle U, AV + VA^* \rangle$. This proves the second part of the lemma. □

We are now ready to prove the following convergence result for iteration (1.7).

Theorem 2.8. *Assume that there exist a solution $\hat{X}$ to $\mathcal{R}(X) \geq 0$ and a Hermitian matrix X_0 such that $\mathcal{R}(X_0) \leq 0$ and $\mathcal{R}'_{X_0}$ is stable. Then the iteration (1.7) defines a sequence $\{X_k\}$ such that*

1. $X_k \geq X_{k+1}$, $X_k \geq \hat{X}$, $\mathcal{R}(X_k) \leq 0$, $k \geq 0$.
2. $\sigma(\mathcal{R}'_{X_k}) \subset \mathbb{C}_<$, $k \geq 0$.
3. $\lim_{k \to \infty} X_k = X_+$, the maximal solution of (1.1).
4. $\sigma(\mathcal{R}'_{X_+}) \subset \mathbb{C}_\leq$.

Proof. By Theorem 1.3, $X_1^N = X_0 - (\mathcal{R}'_{X_0})^{-1}\mathcal{R}(X_0) \geq \hat{X}$. Since $\mathcal{R}(X_0) \leq 0$ and $-\mathcal{R}'_{X_0}$ is inverse positive by Theorem 2.5 and Lemma 2.7, we also have $X_0 \geq X_1^N$. Thus, $X_0 \geq \hat{X}$ is necessarily true. Since $\mathcal{R}'_{X_0}$ is stable, we know from Theorem 2.5 that the operator $\mathcal{L}$ given by

$$\mathcal{L}(H) = (A - DX_0)^*H + H(A - DX_0)$$

is also stable. Thus, $A - DX_0$ is a stable matrix. Therefore, all the conclusions of Theorem 2.2 are true. Since $\lim_{k \to \infty} X_k = \tilde{X} \geq \hat{X}$ and $\hat{X}$ can be taken to be any solution of (1.1), we have $\tilde{X} = X_+$. We have thus proved items 1 and 3 of the theorem. Since item 4 follows from item 2, we need only to prove item 2. Assume that $\mathcal{R}'_{X_k}$ is stable for some $k \geq 0$. We need to prove that $\mathcal{R}'_{X_{k+1}}$ is also stable. If $\mathcal{R}'_{X_{k+1}}$ is not stable, we know from Theorem 2.6 and the note that follows it that $(\mathcal{R}'_{X_{k+1}})^*(V) = \beta V$ for some nonzero $V \geq 0$ and some number $\beta \geq 0$. Therefore,

$$\langle V, \mathcal{R}'_{X_{k+1}}(X_{k+1} - \hat{X}) \rangle = \langle \beta V, X_{k+1} - \hat{X} \rangle \geq 0.$$

On the other hand, we have by (2.3) that

$$\begin{aligned}
&\mathcal{R}'_{X_{k+1}}(X_{k+1} - \hat{X}) \\
&= (A - DX_{k+1})^*(X_{k+1} - \hat{X}) + (X_{k+1} - \hat{X})(A - DX_{k+1}) + \Pi(X_{k+1} - \hat{X}) \\
&\leq -\Pi(X_k - X_{k+1}) - (X_{k+1} - \hat{X})D(X_{k+1} - \hat{X}) \\
&\quad -(X_k - X_{k+1})D(X_k - X_{k+1}) \\
&\leq -(X_k - X_{k+1})D(X_k - X_{k+1}).
\end{aligned}$$

Therefore,
$$\langle V, (X_k - X_{k+1})D(X_k - X_{k+1})\rangle = 0.$$
So, $\text{trace}\big(V^{1/2}(X_k - X_{k+1})D^{1/2}D^{1/2}(X_k - X_{k+1})V^{1/2}\big) = 0.$

It follows that $D^{1/2}(X_k - X_{k+1})V^{1/2} = 0$ and thus $D(X_k - X_{k+1})V = 0$.
Now, by Lemma 2.7,

$$
\begin{aligned}
(\mathcal{R}'_{X_k})^*(V) &= (A - DX_k)V + V(A - DX_k)^* + \Pi^*(V) \\
&= (\mathcal{R}'_{X_{k+1}})^*(V) + D(X_{k+1} - X_k)V + V(X_{k+1} - X_k)D \\
&= (\mathcal{R}'_{X_{k+1}})^*(V) = \beta V,
\end{aligned}
$$

which is contradictory to the stability of $\mathcal{R}'_{X_k}$.   □

We will now make a comparison between Theorem 1.3 and Theorem 2.8. Note first that we need to assume $\mathcal{R}(X_0) \leq 0$ in Theorem 2.8. The Newton iteration does not need this assumption and $\mathcal{R}(X_1) \leq 0$ is necessarily true. The conclusions in Theorem 1.3 and Theorem 2.8 are almost the same. The only difference is that the first conclusion is generally not true for $k = 0$ in Theorem 1.3, since $\mathcal{R}(X_0) \leq 0$ is not assumed there. But that conclusion will be true for $k = 0$ if we also assume $\mathcal{R}(X_0) \leq 0$ in Theorem 1.3. If it is difficult to choose an X_0 with $\mathcal{R}'_{X_0}$ stable and $\mathcal{R}(X_0) \leq 0$, we may get such an X_0 by applying one Newton iteration on a Hermitian matrix Y_0 stabilizing for $\mathcal{R}$.

From the above discussions, the following conclusions can be made.

- Under the conditions of Theorem 2.8, the four conclusions in the theorem would remain valid if the sequence $\{X_k\}_{k=1}^{\infty}$ were obtained by using Newton's method and iteration (1.7) in an arbitrary combination.

- Under the conditions of Theorem 1.3, the four conclusions in the theorem would remain valid if, after X_1 has been obtained by Newton's method, the sequence $\{X_k\}_{k=2}^{\infty}$ were obtained by using Newton's method and iteration (1.7) in an arbitrary combination.

Before we can determine a good combination of the Newton iteration and the iteration (1.7), we need to have some idea about the convergence rates of these two iterations.

3. Convergence rates of the two iterations

We start with a result on the convergence rate of the Newton iteration.

Theorem 3.1. *If $\mathcal{R}'_{X_+}$ is stable in Theorem 1.3, then the convergence of Newton's method is quadratic.*

The above result was proved in [2]. It also follows directly from Theorem 1.3 and a result on the local quadratic convergence of Newton's method in general Banach spaces (see [7], for example).

For iteration (1.7), linear convergence can be guaranteed when $\mathcal{R}'_{X_+}$ is stable. This will be a consequence of the following general result.

Theorem 3.2. (cf. [8, p. 21]) *Let T be a (nonlinear) operator from a Banach space E into itself and $x^* \in E$ be a solution of $x = Tx$. If T is Fréchet differentiable at x^* with $\rho(T'_{x^*}) < 1$, then the iterates $x_{n+1} = Tx_n$ $(n = 0, 1, \ldots)$ converge to x^*, provided that x_0 is sufficiently close to x^*. Moreover, for any $\epsilon > 0$,*

$$\|x_n - x^*\| \le c(x_0; \epsilon)\big(\rho(T'_{x^*}) + \epsilon\big)^n,$$

where $\|\cdot\|$ is the norm in E and $c(x_0; \epsilon)$ is a constant independent of n.

Theorem 3.3. *Let the sequence $\{X_k\}$ be as in Theorem 2.8. If $\mathcal{R}'_{X_+}$ is stable, then*

$$\limsup_{k \to \infty} \sqrt[k]{\|X_k - X_+\|} \le \rho\big((\mathcal{L}_{X_+})^{-1}\Pi\big) < 1,$$

where $\|\cdot\|$ is any matrix norm and the operator $\mathcal{L}_X$ is defined by

$$\mathcal{L}_X(H) = (A - DX)^*H + H(A - DX).$$

Proof. The iteration (1.7) can be written as $X_{k+1} = G(X_k)$ with

$$G(X) = (\mathcal{L}_X)^{-1}(-\Pi(X) - XDX - C).$$

It can easily be shown that

$$G(X_+ + H) - G(X_+) = -(\mathcal{L}_{X_+})^{-1}\Pi(H) + o(H),$$

where $o(H)$ denotes some matrix $W(H)$ with $\lim_{\|H\| \to 0} \frac{\|W(H)\|}{\|H\|} = 0$. Therefore, the Fréchet derivative of G at the matrix X_+ is $G'_{X_+} = -(\mathcal{L}_{X_+})^{-1}\Pi$. Since $\mathcal{R}'_{X_+}$ is stable, we have $\rho\big((\mathcal{L}_{X_+})^{-1}\Pi\big) < 1$ by Theorem 2.5. Therefore,

$$\limsup_{k \to \infty} \sqrt[k]{\|X_k - X_+\|} \le \rho\big((\mathcal{L}_{X_+})^{-1}\Pi\big) < 1$$

by Theorems 2.8 and 3.2. $\qquad\square$

Therefore, when $\mathcal{R}'_{X_+}$ has no eigenvalues on the imaginary axis, the convergence of iteration(1.7) is linear while the convergence of the Newton iteration is quadratic. Next we will examine the convergence rates of the two iterations when $\mathcal{R}'_{X_+}$ has some eigenvalues on the imaginary axis.

In the case of $\Pi = 0$, the two iterations are identical and $\mathcal{R}'_{X_+}$ has eigenvalues on the imaginary axis if and only if $A - DX_+$ has eigenvalues on the imaginary axis. A convergence rate analysis has been given in [6] when all the eigenvalues of $A - DX_+$ on the imaginary axis are semisimple (i.e., all elementary divisors associated with these eigenvalues are linear). If $A - DX_+$ has non-semisimple eigenvalues on the imaginary axis, the convergence rate analysis remains an open problem.

In general, when $\mathcal{R}'_{X_+}$ has eigenvalues on the imaginary axis, we know from Theorem 2.6 that 0 must be one of these eigenvalues. Therefore, $\mathcal{R}'_{X_+}$ is not invertible. The convergence of Newton's method is typically linear in this case (see

[3], for example). If $\mathcal{L}_{X_+}$ is invertible, then $\rho\left((\mathcal{L}_{X_+})^{-1}\Pi\right) = 1$ and the convergence of iteration (1.7) is expected to be sublinear in view of Theorem 3.3. Here is one example.

Example 3.4. For the Riccati equation (1.1) with $n = 1$ and

$$A = \frac{1}{2}, \quad C = -1, \quad D = 1, \quad \Pi(X) = X,$$

it is clear that $X_+ = 1$ is the unique solution and the conditions in Theorems 1.3 and 2.8 are satisfied for any $X_0 > 1$. The iteration (1.7) is given by

$$X_{k+1} = X_k - \frac{(X_k - 1)^2}{2X_k - 1}.$$

So

$$\frac{X_{k+1} - 1}{X_k - 1} = 1 - \frac{X_k - 1}{2X_k - 1}.$$

Therefore, $\lim_{k\to\infty} \frac{X_{k+1}-1}{X_k-1} = 1$, i.e., the convergence is sublinear. The Newton iteration is given by

$$X_{k+1} = X_k - \frac{1}{2}(X_k - 1).$$

So

$$\frac{X_{k+1} - 1}{X_k - 1} = \frac{1}{2}.$$

Thus, the Newton iteration converges to X_+ linearly with rate $\frac{1}{2}$.

If both $\mathcal{R}'_{X_+}$ and $\mathcal{L}_{X_+}$ are singular (this is not very likely when $\Pi \neq 0$), then the convergence of the iteration (1.7) may be linear. However, the rate of convergence may be very close to 1, as the following example shows:

Example 3.5. Consider the Riccati equation (1.1) with $n = 2$ and

$$A = \begin{pmatrix} 0 & 0 \\ 0 & -1 \end{pmatrix}, \quad C = 0, \quad D = \begin{pmatrix} 1 & 0 \\ 0 & 0 \end{pmatrix},$$

$$\Pi(X) = \begin{pmatrix} 0 & 0 \\ 0 & \epsilon \end{pmatrix} X \begin{pmatrix} 0 & 0 \\ 0 & \epsilon \end{pmatrix},$$

where $0 \leq \epsilon < \sqrt{2}$. The conditions in Theorems 1.3 and 2.8 are satisfied for $X_0 = I$. For iteration (1.7), the iterates are

$$X_k = \begin{pmatrix} \left(\frac{1}{2}\right)^k & 0 \\ 0 & \left(\frac{\epsilon^2}{2}\right)^k \end{pmatrix}, \quad k = 1, 2, \ldots.$$

The convergence to the maximal solution $X_+ = 0$ is thus linear with rate $\max\{\frac{1}{2}, \frac{\epsilon^2}{2}\}$. For the Newton iteration, the iterates are

$$X_k = \begin{pmatrix} \left(\frac{1}{2}\right)^k & 0 \\ 0 & 0 \end{pmatrix}, \quad k = 1, 2, \ldots.$$

The convergence to the maximal solution is thus linear with rate $\frac{1}{2}$.

In summary, when $\mathcal{R}'_{X_+}$ is invertible, the convergence of iteration (1.7) is linear and the convergence of Newton's method is quadratic; when $\mathcal{R}'_{X_+}$ is not invertible, the convergence of iteration (1.7) is typically sublinear and the convergence of Newton's method is typically linear. Therefore, we should start with the much less expensive iteration (1.7) (as long as an initial guess X_0 satisfying the conditions of Theorem 2.8 is available) and switch to the Newton iteration at a later stage if the convergence of iteration (1.7) is detected to be too slow or if a very high precision is required of the approximate maximal solution. Note that *one* step of Newton iteration may be needed to find an X_0 for use with iteration (1.7), from a Hermitian matrix Y_0 such that $\mathcal{R}'_{Y_0}$ is stable. The matrix X_0 so obtained may be far away from the maximal solution X_+ and the Newton iteration could take many steps before fast convergence sets in. This makes the use of the iteration (1.7) (after one Newton iteration) particularly important.

4. Improvement of Newton's method in the singular case

From the above discussions, we can see that the Newton iteration is most useful when the convergence of iteration (1.7) is too slow, particularly when the convergence of iteration (1.7) is sublinear. However, when the convergence of iteration (1.7) is sublinear, $\mathcal{R}'_{X_+}$ is singular and the convergence of Newton's method is typically linear. Linear convergence alone is not satisfactory since the method requires a lot of computational work in each iteration. As in [6], we will show that a simple modification can improve the performance of Newton's method significantly in many cases.

We let $\mathcal{N}$ be the null space of $\mathcal{R}'_{X_+}$ and $\mathcal{M}$ be its orthogonal complement in $\mathcal{H}$. Let $P_{\mathcal{N}}$ and $P_{\mathcal{M}}$ be the orthogonal projections onto $\mathcal{N}$ and $\mathcal{M}$, respectively.

Theorem 4.1. *Let the sequence $\{X_k\}$ be as in Theorem 1.3 and, for any fixed $\theta > 0$, let*

$$Q = \{k : \|P_{\mathcal{M}}(X_k - X_+)\| > \theta\|P_{\mathcal{N}}(X_k - X_+)\|\}.$$

Then there is a constant $c > 0$ such that $\|X_k - X_+\| \leq c\|X_{k-1} - X_+\|^2$ for all sufficiently large $k \in Q$.

Proof. Let $\tilde{X}_k = X_k - X_+$. Using Taylor's Theorem with (1.4) and the fact that $\mathcal{R}'_{X_+}(P_{\mathcal{N}}\tilde{X}_k) = 0$,

$$\mathcal{R}(X_k) = \mathcal{R}(X_+) + \mathcal{R}'_{X_+}(\tilde{X}_k) + \frac{1}{2}\mathcal{R}''_{X_+}(\tilde{X}_k, \tilde{X}_k) = \mathcal{R}'_{X_+}(P_{\mathcal{M}}\tilde{X}_k) - \tilde{X}_k D\tilde{X}_k. \quad (4.1)$$

For $k \in Q$, we have $\|\tilde{X}_k\| \leq \|P_{\mathcal{M}}\tilde{X}_k\| + \|P_{\mathcal{N}}\tilde{X}_k\| \leq (\theta^{-1} + 1)\|P_{\mathcal{M}}\tilde{X}_k\|$. Since $\|\mathcal{R}'_{X_+}(P_{\mathcal{M}}\tilde{X}_k)\| \geq c_1\|P_{\mathcal{M}}\tilde{X}_k\|$ for some constant $c_1 > 0$, we have by (4.1)

$$\|\mathcal{R}(X_k)\| \geq c_1\|P_{\mathcal{M}}\tilde{X}_k\| - c_2\|\tilde{X}_k\|^2 \geq \left(c_1(\theta^{-1} + 1)^{-1} - c_2\|\tilde{X}_k\|\right)\|\tilde{X}_k\|. \quad (4.2)$$

On the other hand, we have by (1.6)

$$(A - DX_{k-1})^* X_k + X_k(A - DX_{k-1}) + \Pi(X_k) = -X_{k-1}DX_{k-1} - C,$$

and obviously,

$$(A - DX_+)^* X_+ + X_+ (A - DX_+) + \Pi(X_+) = -X_+ DX_+ - C.$$

By subtraction, we obtain after some manipulations

$$(A - DX_{k-1})^* \tilde{X}_k + \tilde{X}_k (A - DX_{k-1}) + \Pi(\tilde{X}_k) = -\tilde{X}_{k-1} D\tilde{X}_{k-1}.$$

Writing $X_+ = X_{k-1} - \tilde{X}_{k-1}$ in (4.1) and using the last equation it is found that

$$
\begin{aligned}
\mathcal{R}(X_k) &= \left((A - DX_{k-1}) + D\tilde{X}_{k-1}\right)^* \tilde{X}_k + \tilde{X}_k \left((A - DX_{k-1}) + D\tilde{X}_{k-1}\right) \\
&\quad + \Pi(\tilde{X}_k) - \tilde{X}_k D\tilde{X}_k \\
&= -\tilde{X}_{k-1} D\tilde{X}_{k-1} + \tilde{X}_{k-1} D\tilde{X}_k + \tilde{X}_k D\tilde{X}_{k-1} - \tilde{X}_k D\tilde{X}_k.
\end{aligned}
$$

Thus,

$$\|\mathcal{R}(X_k)\| \le c_3 \|\tilde{X}_k\|^2 + c_4 \|\tilde{X}_k\| \|\tilde{X}_{k-1}\| + c_5 \|\tilde{X}_{k-1}\|^2. \tag{4.3}$$

In view of (4.2) and the fact that $X_k \ne X_+$ for any k, we have

$$c_1(\theta^{-1} + 1)^{-1} - c_2 \|\tilde{X}_k\| \le c_3 \|\tilde{X}_k\| + c_4 \|\tilde{X}_{k-1}\| + c_5 \|\tilde{X}_{k-1}\|^2 / \|\tilde{X}_k\|.$$

Since $\tilde{X}_k \to 0$ by Theorem 1.3, $\|\tilde{X}_k\| \le c \|\tilde{X}_{k-1}\|^2$ for all sufficiently large $k \in Q$. □

Corollary 4.2. *Assume that, for given $\theta > 0$,*

$$\|P_{\mathcal{M}}(X_k - X_+)\| > \theta \|P_{\mathcal{N}}(X_k - X_+)\|$$

for all k large enough. Then $X_k \to X_+$ quadratically.

From the corollary, we see that the error will be dominated by the null space component at some stage if the convergence of Newton's method is not quadratic (no examples of quadratic convergence for Newton's method in the singular case have been found for the Riccati equation). We will now examine what will happen if the error is precisely in the null space.

Theorem 4.3. *Let the sequence $\{X_k\}$ be as in Theorem 1.3. If $\mathcal{R}'_{X_+}$ is singular and $X_k - X_+ \in \mathcal{N}$, then*

1. $X_{k+1} - X_+ = \frac{1}{2}(X_k - X_+).$
2. $X_+ = X_k - 2(\mathcal{R}'_{X_k})^{-1} \mathcal{R}(X_k).$

Proof. By Taylor's Theorem,

$$\mathcal{R}'_{X_k}(X_k - X_+) = \mathcal{R}'_{X_+}(X_k - X_+) + \mathcal{R}''_{X_+}(X_k - X_+, X_k - X_+).$$

Since $\mathcal{R}(X_+) = 0$ and $\mathcal{R}'_{X_+}(X_k - X_+) = 0$, we may also write

$$
\begin{aligned}
&\mathcal{R}'_{X_k}(X_k - X_+) \\
&= 2\{\mathcal{R}(X_+) + \mathcal{R}'_{X_+}(X_k - X_+) + \frac{1}{2}\mathcal{R}''_{X_+}(X_k - X_+, X_k - X_+)\} \\
&= 2\mathcal{R}(X_k).
\end{aligned}
$$

The second part of the theorem follows immediately. The first part follows easily from (1.5) and the second part. □

From this result, we know that it is possible to get a better approximation to the maximal solution by using a double Newton step when X_k approaches X_+ slowly but the error $X_k - X_+$ is rapidly dominated by its null space component.

The following example illustrates this point.

Example 4.4. Consider the Riccati equation (1.1) with $n = 2$,

$$A = \begin{pmatrix} 1 & 1 \\ 2 & -1 \end{pmatrix}, \quad C = \begin{pmatrix} -2 & -4 \\ -4 & -3 \end{pmatrix}, \quad D = \begin{pmatrix} 1 & 1 \\ 1 & 1 \end{pmatrix},$$

and

$$\Pi(X) = \begin{pmatrix} 1 & 0 \\ 1 & 2 \end{pmatrix} X \begin{pmatrix} 1 & 1 \\ 0 & 2 \end{pmatrix}.$$

It can be verified that the maximal solution of the equation is

$$X_+ = \begin{pmatrix} 2 & 0 \\ 0 & 1 \end{pmatrix}.$$

The conditions in Theorem 2.8 are satisfied for $X_0 = 10I$. For this example, $\rho((\mathcal{L}_{X_+})^{-1}\Pi) = 1$ and the convergence of iteration (1.7) is indeed sublinear. After 10000 iterations, we get an approximate maximal solution X_{10000} with

$$X_{10000} - X_+ = \begin{pmatrix} 0 & 2.9596 \times 10^{-16} \\ 2.9596 \times 10^{-16} & 4.0049 \times 10^{-4} \end{pmatrix}.$$

We could have switched to Newton's method much earlier. For example, after 40 iterations, we get X_{40} with

$$X_{40} - X_+ = \begin{pmatrix} 5.0493 \times 10^{-12} & 1.4938 \times 10^{-8} \\ 1.4938 \times 10^{-8} & 1.1968 \times 10^{-1} \end{pmatrix}.$$

If we use $X_0^N = X_{40}$ as the initial guess for the Newton iteration, we get X_{20}^N after 20 iterations with

$$X_{20}^N - X_+ = \begin{pmatrix} 0 & 0 \\ 0 & 1.1516 \times 10^{-7} \end{pmatrix}.$$

However, the double Newton step can be used to great advantage. For example, we can get X_2^N after only two Newton iterations with

$$X_2^N - X_+ = \begin{pmatrix} 0 & 2.2547 \times 10^{-11} \\ 2.2547 \times 10^{-11} & 2.9921 \times 10^{-2} \end{pmatrix}$$

and apply a double Newton step on X_2^N to get X_3^{DN} with

$$X_3^{DN} - X_+ = \begin{pmatrix} 1.7764 \times 10^{-15} & -2.1889 \times 10^{-11} \\ -2.1889 \times 10^{-11} & 6.5890 \times 10^{-11} \end{pmatrix}.$$

Note that $\mathcal{N} = \{\mathrm{diag}(0, a) : a \in \mathbb{R}\}$ for this example.

Acknowledgements. This work was supported in part by a grant from the Natural Sciences and Engineering Research Council of Canada. The author also wishes to thank Tobias Damm for sending him the reference [2].

References

[1] Bartels, R. H., Stewart, G. W., Solution of the matrix equation $AX + XB = C$, *Comm. ACM* **15** (1972), 820–826.

[2] Damm, T., Hinrichsen, D., Newton's method for a rational matrix equation occurring in stochastic control, Preprint, 1999.

[3] Decker, D. W., Keller, H. B., Kelley, C. T., Convergence rates for Newton's method at singular points, *SIAM J. Numer. Anal.* **20** (1983), 296–314.

[4] Freiling, G., Jank, G., Existence and comparison theorems for algebraic Riccati equations and Riccati differential and difference equations, *J. Dynam. Control Systems* **2** (1996), 529–547.

[5] Gohberg, I., Lancaster, P., Rodman, L., On Hermitian solutions of the symmetric algebraic Riccati equation, *SIAM J. Control Optimization* **24** (1986), 1323–1334.

[6] Guo, C.-H., Lancaster, P., Analysis and modification of Newton's method for algebraic Riccati equations, *Math. Comp.* **67** (1998), 1089–1105.

[7] Kantorovich, L. V., Akilov, G. P., *Functional Analysis in Normed Spaces*, Pergamon, New York, 1964.

[8] Krasnoselskii, M. A., Vainikko, G. M., Zabreiko, P. P., Rutitskii, Ya. B., Stetsenko, V. Ya., *Approximate Solution of Operator Equations*, Wolters-Noordhoff Publishing, Groningen, 1972.

[9] Lancaster, P., Rodman, L., *Algebraic Riccati Equations*, Clarendon Press, Oxford, 1995.

[10] Mehrmann, V. L., *The Autonomous Linear Quadratic Control Problem*, Lecture Notes in Control and Information Sciences, Vol. 163, Springer Verlag, Berlin, 1991.

[11] Wonham, W. M., On a matrix Riccati equation of stochastic control, *SIAM J. Control* **6** (1968), 681–697.

Department of Mathematics and Statistics
University of Regina
Regina, SK S4S 0A2
Canada

1991 Mathematics Subject Classification.
Primary 15A24, 93B40; Secondary 47B60, 65H10

Received March 28, 2000

Operator Theory:
Advances and Applications, Vol. 130, 223–233
© 2001 Birkhäuser Verlag Basel/Switzerland

Lyapunov Functions and Solutions of the Lyapunov Matrix Equation for Marginally Stable Systems

Wolfhard Kliem and Christian Pommer

Dedicated to Peter Lancaster

Abstract. We consider linear systems of differential equations $I\ddot{x} + B\dot{x} + Cx = 0$ where I is the identity matrix and B and C are general complex n x n matrices. Our main interest is to determine conditions for complete marginal stability of these systems. To this end we find solutions of the Lyapunov matrix equation and characterize the set of matrices (B, C) which guarantees marginal stability. The theory is applied to gyroscopic systems, to indefinite damped systems, and to circulatory systems, showing how to choose certain parameter matrices to get sufficient conditions for marginal stability. Comparison is made with some known results for equations with real system matrices. Moreover more general cases are investigated and several examples are given.

1. Introduction

Investigations of certain marginally stable systems have been studied for more than 100 years. W. THOMSON and P.G. TAIT stated in 1879 [1] that an unstable conservative system $M\ddot{x} + Kx = 0$ with real symmetric matrices $M > 0$ (positive definite) and $K \not\geq 0$ (not positive semi-definite) can be marginally stabilized by suitable gyroscopic forces $G\dot{x}$, if and only if the number of unstable degrees of freedom is even (det $K > 0$). Besides the non-spectacular stability behavior of normal conservative systems $M\ddot{x} + Kx = 0$, $M > 0, K > 0$, these gyroscopic systems with one or several parameters have for a long time been the main examples of marginally stable systems. The literature on gyroscopic systems and on their special application to rotor dynamics is numerous. In many textbooks on stability and vibrating systems, e.g. by P. LANCASTER [2], P.C. MÜLLER [3], K. HUSEYIN [4], and D.R. MERKIN [5], gyroscopic systems are standard examples, and in the special literature on the subject, e.g. by P. HAGEDORN [6], L. BARKWELL and P. LANCASTER [7], P. LANCASTER and P. ZIZLER [8], J.A. WALKER [9], R. HRYNIV et al. [10], A.P. SEYRANIAN et al. [11], and K. VESELIC [12], extensive lists of references are available.

Slightly more sophisticated than that of conservative systems is the behavior of circulatory systems $M\ddot{x} + Kx = 0$, $M > 0$, K non-symmetric, for which stability is marginal as well. A subclass consists of the so-called pseudo-conservative systems, see K. HUSEYIN [4]. Recently, a new case of marginally stable systems appeared in the literature, namely indefinite damped systems, see P. FREITAS [13]. It is therefore natural to investigate marginally stable systems in a more general sense.

Let us consider matrix differential equations of the form

$$M\ddot{x} + B\dot{x} + Cx = 0, \tag{1.1}$$

where M, B, and C are general complex n x n matrices (or linear differential operators), the vector x represents some generalized coordinates (or a deflection function), and a dot means differentiation with respect to time. Such linearized time-invariant systems are important mathematical models in e.g. solid mechanics and fluid dynamics, and in these applications M is symmetric (Hermitian) and for the most part $M > 0$, such that with only little loss of generality we will assume $M = I$, the identity matrix. It is well-known that the stability of such a system is characterized by the position of the eigenvalues λ in the complex plane, where λ are the roots of the characteristic polynomial $\det(\lambda^2 I + \lambda B + C)$. We call the system (1.1) *marginally stable* if *all* eigenvalues λ lie on the imaginary axis and are semi-simple. This case is in the literature often denoted by 'completely' (or totally) marginally stable. For simplicity we drop the word 'completely'. Marginal stability basically means that all solutions are superpositions of harmonic oscillations.

In the following stability investigations we will concentrate on Lyapunov functions and the Lyapunov matrix equation, which in essence is a generalization of energy considerations for mechanical systems. The properties of the system matrices B and C play an important role in this approach, and we avoid the cumbersome procedures connected to the computation of the characteristic polynomial and its roots. By solving the Lyapunov matrix equation, we will be able to give a complete characterization of the set of matrices (B, C) which guarantees marginal stability.

2. Lyapunov functions and the Lyapunov matrix equation

System (1.1) with $M = I$ is equivalent to the first order system

$$\dot{z} = L z, \quad L = \begin{bmatrix} 0 & I \\ -C & -B \end{bmatrix}, \quad z = \begin{bmatrix} x \\ \dot{x} \end{bmatrix}. \tag{2.1}$$

A function $V(t)$ is called a *Lyapunov function* for system (2.1) if $V(t) > 0$ and the time derivative $\dot{V}(t) \leq 0$ on all solutions of (2.1). It is obvious that Lyapunov functions can be regarded as suitable test functions for stability investigations, since the existence of such a function simply implies stability for (2.1), see e.g. [3], [4], [5] and P. LANCASTER and M. TISMENETSKY [14].

For linear mechanical systems the energy is a quadratic form in the velocities $\dot{x}$ (kinetic energy) and in the coordinates x (potential energy). Since Lyapunov functions can be considered as generalized energy expressions, it makes sense to look for $V(t)$ as a quadratic form

$$V(t) = z^* P z \tag{2.2}$$

with a Hermitian matrix $P > 0$. On the trajectories of (2.1) we then have

$$\dot{V}(t) = z^*(L^* P + PL)z, \tag{2.3}$$

such that $\dot{V}(t) \leq 0$ is expressed by the so-called *Lyapunov matrix equation*

$$L^* P + PL = -Q, \quad Q = Q^* \geq 0.$$

For further details and other stability criteria which issue from this direct method of Lyapunov, see the already mentioned literature.

3. Conditions for marginal stability

A well-known result about marginal stability, see e.g. P.C. MÜLLER [3], can now be formulated.

Theorem 3.1. *System* (2.1) *is marginally stable if and only if the homogeneous Lyapunov matrix equation*

$$L^* P + PL = 0 \tag{3.1}$$

has a Hermitian, positive definite solution $P = P^* > 0$.

According to (2.3) this means that the generalized energy function neither increases nor decreases, such that marginal stability is obvious.

Seeking solutions $P > 0$ of (3.1) we assume the form

$$P = \begin{bmatrix} R & S \\ S^* & W \end{bmatrix} \tag{3.2}$$

where $R = R^*$, $W = W^*$, and S are complex n x n matrices. Matrix P is positive definite, if and only if the Schur condition

$$R > 0 \quad \text{and} \quad W - S^* R^{-1} S > 0 \tag{3.3}$$

is satisfied.

Substituting (3.2) and (3.3) into (3.1) we get

$$
\begin{aligned}
C^* S^* + SC &= 0, & (3.4) \\
R &= SB + C^* W > 0, & (3.5) \\
WB + B^* W &= S + S^*, & (3.6) \\
W &> S^* R^{-1} S. & (3.7)
\end{aligned}
$$

Note that the matrices B and C cannot be chosen arbitrarily. The question is therefore how to characterize a set (B, C) of matrices for which the system (2.1) is marginally stable. For a non-singular matrix C, (3.4) shows that S must have the form

$$S = C^*A, \quad A = -A^*, \quad \det(C) \neq 0, \tag{3.8}$$

where A is skew-Hermitian. To simplify (3.5) - (3.7) we can write R as

$$R = C^*HC, \quad H = H^* > 0,$$

without loss of generality. Then (3.5) and (3.6) become

$$W = HC - AB, \tag{3.9}$$
$$B^*HC + C^*HB = C^*A - AC. \tag{3.10}$$

Now we want to get explicit expressions for B and C. Rewriting (3.10) as

$$C^*(HB - A) = -(B^*H + A)C, \tag{3.11}$$

we see that (3.11) can only be satisfied if there exists a skew-Hermitian matrix A_1 such that

$$HB - A = A_1C, \quad A_1 = -A_1^* = 2E \tag{3.12}$$

Solving (3.12) with respect to B results in

$$B = H^{-1}(A + A_1C). \tag{3.13}$$

The Schur condition (3.7) is then expressed by

$$W = HC - AB = (H - AH^{-1}A_1)C - AH^{-1}A > A^*H^{-1}A > 0$$

which is satisfied if and only if

$$(H - AH^{-1}A_1)C > 0.$$

From this we conclude that there must exist a positive Hermitian matrix H_1 such that

$$(H - AH^{-1}A_1) = C^*H_1, \quad H_1 = H_1^* > 0. \tag{3.14}$$

With these results we can express the matrix P in (3.2) as

$$R = C^*HC, \qquad H = H^* > 0, \quad \det(C) \neq 0, \tag{3.15}$$
$$W = C^*H_1C + A^*H^{-1}A, \quad H_1 = H_1^* > 0, \tag{3.16}$$
$$S = C^*A, \qquad A = -A^*. \tag{3.17}$$

Moreover, (3.13) and (3.14) characterize the set (B, C) of matrices for which system (2.1) is marginally stable. We formulate this in the following theorem.

Theorem 3.2. *If C is non-singular then the system* (2.1) *is marginally stable if and only if there exist two positive definite Hermitian matrices H and H_1 and two skew-Hermitian matrices A and A_1 such that B and C can be expressed as*

$$B = H^{-1}(A + A_1C), \qquad H = H^* > 0, \quad H_1 = H_1^* > 0, \tag{3.18}$$
$$C = H_1^{-1}(H - A_1H^{-1}A), \quad A = -A^*, \qquad A_1 = -A_1^*. \tag{3.19}$$

One way is simple: for any choice of H, H_1, A, A_1 with the mentioned properties, theorem 3.2 will guarantee marginal stability. The other way, how to determine H, H_1, A, A_1 for a given set (B, C) in a marginally stable system, will be demonstrated in the next sections. To this purpose it may be convenient to introduce a skew-Hermitian matrix A_2 by $A_1 = HA_2H$ and a Hermitian matrix H_2 by $H_1 = HH_2^{-1}H$ and rewrite (3.18) and (3.19) as

$$B = H^{-1}A + A_2H_2(I - A_2A), \qquad (3.20)$$
$$C = H^{-1}H_2(I - A_2A). \qquad (3.21)$$

4. Gyroscopic systems

We want to study gyroscopic systems where the matrix C is Hermitian and B is skew-Hermitian. If $C > 0$, then the system (2.1) is always stable independently of B. A special case of interest, appearing e.g. in rotor dynamics, is therefore an unstable system with $C < 0$ and $\det(C) > 0$ which we wish to stabilize by a matrix $B = -B^*$, $\det(B) \neq 0$. According to theorem 3.2 we have to find suitable matrices H, H_1, A and A_1 satisfying (3.18) and (3.19) to ensure marginal stability. It should be mentioned that stability for gyroscopic systems always means marginal stability since together with an eigenvalue λ also $-\bar{\lambda}$ is eigenvalue.

In a first attempt we choose $H_1 = I$, (identity matrix), and $A_1 = -\beta A$, A non-singular, $\beta > 0$. Then C is Hermitian and $C < 0$ leads according to (3.19) to the condition

$$H < \beta A^* H^{-1} A. \qquad (4.1)$$

Matrix B is defined by (3.18) as

$$B = H^{-1}A(I - \beta C) \qquad (4.2)$$

which is skew-Hermitian if we choose $H = (I - \beta C)^{-1}$. Then $B = H^{-1}AH^{-1}$.

If we finally insert equation (4.2) into inequality (4.1) and multiply with H^{-2} we get the following sufficient condition for marginal stability.

Theorem 4.1. *A gyroscopic system (2.1) with $C = C^* < 0$, $\det(C) > 0$ and a non-singular $B = -B^*$ is marginally stable if there exists a number $\beta > 0$ such that*

$$(I - \beta C) < \beta B^* (I - \beta C)^{-1} B. \qquad (4.3)$$

To our knowledge condition (4.3) was for the first time introduced by A.P. SEYRANIAN et al. [11] for real system matrices in a slightly different setting and without derivation. In R. HRYNIV et al. [10], inequality (4.3) was derived directly from the differential equation and used to predict a precise bound for gyroscopic stabilization. Several numerical examples to theorem 4.1 with real system matrices can be found there as well.

The following example on the stability of rotor systems shows the applicability of theorem 4.1.

Example 4.2. A non-symmetrical shaft rotates with angular velocity ω. The linearized equations of motion in a rotating system can be written as a gyroscopic system (1.1) with real system matrices which are defined as

$$M = I, \quad B = 2\omega J, \quad C = U - \omega^2 I, \quad J = \begin{bmatrix} 0 & -I \\ I & 0 \end{bmatrix}, \qquad (4.4)$$

$$U = \begin{bmatrix} U_{11} & U_{12} \\ U_{21} & U_{22} \end{bmatrix}$$

where the symmetric matrix U is not necessarily positive definite. The appearance of the parameter ω in both system matrices B and C shall be noted.

For the gyroscopic system with the matrices (4.4) we apply theorem 4.1 and assume $\beta = 1/\omega^2$. Multiplying (4.3) from left by J and from right by J^* we obtain after some calculations the following expression for the inequality

$$U + J^*UJ > -\beta U^2 (I - \beta C)^{-1}.$$

For large values of ω this condition implies marginal stability of the rotor system if

$$U + J^*UJ > 0.$$

This is in agreement with a previous result of K. VESELIC [12].

Another interesting possibility in the present gyroscopic case with $C < 0$ is the following. Consider the alternative formulation (3.20) and (3.21) and assume

$$A_2 = A^* = -A$$

and

$$H_2 = -H^{-1}(I - A^*A)^{-1}.$$

Since $H_2 > 0$ and $H > 0$, this requires

$$\begin{aligned} A^*A &> I, \\ (A^*A)H &= H(A^*A). \end{aligned}$$

For real systems the order of the matrices has to be even, since A must be non-singular.

Then (3.20) and (3.21) can be written as

$$\begin{aligned} B &= H^{-1}A + AH^{-1}, \\ C &= -H^{-2} < 0 \end{aligned}$$

which implies a stabilizing gyroscopic matrix

$$B = (-C)^{1/2}A + A(-C)^{1/2}.$$

Theorem 4.3. *Consider a gyroscopic system* (B, C) *(of even order in the case of real systems) with* $B = -B^*$ *and* $C = C^* < 0$. *Let* A *be any skew-Hermitian matrix such that*

$$A^* A > I$$

and

$$(A^* A)(-C)^{1/2} = (-C)^{1/2}(A^* A) . \qquad (4.5)$$

Then the system is marginally stable if

$$B = (-C)^{1/2} A + A(-C)^{1/2}. \qquad (4.6)$$

This result is an extension of a theorem by A.P. SEYRANIAN et al. [11], where A was chosen as a skew-Hermitian permutation matrix. Therefore the stabilizing matrix B coupled the degrees of freedom pairwise in a perfect matching, whereas our choice of matrix A opens for further possibilities.

Especially if C is chosen diagonal with distinct entries, the condition (4.5) implies that the product $A^* A$ has to be diagonal as well. This can only be fulfilled if the columns of A are orthogonal to each other. We can e.g. choose $A = \epsilon U$ where $|\epsilon| > 1$ and U is any skew-Hermitian unitary matrix, see the following example.

Example 4.4. Let $C = \mathrm{diag}[-1^2, -2^2, -3^2, -4^2]$, $A = \epsilon U$ and let the skew-Hermitian unitary matrix U be given as

$$U = \frac{1}{\sqrt{3}} \begin{bmatrix} 0 & 1 & 1 & 1 \\ -1 & 0 & -1 & 1 \\ -1 & 1 & 0 & -1 \\ -1 & -1 & 1 & 0 \end{bmatrix} .$$

Then equation (4.6) determines the stabilizing matrix

$$B = \epsilon \frac{1}{\sqrt{3}} \begin{bmatrix} 0 & 3 & 4 & 5 \\ -3 & 0 & -5 & 6 \\ -4 & 5 & 0 & -7 \\ -5 & -6 & 7 & 0 \end{bmatrix}, \quad |\epsilon| > 1.$$

5. Indefinite damped systems

We consider systems (1.1) with $M = I$ where both matrices B and C are Hermitian. It is well-known that if C is indefinte then the system is unstable. Therefore we assume $C > 0$ in this section. Consequently the system with $B = 0$ is marginally stable. But can marginal stability be the case as well if $B \neq 0$? This question was posed by P. FREITAS [13]. His investigation shows that if a real symmetric matrix C possesses only distinct positive eigenvalues and the real symmetric matrix B belongs to a certain class of indefinite matrices then there exists a positive number ϵ_o such that all pairs $(\epsilon B, C)$ result in marginal stability if $0 < \epsilon < \epsilon_o$. Nevertheless, the number ϵ_o was not determined.

Our investigation will be extended to complex system matrices. By using theorem 3.2 we want to determine a value ϵ_o for a sub class B of matrices described by P. FREITAS [13]. To this end we consider the alternative formulation (3.20) and (3.21) and assume

$$A_2 = A^* = -A$$

and

$$H_2 = H^{-1}(I - A^*A)^{-1}$$

which now requires

$$A^*A < I,$$
$$(A^*A)H = H(A^*A).$$

Then (3.20) and (3.21) can be written as

$$B = H^{-1}A - AH^{-1},$$
$$C = H^{-2} > 0$$

which implies a stabilizing indefinite Hermitian matrix

$$B = C^{1/2}A - AC^{1/2}.$$

Because trace(B) = 0, B is indefinite. Furthermore we observe using $A^2H = HA^2$ that

$$AB = -BA .$$

If A is non-singular we then have

$$B = -A^{-1}BA ,$$

which means that the spectrum of B must be Hamiltonian.

Theorem 5.1. *Consider a system* (B, C) *with* $C = C^* > 0$ *and an indefinite damping matrix* $B = B^*$. *Let* A *be any skew-Hermitian matrix such that*

$$A^*A < I$$

and

$$(A^*A)C^{1/2} = C^{1/2}(A^*A). \tag{5.1}$$

Then the system is marginally stable if

$$B = C^{1/2}A - AC^{1/2} . \tag{5.2}$$

If A *is non-singular then the spectrum of* B *is Hamiltonian.*

The similarity of this theorem with theorem 4.3 is obvious. Again the choice $A = \epsilon U$ where U is a skew-Hermitian unitary matrix and now $|\epsilon| < \epsilon_o = 1$ is especially convenient.

Example 5.2. Let $C = \text{diag}[1^2, 2^2, 3^2, 4^2]$ and assume the same matrix A as in example 4.4. Then (5.2) determines a stabilizing matrix

$$B = \epsilon \frac{1}{\sqrt{3}} \begin{bmatrix} 0 & -1 & -2 & -3 \\ -1 & 0 & 1 & -2 \\ -2 & 1 & 0 & 1 \\ -3 & -2 & 1 & 0 \end{bmatrix}, \quad |\epsilon| < \epsilon_o = 1.$$

6. Circulatory and pseudo-conservative systems

50 years ago people started to model certain problems in elasticity by circulatory systems $M\ddot{x}+Cx=0$, $M>0$, C real but non-symmetric, see e.g. H. ZIEGLER [15]. For circulatory systems stability always means marginal stability (this follows simply from the Hamiltonian symmetry of the eigenvalues λ) and a circulatory system is marginally stable if and only if $M^{-1}C$ is similar to a positive definite symmetric matrix, see e.g. P.C. MÜLLER [3]. One part of this result - the sufficiency - follows from theorem 4.1 if we choose $A=A_1=0$. (We have $M=I$, which is no loss of generality). Then $B=0$ and $C=H_1^{-1}H$ with $H_1^{-1}>0, H>0$, which implies that C is symmetrizable, that is similar to a symmetric matrix and this symmetric matrix is positive definite matrix, see O. TAUSSKY [16]. The necessity of the condition for $M^{-1}C$ follows directly from (3.5).

In the case C is symmetrizable, circulatory systems are often called pseudo-conservative, see K. HUSEYIN [4], since they behave very much like conservative systems (no flutter instability, loss of stability under a change of parameters can occur by divergence only).

7. More general systems

In sections 4 and 5 we assumed the matrix C to be Hermitian. More generally we can consider the case where C is normalizable which means that C is similar to a normal matrix N, $NN^*=N^*N$. This property is equivalent with (see e.g. C. POMMER and W. KLIEM [17])

$$C=PQ, \qquad P=P^*>0, \qquad Q^*PQ=QPQ^*.$$

Certain classes of normalizable matrices consist of the Hermitizable matrices (normalizable matrices with only real eigenvalues) and of the real symmetrizable matrices. For both these classes is $N=N^*$ and $Q=Q^*$. Now it follows directly from (3.19) that if A and A_1 are skew-Hermitian matrices and $A_1=\beta A$, then C is Hermitizable for any Hermitian matrices $H>0,\ \ H_1>0$. Moreover, (3.18) determines a matrix B such that the set (B,C) implies marginal stability of system (2.1). B will in general be the sum of a Hermitian and of a skew-Hermitian matrix.

Example 7.1. Let

$$A=\begin{bmatrix} i & 1+i \\ -1+i & 2i \end{bmatrix}, \quad H_1=\begin{bmatrix} 3 & 1-i \\ 1+i & 2 \end{bmatrix}, \quad H=\begin{bmatrix} 4 & 2+2i \\ 2-2i & 6 \end{bmatrix},$$

$A_1=-2A$. Then (3.18) and (3.19) determine a marginally stable system with matrices

$$B=\frac{1}{16}\begin{bmatrix} -6+43i & -15-3i \\ -35-15i & 6-12i \end{bmatrix}, \quad C=\frac{1}{8}\begin{bmatrix} 16+8i & -4+20i \\ -3-27i & 22-8i \end{bmatrix}.$$

Here C is Hermitizable, since its eigenvalues are real.

Example 7.2. Using (3.20) and (3.21) with

$$H = \begin{bmatrix} 4 & 2+2i \\ 2-2i & 6 \end{bmatrix} > 0, \; H_2 = \begin{bmatrix} 3 & 1-i \\ 1+i & 2 \end{bmatrix} > 0,$$

$$A = \begin{bmatrix} 5i & 1+i \\ -1+i & 4i \end{bmatrix},$$

and taking $A_2 = -A$ we can find a marginally stable system (B, C) with

$$B = \frac{1}{8} \begin{bmatrix} 2+4559i & 2183+2175i \\ -2183+2173i & -2+2376i \end{bmatrix},$$

$$C = \begin{bmatrix} -36+11i & -10+23i \\ 1-23i & -13-11i \end{bmatrix}.$$

HC is Hermitian and negative definite and commutes with A^*A. Because $H_2 > 0$ we get from (3.21) that $I - A^*A$ must be negative definite.

We now take B and C as constant matrices and consider a new system $(\epsilon B, C)$, where ϵ is a real parameter. This system can be defined from (3.20) and (3.21) if we replace A by ϵA, A_2 by $-\epsilon A$ and keep H unchanged. Then we can always find a Hermitian and positive definite matrix H_2 from (3.21) if $I - \epsilon^2 A^*A < 0$. The system is then marginally stable if $|\epsilon| > 1/3$. A numerical calculation shows that the system is marginally stable even if $|\epsilon| > 0.12965$. This limit cannot be obtained by the actual choice of H and A.

8. Conclusions

Conditions for (complete) marginal stability of linear systems with complex system matrices have been investigated. The analysis was carried out by Lyapunov's direct method, namely an inspection of the solutions of the Lyapunov matrix equation. The result was formulated in a way that comparison could be made with known criteria for gyroscopic systems with real system matrices and for rotor systems. For classical systems with positive definite stiffness matrices the theory enabled us to determine bounds for the damping matrices such that certain classes of indefinite damped systems are marginally stable. A result for circulatory systems was confirmed as well, and finally systems where the stiffness is characterized by normal matrices were considered.

Apart from gyroscopic systems, the literature on marginally stable systems is not very comprehensive. The present work wants to draw attention to several more general aspects of marginal stability. There still remain open questions, e.g. the necessity of the conditions in theorem 4.1 and in theorem 4.3. Work to clarify these problems is in progress.

Acknowledgements. The authors are indebted to Peter Lancaster for fruitful discussions about stability properties of gyroscopic systems.

References

[1] Thomson, W., Tait, P.G., *Treatise on Natural Philosophy, Vol. I*, Cambridge University Press 1879.

[2] Lancaster, P., *Lambda-matrices and Vibrating Systems*, Pergamon Press 1966.

[3] Müller, P.C., *Stabilität und Matrizen*, Springer-Verlag 1977.

[4] Huseyin, K., *Vibrations and stability of multiple parameter systems*, Noordhoff International publishing, Alphen aan den Rijn 1978.

[5] Merkin, D.R., *Introduction to the theory of stability of motion*, Nauka, Moscow 1987.

[6] Hagedorn, P., Über die Instabilität konservativer Systeme mit gyroskopischen Kräften, *Arch. Rat. Mech. Anal.* **58** (1975), 1-9.

[7] Barkwell, L., Lancaster, P., Overdamped and Gyroscopic Vibrating Systems, *ASME J. Appl. Mech.* **59** (1992), 176-181.

[8] Lancaster, P., Zizler, P., On the Stability of gyroscopic Systems, *ASME J. Appl. Mech.* **65** (1998), 519-522.

[9] Walker, J.A., Stability of linear conservative gyroscopic systems, *ASME J. Appl. Mech.* **58** (1991), 229-232.

[10] Hryniv, R., Kliem, W., Lancaster, P., Pommer, C., A Precise Bound for Gyroscopic Stabilization, *Z. Angew. Math. Mech., ZAMM* (2000), to appear.

[11] Seyranian, A.P., Stoustrup, J., Kliem, W., On gyroscopic stabilization, *Z. angew. Math. Phys., ZAMP* **46** (1995), 255-267.

[12] Veselic, K., On the Stability of Rotating Systems, *Z. angew. Math. Mech., ZAMM* **75** (1995), 325-328.

[13] Freitas, P., Quadratic matrix polynomials with Hamiltonian spectrum and oscillatory damped systems, *Z. angew. Math. Phys., ZAMP* **50** (1999), 64-81.

[14] Lancaster, P., Tismenetsky, M., *The Theory of Matrices*, Academic Press 1985.

[15] Ziegler, H., Linear Elastic Stability, *Z. angew. Math. Phys., ZAMP* **4** (1953), 89-121.

[16] Taussky, O., The role of symmetric matrices, *Linear Algebra Appl.* **5** (1972), 147-154.

[17] Pommer, C., Kliem, W., Simultaneously Normalizable Matrices, *Linear Algebra Appl.* **94** (1987), 113-125.

Technical University of Denmark
Department of Mathematics
Matematiktorvet, Bldg. 303
DK-2800 Kgs. Lyngby
Denmark

1991 Mathematics Subject Classification. 34A30, 34D20

Operator Theory:
Advances and Applications, Vol. 130, 235–254
© 2001 Birkhäuser Verlag Basel/Switzerland

Invariant Subspaces of Infinite Dimensional Hamiltonians and Solutions of the Corresponding Riccati Equations

H. Langer, A.C.M. Ran and B.A. van de Rotten

To our friend, colleague and teacher Peter Lancaster

Abstract. We consider an infinite dimensional algebraic Riccati equation which arises in systems theory. Using a dichotomy property of the corresponding Hamiltonian and results on invariant subspaces of operators in spaces with an indefinite inner product we show the existence of bounded and unbounded solutions of this Riccati equation.

1. Introduction

In this paper we consider the algebraic Riccati equation

$$\Pi D \Pi - A^* \Pi - \Pi A - Q = 0 \tag{1.1}$$

with an unbounded and e.g. boundedly invertible and sectorial operator A and bounded nonnegative operators D and Q in some Hilbert space $\mathcal{H}$. In the monograph [CZ, Chapter 6], this equation, which corresponds to a control problem on an infinite time interval, was solved by an approximation procedure using solutions for finite time intervals. Here we show the existence of suitable solutions Π by a method which in another situation was exploited in [LT], and which is close to the method used in the finite–dimensional case, see, e.g. [LR]. We consider the so–called Hamiltonian of the equation (1.1)

$$\widetilde{A} := \begin{pmatrix} A & -D \\ -Q & -A^* \end{pmatrix}$$

in the Hilbert space $\widetilde{\mathcal{H}} := \mathcal{H} \oplus \mathcal{H}$ and show that this Hamiltonian has certain invariant subspaces which admit an angular operator representation. It is well–known that these angular operators turn out to be the solutions of a Riccati equation, which here needs specification since in general the entries A and A^* of the Hamiltonian can be unbounded.

In the following Section 2 we adapt a result from [GGK] about the dichotomy of an operator, which is applied in Section 4 in order to show that under our assumptions the Hamiltonian $\widetilde{A}$ is dichotomous. In Section 3 indefinite inner products and operators which are accretive with respect to an indefinite inner product are considered. Two indefinite inner products are introduced in the space $\widetilde{\mathcal{H}}$ in Section 5. In one of these inner products the Hamiltonian $\widetilde{A}$ is accretive, in the other inner product the operator $i\widetilde{A}$ is self–adjoint. These properties, together with the dichotomy of $\widetilde{A}$ lead to the existence of two invariant subspaces of $\widetilde{A}$ with angular operator representations, and it is shown that the corresponding angular operators are selfadjoint if the pair (A, D) is approximately controllable and the pair (Q, A) is approximately observable. Starting from Section 6 we suppose that A is maximal and uniformly accretive, which implies that one of these angular operators is bounded. Finally, in Section 7 solutions of the corresponding Riccati equation are considered.

There is a large and expanding literature concerning the algebraic Riccati equation for several classes of infinite dimensional systems, see [CZ, Section 6.4] and also e.g. [PS], [S], [W], [WW], [CDW]. It is the aim of the present paper to provide an approach to this problem using invariant subspaces of the Hamiltonian. To this end, in order to avoid more technical complications, we suppose that the operators D and Q are bounded. In the papers quoted above these assumptions are sometimes weaker.

2. Dichotomy

Let S be a densely defined closed operator in some Banach space $\mathcal{X}$. As usual, we denote by $\rho(S)$ the resolvent set of S, and by $\sigma(S)$ its complement in the complex plane. Also, by $\mathcal{D}(S)$ we denote the domain of S. We call S *dichotomous* if for some $\omega_-, \omega_+ > 0$

$$\{z : -\omega_- < \Re z < \omega_+\} \subset \rho(S) \tag{2.1}$$

and the space $\mathcal{X}$ is the direct sum of two subspaces $\mathcal{X}_+$, $\mathcal{X}_-$ such that $\mathcal{D}(S) \cap \mathcal{X}_\pm$ is dense in $\mathcal{X}_\pm$, the restriction $S_\pm := S|_{\mathcal{D}(S) \cap \mathcal{X}_\pm}$ maps $\mathcal{D}(S) \cap \mathcal{X}_\pm$ into $\mathcal{X}_\pm$ and

$$\inf_{z \in \sigma(S_+)} \Re z \geq \omega_+, \qquad \sup_{z \in \sigma(S_-)} \Re z \leq -\omega_-.$$

The operator S is called *exponentially dichotomous* if it is dichotomous and for the resolvent operators of $S_\pm$ the estimations

$$\begin{aligned} \|(S_+ - z)^{-1}\| &\leq \frac{\gamma_+}{\omega_+ - \Re z}, \quad \Re z < \omega_+, \\ \|(S_- - z)^{-1}\| &\leq \frac{\gamma_-}{\Re z - \omega_-}, \quad \Re z > \omega_-, \end{aligned} \tag{2.2}$$

hold with positive numbers $\gamma_\pm$.

Suppose now that for the closed operator S the inclusion (2.1) holds and that the resolvent operator $\|(S - z)^{-1}\|$ is uniformly bounded on a strip around the imaginary axis:

$$\sup_{|\Re z| \leq \omega} \|(S - z)^{-1}\| < \infty \quad \text{for some } \omega > 0. \tag{2.3}$$

It follows from [GGK, Theorem XV.3.1] that the operator S is dichotomous if and only if for some (and hence for all) α, $0 < \alpha < \omega$, the integral

$$\frac{1}{2\pi i} \int_{\alpha - i\infty}^{\alpha + i\infty} z^{-2}(S - z)^{-1}S^2 x \, dz, \quad x \in \mathcal{D}(S^2), \tag{2.4}$$

defines a bounded linear operator in $\mathcal{X}$. In this case the closure of this operator is the projection P_+ onto $\mathcal{X}_+$ along $\mathcal{X}_-$.

In the following we sharpen the condition (2.3) to the condition that there exist positive numbers β and γ such that for some $\omega > 0$ and real s

$$\sup_{|s| \leq \omega} \|(S - s - it)^{-1}\| \leq \frac{\gamma}{1 + |t|^\beta}, \quad t \in \mathbb{R}. \tag{2.5}$$

Theorem 2.1. *If the condition (2.5) is satisfied then the operator S in the Banach space $\mathcal{X}$ is dichotomous if and only if for some (and hence for all) α, $0 < \alpha < \omega$, the integral*

$$\frac{1}{2\pi i} \int_{\alpha - i\infty}^{\alpha + i\infty} z^{-2}S(S - z)^{-1}S x \, dz, \quad x \in \mathcal{D}(S), \tag{2.6}$$

defines a bounded linear operator in $\mathcal{X}$. In this case the closure of this operator is the projection P_+ onto $\mathcal{X}_+$ along $\mathcal{X}_-$.

Proof. Suppose that S is dichotomous. For each $x \in \mathcal{D}(S)$ there exists a sequence of elements x_n in $\mathcal{D}(S^2)$, $n = 1, 2, \ldots$, such that $x_n \to x$ and $Sx_n \to Sx$ if $n \to \infty$. Then

$$\left\| \int_{\alpha - i\infty}^{\alpha + i\infty} z^{-2}S(S - z)^{-1}Sx \, dz - \int_{\alpha - i\infty}^{\alpha + i\infty} z^{-2}S(S - z)^{-1}Sx_n \, dz \right\|$$

$$\leq \|Sx - Sx_n\| \int_{\alpha - i\infty}^{\alpha + i\infty} \frac{1}{|z|^2} \|I + z(S - z)^{-1}\| \, |dz|$$

$$\leq \|Sx - Sx_n\| \int_{\alpha - i\infty}^{\alpha + i\infty} \frac{1}{|z|^2} \left(1 + \gamma \frac{|z|}{1 + |\Im z|^\beta}\right) |dz| \to 0 \text{ if } n \to \infty.$$

Since, on the other hand, $P_+ x_n \to P_+ x$ if $n \to \infty$, it follows that

$$P_+ x = \frac{1}{2\pi i} \int_{\alpha - i\infty}^{\alpha + i\infty} z^{-2}S(S - z)^{-1}Sx \, dz, \quad x \in \mathcal{D}(S).$$

Conversely, if the integral in (2.6) defines a bounded linear operator on $\mathcal{D}(S)$ then this is trivially true for the integral in (2.4) on $\mathcal{D}(S^2)$. $\qquad \square$

Our next result is close to the perturbation result in [GGK, Theorem XV.4.1]. We strengthen the condition about the resolvent of the unperturbed operator in order to avoid the assumption $\mathcal{D}(S^2) \subset \mathcal{D}(S_0^2)$ for the perturbed operator S which was imposed there.

Theorem 2.2. *Let S_0 be a dichotomous operator in $\mathcal{X}$ such that for some $\omega > 0$ the strip $\{z : |\Re z| \leq \omega\}$ belongs to the resolvent set $\rho(S_0)$, and*

$$\|(S_0 - z)^{-1}\| < \frac{\gamma_0}{1 + |z|^\beta}, \quad |\Re z| \leq \omega, \tag{2.7}$$

for some $\beta > 1/2$ and $\gamma_0 > 0$. If B is a bounded operator and for the operator $S := S_0 + B$ the strip $\{z : |\Re z| \leq \omega\}$ is in the resolvent set $\rho(S)$ then the operator S is dichotomous as well.

Proof. The relation $(S - z)^{-1} = (S_0 - z)^{-1} \left(I + B(S_0 - z)^{-1}\right)^{-1}$ easily implies that the resolvent of S satisfies in the strip $\{z : |\Re z| \leq \omega\}$ an estimation

$$\sup_{|s| \leq \omega} \|(S - s - it)^{-1}\| < \frac{\gamma}{1 + |t|^\beta}$$

with some $\gamma > 0$ and the same $\beta > 1/2$ as in (2.7). Choose α, $0 < \alpha < \omega$, and define

$$L_0 x := \frac{1}{2\pi i} \int_{\alpha - i\infty}^{\alpha + i\infty} z^{-2} S_0 (S_0 - z)^{-1} S_0 x \, dz, \quad x \in \mathcal{D}(S_0),$$

and an operator L by the analogous formula with S_0 replaced by S. Since S_0 is dichotomous, by Theorem 2.1 the operator L_0 extends to a bounded linear operator on $\mathcal{X}$. Hence, in order to prove the theorem it suffices to show that there exists a constant c such that

$$\|Lx - L_0 x\| \leq c \|x\|, \quad x \in \mathcal{D}(S_0). \tag{2.8}$$

To this end we observe the identity

$$z^{-2} \left(S(S - z)^{-1} Sx - S_0(S_0 - z)^{-1} S_0 x\right)$$
$$= z^{-2}(Sx + z(S - z)^{-1} Sx - S_0 x - z(S_0 - z)^{-1} S_0 x)$$
$$= z^{-2}(Sx - S_0 x + z((S - z)^{-1} - (S_0 - z)^{-1})Sx + z(S_0 - z)^{-1}(Sx - S_0 x))$$

which implies the estimation

$$\|z^{-2} \left(S(S - z)^{-1} Sx - S_0(S_0 - z)^{-1} S_0 x\right)\|$$

$$\leq |z|^{-2} \|B\| \|x\| + |z|^{-1} \|(S_0 - z)^{-1}\| \|B\| \|(S - z)^{-1} S\| \|x\|$$

$$+ |z|^{-1} \|(S_0 - z)^{-1}\| \|B\| \|x\|$$

$$\leq |z|^{-2} \|B\| \|x\| + \frac{1}{|z|} \frac{\gamma_0}{1 + |z|^\beta} \|B\| \left(1 + \frac{|z| \gamma}{1 + |z|^\beta}\right) \|x\| + \frac{1}{|z|} \frac{\gamma_0}{1 + |z|^\beta} \|B\| \|x\|.$$

Since $\beta > 1/2$ the expression on the right hand side is summable along the line $z = \alpha + it$, $-\infty < t < +\infty$, and hence the relation (2.8) follows. □

3. Indefinite inner products

Now we consider the case that $\mathcal{X}$ is a Hilbert space $\mathcal{H}$ with inner product $(\cdot, \cdot)$. Recall that a densely defined operator B in a Hilbert space $\mathcal{H}$ is called *accretive* if $\Re(Bx, x) \geq 0$ for $x \in \mathcal{D}(B)$. Suppose that in $\mathcal{H}$ a fundamental symmetry J is given, that means $J = P_+ - P_-$ with two orthogonal projections $P_\pm$ in $\mathcal{H}$ such that $P_+ + P_- = I$. Equipped with the inner product $[x, y]_J := (Jx, y)$, $x, y \in \mathcal{H}$, the space $\mathcal{H}$ becomes a Krein space, which we denote by $\mathcal{H}_J$. The operator S in $\mathcal{H}$ is called J–*accretive* if $\Re[Sx, x]_J \geq 0$ for all $x \in \mathcal{D}(S)$, *strictly J–accretive* if $\Re[Sx, x]_J > 0$ for all $x \in \mathcal{D}(S), x \neq 0$, and *uniformly J–accretive* if for some $\gamma > 0$

$$\Re[Sx, x]_J \geq \gamma \|x\|^2, \quad x \in \mathcal{D}(S). \tag{3.1}$$

For the next theorem compare [LT, Theorem 1.4].

Theorem 3.1. *Let S be a dichotomous operator in $\mathcal{H}$ such that for some $\omega > 0$ the assumption*

$$\lim_{t \to \infty} \sup_{|s| \leq \omega} \|(S - s - it)^{-1}\| = 0 \tag{3.2}$$

is satisfied. Further, suppose that there exists a fundamental symmetry J in $\mathcal{H}$ such that S is J–accretive. Then the invariant subspace $\mathcal{H}_+ \subset \mathcal{H}$ of S, corresponding to its spectrum in the right half plane, is maximal J–nonnegative, and the invariant subspace $\mathcal{H}_- \subset \mathcal{H}$ of S, corresponding to its spectrum in the left half plane, is maximal J–nonpositive. If S is strictly J–accretive (bounded and uniformly J–accretive, respectively) then $\mathcal{H}_+$ is maximal J–positive (uniformly maximal J–positive, respectively) and $\mathcal{H}_-$ is maximal J–negative (uniformly maximal J–negative, respectively).

Proof. Choose α such that $0 < \alpha < \omega$. Since S is dichotomous the integral (2.4) exists and coincides with $P_+ x$ where P_+ is the projection onto $\mathcal{H}_+$ along $\mathcal{H}_-$. The relation

$$z^{-2}(S - z)^{-1}S^2 = (S - z)^{-1} + z^{-1} + z^{-2}S$$

yields for $x \in \mathcal{D}(S^2)$

$$P_+ x = \frac{1}{2\pi i} \int_{\alpha + i\mathbb{R}} z^{-2}(S - z)^{-1}S^2 x \, dz = \frac{1}{2\pi i} \int_{\alpha + i\mathbb{R}}' (S - z)^{-1}x \, dz \tag{3.3}$$
$$+ \frac{1}{2\pi i} \int_{\alpha + i\mathbb{R}}' z^{-1}dz \, x + \frac{1}{2\pi i} \int_{\alpha + i\mathbb{R}} z^{-2} \, dz \, Sx.$$

where the prime at the integrals denotes the Cauchy principal value at ∞. It is easy to see that the last integral vanishes, and that the second last integral equals $x/2$. Further, for $\nu > 0$ we have

$$\frac{1}{2\pi} \int_{-\nu}^{\nu} (S - it)^{-1}dt = \frac{1}{2\pi} \int_{-\nu}^{\nu} (S - (\alpha + it))^{-1}dt$$
$$- \frac{1}{2\pi i} \int_0^{\alpha} (S - (i\nu + s))^{-1}ds - \frac{1}{2\pi i} \int_{\alpha}^0 (S - (-i\nu + s))^{-1}ds.$$

The assumption (3.2) implies that for $\nu \to \infty$ the last two integrals tend to zero in the uniform operator topology. Because of (3.3), for $x \in \mathcal{D}(S^2)$ the integral $\dfrac{1}{2\pi i}\displaystyle\int_{i\mathbb{R}}'(S-z)^{-1}x\,dz$ exists in the strong operator topology and coincides with $\dfrac{1}{2\pi i}\displaystyle\int_{\alpha+i\mathbb{R}}'(S-z)^{-1}x\,dz$. Finally, the relation (3.3) yields for $x \in \mathcal{D}(S^2)$

$$\frac{1}{2\pi i}\int_{i\mathbb{R}}'(S-z)^{-1}x\,dz = \frac{1}{2}(P_+ - P_-)x.$$

On the other hand, $\mathcal{D}(S^2) \cap \mathcal{H}_+$ is dense in $\mathcal{H}_+$, and for $x \in \mathcal{D}(S^2) \cap \mathcal{H}_+$ we obtain

$$
\begin{aligned}
[x,x]_J &= [P_+x,x]_J = \Re\,[P_+x,x]_J = \Re\,[(P_+ - P_-)x,x]_J\\
&= \Re\left(\frac{1}{\pi i}\int_{i\mathbb{R}}'[(S-z)^{-1}x,x]_J\,dz\right)\\
&= \frac{1}{\pi}\int_{-\infty}^{'+\infty}\Re\Big([S(S-it)^{-1}x,(S-it)^{-1}x]_J\Big)dt.
\end{aligned}
$$

Evidently, for $x \neq 0$ the last integral is nonnegative (positive, respectively) if S is J–accretive (strictly J–accretive, respectively). Therefore $\mathcal{H}_+$ is a J–nonnegative (J–positive, respectively) subspace of $\mathcal{H}$, and, analogously, $\mathcal{H}_-$ is a J–nonpositive (J–negative, respectively) subspace of $\mathcal{H}$. If S is uniformly accretive and bounded, we consider an arbitrary nonempty open bounded real interval Δ with a positive distance from zero. Then with γ from (3.1) the integral in the last relation is greater than or equal to

$$\frac{1}{\pi}\int_\Delta \gamma\|(S-it)^{-1}x\|^2\,dt \geq \frac{\gamma}{\pi}l(\Delta)\left(\max_{t\in\Delta}\|(S-it)\|\right)^{-2}\|x\|^2,$$

where $l(\Delta)$ is the length of Δ. That the subspace $\mathcal{H}_+$ ($\mathcal{H}_-$, respectively) is maximal J–nonnegative (maximal J– nonpositive, respectively) follows from the decomposition $\mathcal{H} = \mathcal{H}_+ \dot{+} \mathcal{H}_-$ and [AI, I.1.25°]. □

In this direction we also have the following result.

Theorem 3.2. *Let S be an exponentially dichotomous operator in $\mathcal{H}$. Suppose that there exists a fundamental symmetry J in $\mathcal{H}$ such that S is J–accretive. Then the invariant subspace $\mathcal{H}_+ \subset \mathcal{H}$ of S, corresponding to its spectrum in the right half plane, is maximal J–nonnegative, and the invariant subspace $\mathcal{H}_- \subset \mathcal{H}$ of S, corresponding to its spectrum in the left half plane, is maximal J–nonpositive.*

Proof. For $x \in \mathcal{H}_+$ we consider the function

$$\varphi_x(z) := \big[(S-z)^{-1}x,x\big]_J, \quad z \in \rho(S|\mathcal{H}_+).$$

It is analytic in the closed left half plane, and on the imaginary axis it has the property

$$\Re\varphi_x(it) = \Re\left[S(S-it)^{-1}x,(S-it)^{-1}x\right]_J \geq 0. \tag{3.4}$$

Consider the fractional linear mapping

$$\zeta(z) = \frac{z+1}{z-1}.$$

It maps the left half plane of the z–plane onto the unit disc of the ζ–plane. Under this transformation the function φ_x becomes

$$\widehat{\varphi}_x(\zeta) := \varphi_x\left(\frac{\zeta+1}{\zeta-1}\right),$$

which we consider on the open unit disc. Evidently, for $|\zeta| = 1$, $\zeta \neq 1$, the boundary values $\widehat{\varphi}_x(\zeta)$ exist and by (3.4) they have a nonnegative real part. We show that this function $\widehat{\varphi}_x$ has also a nontangential boundary value in $\zeta = 1$ which is zero. To this end we observe the relation

$$\lim_{\zeta \widehat{\rightarrow} 1} \widehat{\varphi}_x(\zeta) = \lim_{z \widehat{\rightarrow} \infty} \varphi_x(z) \tag{3.5}$$

where $\widehat{\rightarrow}$ denotes the nontangential boundary value, which for $z \widehat{\rightarrow} \infty$ means that z tends to ∞ within an arbitrary angle $\pi - \alpha \leq \arg z \leq \pi + \alpha$, where $0 < \alpha < \pi/2$. That the limit on the right hand side of (3.5) exists and equals zero follows immediately from the first inequality in (2.2).

The real part of the function $\widehat{\varphi}_x$ is a harmonic function having nonnegative boundary values on the unit cirle. Therefore, by the maximum principle for harmonic functions, the real part of $\widehat{\varphi}_x$ is nonnegative on the whole unit disc, and hence the real part of the function φ_x is nonnegative on the whole left half plane. With $z = u < 0$ we obtain

$$[x, x]_J = - \lim_{u \to -\infty} \Re u[(S - u)^{-1}x, x)]_J \geq 0. \qquad \square$$

4. The Hamiltonian

Let A be a closed linear operator in the Hilbert space $\mathcal{H}$ with the property that for some $\omega_A > 0$ the strip $\{z : |\Re z| < \omega_A\}$ belongs to $\rho(A)$, and let D and Q be bounded and nonnegative selfadjoint operators in $\mathcal{H}$. Here an operator B is called *nonnegative* (*positive*, etc., respectively) if $(Bx, x) \geq 0$, $((Bx, x) > 0$, etc., respectively) for all $x \in \mathcal{D}(B)$, $x \neq 0$. In the space $\widetilde{\mathcal{H}} := \mathcal{H} \oplus \mathcal{H}$ we consider the operators

$$\widetilde{A}_0 = \begin{pmatrix} A & 0 \\ 0 & -A^* \end{pmatrix} \tag{4.1}$$

and

$$\widetilde{A} := \begin{pmatrix} A & -D \\ -Q & -A^* \end{pmatrix}. \tag{4.2}$$

Evidently, $\widetilde{A}_0$ and $\widetilde{A}$ are closed operators with

$$\mathcal{D}(\widetilde{A}) = \mathcal{D}(\widetilde{A}_0) = \mathcal{D}(A) \oplus \mathcal{D}(A^*),$$

and the strip $\{z : |\Re z| < \omega_A\}$ belongs to the resolvent set of $\widetilde{A}_0$.

The operator $\widetilde{A}$ is also called a *Hamiltonian*. It is the aim of this section to find conditions under which $\widetilde{A}$ is dichotomous.

Lemma 4.1. *Suppose that the operator A satisfies the condition*

$$\sup_{|\Re z| \leq \omega_0} \|(A - z)^{-1}\| \leq c < \infty \tag{4.3}$$

with positive numbers c, ω_0, and that D, Q are bounded nonnegative operators in $\mathcal{H}$. Then there exists an ω, $0 < \omega < \omega_A$, such that for the operator $\widetilde{A}$ from (4.2) it holds

$$\{z : |\Re z| \leq \omega\} \subset \rho(\widetilde{A}), \qquad \sup_{|\Re z| \leq \omega} \|(\widetilde{A} - z)^{-1}\| \leq c. \tag{4.4}$$

Proof. We choose $\omega > 0$ and such that

$$\omega + \max(\|D\|^{1/2}, \|Q\|^{1/2})\sqrt{\omega} < \frac{1}{c\sqrt{2}}. \tag{4.5}$$

First we show that for the strip $\{z : |\Re z| < \omega\}$ there exists a positive number d such that for all $\widetilde{x} \in \widetilde{\mathcal{H}}$, $\|\widetilde{x}\| = 1$,

$$\|(\widetilde{A} - z)\widetilde{x}\| \geq d \text{ if } |\Re z| \leq \omega. \tag{4.6}$$

Assume to the contrary that there exist sequences of elements $x_n, y_n \in \mathcal{H}$ and a sequence of numbers $z_n = s_n + it_n$, $|s_n| < \omega$, $t_n \in \mathbb{R}$, $n = 1, 2, \ldots$, such that $\|x_n\|^2 + \|y_n\|^2 = 1$ and

$$(A - it_n - s_n)x_n - Dy_n \to 0, \quad -Qx_n - (A^* + it_n + s_n)y_n \to 0 \tag{4.7}$$

if $n \to \infty$. Without loss of generality we can suppose that

$$\underline{\lim}_{n\to\infty} \|x_n\|^2 \geq 1/2. \tag{4.8}$$

If we take the inner product of the first relation in (4.7) with y_n and of the second relation in (4.7) with x_n, add the resulting relations and let $n \to \infty$ it follows that

$$2s_n(x_n, y_n) + (Dy_n, y_n) + (Qx_n, x_n) \to 0$$

and

$$
\begin{aligned}
0 &\leq \|D^{1/2}y_n\|^2 + \|Q^{1/2}x_n\|^2 &\leq 2|s_n|\,|(x_n, y_n)| + o(1) \\
&\leq |s_n|(\|x_n\|^2 + \|y_n\|^2) + o(1) &= |s_n| + o(1).
\end{aligned} \tag{4.9}
$$

Now from (4.8), (4.3), (4.7) and (4.9) we obtain the following chain of estimations:

$$
\begin{aligned}
\frac{1}{c\sqrt{2}} &\leq \underline{\lim}_{n\to\infty}\frac{1}{c}\|x_n\| \\
&\leq \underline{\lim}_{n\to\infty}\|(A - it_n)x_n\| \leq \underline{\lim}_{n\to\infty}(\|s_n x_n\| + \|Dy_n\|) \\
&\leq \underline{\lim}_{n\to\infty}\left(|s_n|\,\|x_n\| + \|D^{1/2}\|\,\|D^{1/2}y_n\|\right) \\
&\leq \underline{\lim}_{n\to\infty}\left(|s_n| + \|D^{1/2}\|\sqrt{|s_n|}\right) \leq \omega + \|D^{1/2}\|\sqrt{\omega},
\end{aligned}
$$

which gives a contradiction to (4.5).

In the same way, for the operator

$$\widetilde{A}^* = \begin{pmatrix} A^* & -Q \\ -D & -A \end{pmatrix}$$

we obtain for all $\widetilde{x} \in \widetilde{\mathcal{H}}$, $\|\widetilde{x}\| = 1$,

$$\|(\widetilde{A}^* - z)\widetilde{x}\| \geq d, \quad |\Re z| \leq \omega,$$

and (4.4) follows easily. $\qquad \square$

Theorem 4.2. *Let A be a closed operator in $\mathcal{H}$ such that for some $\omega_0 > 0$ the strip $\{z : |\Re z| \leq \omega_0\}$ belongs to $\rho(A)$ and that there exist numbers M and $\beta > 1/2$ such that*

$$\|(A - z)^{-1}\| \leq \frac{M}{1 + |z|^\beta}, \quad |\Re z| \leq \omega_0. \tag{4.10}$$

Further, let D, Q be bounded nonnegative operators in $\mathcal{H}$. Then the operator $\widetilde{A}$ in (4.2) is dichotomous and the relation

$$\lim_{|t| \to \infty} \sup_{|s| \leq \omega} \|(\widetilde{A} - s - it)^{-1}\| = 0 \tag{4.11}$$

holds.

Proof. Lemma 4.1 implies that a strip around the imaginary axis belongs to $\rho(\widetilde{A})$. An application of Theorem 2.2 to $S_0 = \widetilde{A}_0 = \begin{pmatrix} A & 0 \\ 0 & -A^* \end{pmatrix}$ and $B = \begin{pmatrix} 0 & -D \\ -Q & 0 \end{pmatrix}$ yields that $\widetilde{A}$ is dichotomous. The relation (4.11) is now an easy consequence of the identity

$$(\widetilde{A} - z)^{-1} = (\widetilde{A}_0 - z)^{-1}\left(I + (\widetilde{A} - \widetilde{A}_0)(\widetilde{A}_0 - z)^{-1}\right)^{-1},$$
$$z \in \rho(\widetilde{A}_0) \cap \rho(\widetilde{A}). \qquad \square$$

5. Invariant subspaces of the Hamiltonian

In the following we need some simple facts about linear relations in a Hilbert space, see e.g. [DS]. Let $\mathcal{H}$ be a Hilbert space and set $\widetilde{\mathcal{H}} = \mathcal{H} \dotplus \mathcal{H}$. A linear subspace $\widetilde{\mathcal{L}} \subset \widetilde{\mathcal{H}}$ is called a *linear relation in $\mathcal{H}$*. We write the elements of a linear relation $\widetilde{\mathcal{L}}$ as $\begin{pmatrix} x \\ y \end{pmatrix}$, $x, y \in \mathcal{H}$. For the linear relation $\widetilde{\mathcal{L}}$ we define

$$\ker(\widetilde{\mathcal{L}}) := \left\{ x : \begin{pmatrix} x \\ 0 \end{pmatrix} \in \widetilde{\mathcal{L}} \right\}, \quad \mathrm{infty}(\widetilde{\mathcal{L}}) := \left\{ y : \begin{pmatrix} 0 \\ y \end{pmatrix} \in \widetilde{\mathcal{L}} \right\}.$$

If $\mathrm{infty}(\widetilde{\mathcal{L}}) = \{0\}$ then $\widetilde{\mathcal{L}}$ is the graph of a linear operator Π in $\mathcal{H}$:

$$\widetilde{\mathcal{L}} = \left\{ \begin{pmatrix} x \\ y \end{pmatrix} : x \in \mathcal{D}(\Pi), \Pi x = y \right\}.$$

If $\widetilde{\mathcal{L}}$ is a linear relation in $\mathcal{H}$ its *adjoint linear relation* $\widetilde{\mathcal{L}}^*$ is defined as follows:

$$\widetilde{\mathcal{L}}^* := \left\{ \begin{pmatrix} u \\ v \end{pmatrix} : u, v \in \mathcal{H}, (u, y) - (v, x) = 0 \text{ for all } \begin{pmatrix} x \\ y \end{pmatrix} \in \widetilde{\mathcal{L}} \right\}.$$

The linear relation $\widetilde{\mathcal{L}}$ is called *symmetric* if $\widetilde{\mathcal{L}} \subset \widetilde{\mathcal{L}}^*$, and *selfadjoint* if $\widetilde{\mathcal{L}} = \widetilde{\mathcal{L}}^*$. A selfadjoint linear relation $\widetilde{\mathcal{L}}$ with infty$(\widetilde{\mathcal{L}}) = \{0\}$ is the graph of a selfadjoint operator. The symmetric linear relation $\widetilde{\mathcal{L}}$ is called *nonnegative* (*positive* etc., respectively) if $(x, y) \geq 0$ (> 0 etc., respectively) for $\begin{pmatrix} x \\ y \end{pmatrix} \in \mathcal{L}$, $\begin{pmatrix} x \\ y \end{pmatrix} \neq \begin{pmatrix} 0 \\ 0 \end{pmatrix}$.

Now we resume the study of the operator $\widetilde{A}$ in (4.2) under the assumption that its entries A, D, Q satisfy the conditions in Theorem 4.2. We introduce two fundamental symmetries $\widetilde{J}_1, \widetilde{J}_2$ in $\mathcal{H}$:

$$\widetilde{J}_1 := \begin{pmatrix} 0 & -I \\ -I & 0 \end{pmatrix}, \qquad \widetilde{J}_2 := \begin{pmatrix} 0 & iI \\ -iI & 0 \end{pmatrix}.$$

Then the relations

$$\widetilde{J}_1 \widetilde{A} = \begin{pmatrix} Q & -A^* \\ A & D \end{pmatrix}, \qquad \widetilde{J}_2(i\widetilde{A}) = \begin{pmatrix} Q & A^* \\ A & -D \end{pmatrix}.$$

hold. Since D and Q are positive operators and

$$\Re\left(\widetilde{J}_1 \widetilde{A} \right) = \begin{pmatrix} Q & 0 \\ 0 & D \end{pmatrix},$$

the operator $\widetilde{A}$ is $\widetilde{J}_1$–accretive, and, according to Theorem 4.2, it is dichotomous and has the property (4.11). It follows from Theorem 3.1 that the invariant subspace $\widetilde{\mathcal{H}}_+ \subset \widetilde{\mathcal{H}}$ of $\widetilde{A}$, corresponding to its spectrum in the right half plane, is maximal $\widetilde{J}_1$–nonnegative, and the invariant subspace $\widetilde{\mathcal{H}}_- \subset \widetilde{\mathcal{H}}$ of $\widetilde{A}$, corresponding to its spectrum in the left half plane, is maximal $\widetilde{J}_1$–nonpositive. If we observe the form of $\widetilde{J}_1$ it follows that $(x, y) \geq 0$ for $\begin{pmatrix} x \\ y \end{pmatrix} \in \widetilde{\mathcal{H}}_-$ and $(x, y) \leq 0$ for $\begin{pmatrix} x \\ y \end{pmatrix} \in \widetilde{\mathcal{H}}_+$.

Since the operator $i\widetilde{A}$ is $\widetilde{J}_2$–selfadjoint, the spectrum $\sigma(\widetilde{A})$ is symmetric with respect to the imaginary axis, and the invariant subspaces $\widetilde{\mathcal{H}}_+, \widetilde{\mathcal{H}}_-$ are maximal $\widetilde{J}_2$–neutral. These subspaces can be considered as linear relations in $\mathcal{H}$. Then 'maximal $\widetilde{J}_2$–neutral' means that these linear relations are selfadjoint. Summing up, the first claims of the following theorem are proved.

Theorem 5.1. *Let A, D and Q satisfy the assumptions of Theorem 4.2. Then the invariant subspace $\widetilde{\mathcal{H}}_+$ of $\widetilde{A}$, corresponding to its spectrum in the right half plane, is a nonpositive selfadjoint linear relation in $\mathcal{H}$, and the invariant subspace $\widetilde{\mathcal{H}}_-$ of $\widetilde{A}$, corresponding to its spectrum in the left half plane, is a nonnegative selfadjoint linear relation in $\mathcal{H}$. Moreover, if the assumption*

$$\text{c.l.s. } \left\{ (A - z)^{-1} D x : x \in \mathcal{H}, |\Re z| \leq \omega_0 \right\} = \mathcal{H} \tag{5.1}$$

is satisfied, then $\mathrm{infty}(\widetilde{\mathcal{H}}_+) = \{0\}$, *if the assumption*

$$\text{c.l.s. } \{(A^* - z)^{-1}Qx : x \in \mathcal{H}, \, |\Re z| \leq \omega_0\} = \mathcal{H} \tag{5.2}$$

is satisfied, then $\ker(\widetilde{\mathcal{H}}_\pm) = \{0\}$.

Proof. First we mention that the assumption (5.1) is equivalent to the fact that $D(A^* - z)^{-1}y = 0$ for all z with $|\Re z| \leq \omega_0$ implies $y = 0$. Secondly, it is easy to check that for $z \in \rho(\widetilde{A})$ the resolvent $(\widetilde{A} - z)^{-1}$ admits the matrix representation

$$\begin{pmatrix} S_1(z)^{-1} & -S_1(z)^{-1}D(A^* + z)^{-1} \\ -(A^* + z)^{-1}QS_1(z)^{-1} & -S_2(z)^{-1} \end{pmatrix} \tag{5.3}$$

with

$$S_1(z) = \left(I + D(A^* + z)^{-1}Q(A - z)^{-1}\right)(A - z);$$
$$S_2(z) = \left(I + Q(A - z)^{-1}D(A^* + z)^{-1}\right)(A^* + z).$$

Now consider an element $\begin{pmatrix} 0 \\ y \end{pmatrix} \in \widetilde{\mathcal{H}}_-$. Since $\widetilde{\mathcal{H}}_-$ is $\widetilde{J}_2$–neutral and invariant under $(\widetilde{A} - it)^{-1}$ we obtain for real t

$$\begin{aligned} 0 &= \left[\begin{pmatrix} 0 \\ y \end{pmatrix}, (\widetilde{A} - it)^{-1}\begin{pmatrix} 0 \\ y \end{pmatrix}\right]_{\widetilde{J}_2} \\ &= -i\left(y, S_1(it)^{-1}D(A^* + it)^{-1}y\right) \\ &= -i\left(y, \left(A - it + D(A^* + it)^{-1}Q\right)^{-1}D(A^* + it)^{-1}y\right) \\ &= -i\left(u, \left(I + D^{1/2}(A^* + it)^{-1}Q(A - it)^{-1}D^{1/2}\right)^{-1}u\right) \end{aligned}$$

where $u = D^{1/2}(A^* + it)^{-1}y$. It follows that $u = D^{1/2}(A^* + it)^{-1}y = 0$, and the assumption (5.1) implies $y = 0$. The proof of the other claims is similar. ⊔⊓

If the condition (5.1) is satisfied then the pair (A, D) is called *approximately controllable*, if the condition (5.2) is satisfied the pair (Q, A) is called *approximately observable* (comp. [CZ]).

Corollary 5.2. *Under the assumptions of Theorem 4.2 and of (5.1) and (5.2) the linear relations* $\widetilde{\mathcal{H}}_\pm$ *are graph subspaces of possibly unbounded linear operators* Π_+ *and* Π_-, *in fact*

$$\widetilde{\mathcal{H}}_- = \left\{\begin{pmatrix} x \\ \Pi_- x \end{pmatrix} : x \in \mathcal{D}(\Pi_-)\right\}, \qquad \widetilde{\mathcal{H}}_+ = \left\{\begin{pmatrix} x \\ \Pi_+ x \end{pmatrix} : x \in \mathcal{D}(\Pi_+)\right\}.$$

Π_- *is a positive selfadjoint operator,* Π_+ *is a negative selfadjoint operator and* $\ker \Pi_\pm = \{0\}$.

Starting from the next section we strengthen the assumptions about A in order to assure the boundedness of the operator Π_-.

6. Boundedness of the operator Π_-.

We call an operator B in $\mathcal{H}$ mu-*sectorial* if the right half plane belongs to $\rho(B)$ and if there exist numbers θ, $0 < \theta < \pi/2$, and $\beta > 0$ such that

$$\pi/2 + \theta \leq \text{Arg}\,(Bx, x) \leq 3\pi/2 - \theta \text{ and } \Re(Bx, x) \leq -\beta, \quad x \in \mathcal{D}(B).$$

In 'mu-sectorial' m stands for maximal and u for uniformly, comp. [K]. If the operator B is mu-sectorial then its resolvent satisfies an estimate

$$\|(B - z)^{-1}\| \leq \frac{\gamma}{1 + |z|}, \quad \Re z \geq 0, \tag{6.1}$$

for some $\gamma > 0$.

We shall show that the operator Π_- in Corollary 5.2 is bounded if the operator A in the Hamiltonian is mu-sectorial. As a motivation, we consider the simple case where A is a uniformly negative unbounded operator and $D = Q = I$. It is easy to check that in this case

$$\Pi_- = \sqrt{A^2 + I} + A = \left(\sqrt{A^2 + I} - A\right)^{-1}, \quad \Pi_+ = A - \sqrt{A^2 + I},$$

hence Π_- is bounded and positive, but Π_+ is unbounded and uniformly negative.

We start with the following simple lemma. All the operators we consider are supposed to be densely defined.

Lemma 6.1. *Let B be a closed operator in some Banach space $\mathcal{X}$ such that the closed right half plane belongs to $\rho(B)$ and*

$$\|(B - z)^{-1}\| \leq \frac{\gamma}{1 + |z|}, \quad -\pi/2 \leq \varphi \leq \pi/2, \tag{6.2}$$

with some positive number γ. (Here $\varphi = \arg(z)$.) Then

$$-\frac{1}{\pi} \int_{-\infty}^{\prime +\infty} (B - it)^{-1}\,dt = I,$$

where the prime denotes the Cauchy principal value at ∞, and the integral exists in the strong operator topology.

Proof. We start from the relation

$$\frac{1}{\pi} \int_{-R}^{R} (B - it)^{-1}\,dt + \frac{1}{\pi} \int_{+\pi/2}^{-\pi/2} (B - Re^{i\varphi})^{-1}\,Re^{i\varphi}\,d\varphi = 0.$$

and show that the latter integral tends to I if $R \to \infty$. Indeed, if $x \in \mathcal{D}(B)$ we obtain

$$I_R x := \frac{1}{\pi} \int_{+\pi/2}^{-\pi/2} (B - Re^{i\varphi})^{-1}\,Re^{i\varphi}x\,d\varphi = x + \frac{1}{\pi} \int_{+\pi/2}^{-\pi/2} (B - Re^{i\varphi})^{-1}\,Bx\,d\varphi.$$

This integral tends to 0 if $R \to \infty$, hence $\|I_R x - x\| \to 0$ for all x from the dense subset $\mathcal{D}(B)$. On the other hand, the assumption (6.2) also yields that $\|I_R\|$ is uniformly bounded if $R \to \infty$, which implies that $I_R \to I$ strongly. $\qquad\square$

Lemma 6.2. *Let B be a bounded mu-sectorial operator and let Π be a nonnegative selfadjoint operator in the Hilbert space $\mathcal{H}$ such that for some bounded positive operator C in $\mathcal{H}$ the following inequality holds:*

$$-(Bx, \Pi x) - (B^*\Pi x, x) \le (Cx, x), \quad x \in \mathcal{D}(\Pi). \tag{6.3}$$

Then the operator Π is bounded.

Proof. We denote by E the spectral function of Π, choose a bounded interval Δ of the real axis and consider the inequality (6.3) for $x = x_\Delta \in E(\Delta)\mathcal{H}$. Then we obtain, with Π_Δ denoting the bounded operator generated by Π in $E(\Delta)\mathcal{H}$ and $B_\Delta = E(\Delta)B|_{E(\Delta)\mathcal{H}}$, $C_\Delta = E(\Delta)C|_{E(\Delta)\mathcal{H}}$:

$$-((B_\Delta - it)x, \Pi_\Delta x) - ((B_\Delta^* + it)\Pi_\Delta x, x) \le (C_\Delta x, x),$$

or, in $\mathcal{H}_\Delta$ with all the operators being bounded,

$$-(B_\Delta^* + it)^{-1}\Pi_\Delta - \Pi_\Delta(B_\Delta - it)^{-1} \le (B_\Delta^* + it)^{-1}C_\Delta(B_\Delta - it)^{-1}.$$

If we integrate this relation with respect to t from $-\infty$ to $+\infty$ and observe Lemma 6.1 and the relation (6.1) we obtain for $x \in E(\Delta)\mathcal{H}$

$$\begin{aligned}
(x, \Pi_\Delta x) + (\Pi_\Delta x, x) &\le \frac{1}{\pi} \int_{-\infty}^{+\infty} (C_\Delta(B_\Delta - it)^{-1}x, (B_\Delta - it)^{-1}x)\,dt \\
&\le \frac{\|C\|\,\|x\|^2|\gamma|^2}{\pi} \int_{-\infty}^{+\infty} \frac{dt}{(1 + |t|)^2}.
\end{aligned}$$

It follows that $0 \le \Pi_\Delta \le c$ with a positive number c which is independent of Δ. Since Δ is an arbitrary interval we obtain $0 \le \Pi \le c$. $\qquad \Box$

Remark 6.3. *From the proof of Lemma 6.2 it is clear that the bound for the operator Π depends only on the norm of C and the constant γ in the estimation (6.1) but not on the norm of B.*

Corollary 6.4. *Suppose that the operator A is mu-sectorial and bounded, and that D and Q are bounded and nonnegative and such that (5.1) and (5.2) hold. Then the positive self-adjoint operator Π_- in Corollary 5.2 is bounded.*

In order to see this we observe that the invariance of the subspace $\widetilde{\mathcal{H}}_-$ under the bounded operator $\widetilde{A}$ means that for each $x \in \mathcal{D}(\Pi_-)$ there exists an $y \in \mathcal{D}(\Pi_-)$ such that

$$\begin{pmatrix} A & -D \\ -Q & -A^* \end{pmatrix} \begin{pmatrix} x \\ \Pi_- x \end{pmatrix} = \begin{pmatrix} y \\ \Pi_- y \end{pmatrix},$$

which implies

$$-Qx - A^*\Pi_- x = \Pi_- y = \Pi_-(Ax - D\Pi_- x).$$

Taking the inner product with x we obtain the relation

$$-(Ax, \Pi_- x) - (A^*\Pi_- x, x) = (Qx, x) - (D\Pi_- x, \Pi_- x) \le (Qx, x), \quad x \in \mathcal{D}(\Pi_-),$$

and the claim follows from Lemma 6.2.

The next theorem is the main result of this section.

Theorem 6.5. *Suppose that A is a mu-sectorial operator, that D, Q are bounded nonnegative operators such that the relations (5.1) and (5.2) hold, and that*

$$\widetilde{\mathcal{H}}_- = \left\{ \begin{pmatrix} x \\ \Pi_- x \end{pmatrix} : x \in \mathcal{D}(\Pi_-) \right\}, \quad \widetilde{\mathcal{H}}_+ = \left\{ \begin{pmatrix} x \\ \Pi_+ x \end{pmatrix} : x \in \mathcal{D}(\Pi_+) \right\}$$

are the spectral invariant subspaces of $\widetilde{A}$ corresponding to its spectrum in the left and right half plane, respectively, according to Theorem 5.1. Then Π_- is a bounded positive, Π_+ a boundedly invertible negative selfadjoint operator, and the operator $\Pi_- - \Pi_+$ is uniformly positive. Moreover, $\Pi_-\mathcal{D}(A) \subset \mathcal{D}(A^*)$, and

$$\mathcal{D}(\widetilde{A}|_{\widetilde{\mathcal{H}}_-}) = \left\{ \begin{pmatrix} x \\ \Pi_- x \end{pmatrix} : x \in \mathcal{D}(A) \right\}.$$

Proof. For any positive integer n we define the bounded linear operator

$$A_n := -nA(A - n)^{-1}.$$

Then for $x \in \mathcal{D}(A)$ it holds

$$\|A_n x - Ax\| \to 0 \text{ if } n \to \infty. \tag{6.4}$$

In order to see this we observe the relation $A_n x - Ax = -A(A - n)^{-1}Ax$ and the fact that $A(A - n)^{-1}$ tends strongly to zero if $n \to \infty$ since the resolvent of A satisfies an inequality (6.1).

Next, we show that the inequality

$$\|(A_n - it)^{-1}\| \le \frac{\gamma}{1 + |t|} \tag{6.5}$$

holds with a fixed number γ for all real t and positive integers n. Indeed, since the operator A is mu-sectorial there exists a number θ, $0 < \theta < \pi/2$, such that

$$\pi/2 + \theta \le \text{Arg}(Ax, x) \le 3\pi/2 - \theta, \ x \in \mathcal{D}(A).$$

The relation

$$\begin{aligned} (A_n x, x) &= -n(A(A - n)^{-1}x, x) = -n(Ay, (A - n)y) \\ &= -n(Ay, Ay) + n^2(Ay, y) \end{aligned}$$

with $y = (A - n)^{-1}x$ implies that also

$$\pi/2 + \theta \le \text{Arg}(A_n x, x) \le 3\pi/2 - \theta, \ x \in \mathcal{H},$$

and hence (6.5) holds for all t outside any interval around zero. Further, one checks that

$$(A_n - z)^{-1} = -\frac{1}{n + z}\left(I + \frac{-n^2}{n + z}\left(A - \frac{zn}{n + z}\right)^{-1}\right)$$

and it is easy to see that in a neighbourhood of $z = 0$ the norms $\|(A_n - z)^{-1}\|$ are uniformly bounded with respect to n. Therefore (6.5) holds for all real t.

We consider for $n = 1, 2, \ldots$ the Hamiltonians

$$\widetilde{A}_n := \begin{pmatrix} A_n & -D \\ -Q & -A_n^* \end{pmatrix}.$$

The relation (6.4) implies that for $\widetilde{x} \in \mathcal{D}(\widetilde{A})$

$$\|\widetilde{A}_n x - \widetilde{A}x\| \to 0 \text{ if } n \to \infty. \tag{6.6}$$

According to Lemma 4.1 a strip around the imaginary axis belongs to $\rho(\widetilde{A}_n)$ for all n. The reasoning in the proof of Theorem 4.2 yields also an inequality

$$\|(\widetilde{A}_n - z)^{-1}\| \leq \frac{\widetilde{\gamma}}{1 + |z|} \tag{6.7}$$

with a fixed number $\widetilde{\gamma}$ for all z in this strip and all positive integers n.

Now we apply the results of Section 5 to the Hamiltonians $\widetilde{A}_n$. The corresponding nonnegative selfadjoint operators $\Pi_{-,n}$ are bounded according to Corollary 6.4, and even uniformly bounded because of the uniform boundedness (6.5) of the resolvents and of Remark 6.3. It follows that a subsequence of the sequence $(\Pi_{-,n})_{n=1}^{\infty}$ converges in the weak operator topology to some bounded operator Π_0.

On the other hand, the formula

$$(A_n - z)^{-1} - (A - z)^{-1} = -\frac{1}{n+z} A^2 (A - z)^{-1} \left(A - \frac{zn}{n+z} \right)^{-1}$$

implies

$$\|(A_n - z)^{-1} - (A - z)^{-1}\| \longrightarrow 0 \text{ if } n \to \infty,$$

uniformly for all z in the closed right half plane. Now the relation (5.3) yields that for all z in a strip around the imaginary axis

$$\|(\widetilde{A}_n - z)^{-1} - (\widetilde{A} - z)^{-1}\| \longrightarrow 0 \text{ if } n \to \infty.$$

Consequently, for all $\widetilde{x} \in \mathcal{D}(\widetilde{A})$ and sufficiently small $\alpha > 0$ we obtain

$$\left\| \frac{1}{2\pi i} \int_{\alpha - i\infty}^{\alpha + i\infty} \left(\widetilde{A}_n (\widetilde{A}_n - z)^{-1} \frac{\widetilde{A}_n \widetilde{x}}{z^2} - \widetilde{A}(\widetilde{A} - z)^{-1} \frac{\widetilde{A}\widetilde{x}}{z^2} \right) dz \right\|$$

$$\leq \frac{1}{2\pi} \int_{\alpha - i\infty}^{\alpha + i\infty} \left\| (\widetilde{A}_n - z)^{-1} - (\widetilde{A} - z)^{-1} \right\| \frac{\|\widetilde{A}_n \widetilde{x}\|}{|z|} |dz|$$

$$+ \frac{1}{2\pi} \int_{\alpha - i\infty}^{\alpha + i\infty} \left\| \widetilde{A}(\widetilde{A} - z)^{-1} \right\| \frac{\|\widetilde{A}_n \widetilde{x} - \widetilde{A}\widetilde{x}\|}{|z|^2} |dz|.$$

The first integral tends to zero if $n \to \infty$ since the integrands tend pointwise to zero and are bounded by an integrable function, the second integral tends to zero because of the relation (6.6).

As a consequence, with the projection $\widetilde{P}_{-,n}$ onto the invariant subspace of $\widetilde{A}_n$ and $\widetilde{P}_-$ onto the invariant subspace of $\widetilde{A}$, corresponding to the spectrum in the left half plane, we obtain for $\widetilde{x} \in \mathcal{D}(\widetilde{A})$

$$\widetilde{P}_{-,n} \widetilde{x} \longrightarrow \widetilde{P}_- \widetilde{x}, \quad n \to \infty.$$

With $\tilde{x} = \begin{pmatrix} x \\ \Pi_- x \end{pmatrix} \in \tilde{\mathcal{H}}_- \cap \mathcal{D}(\tilde{A})$ this relation can be written as

$$\tilde{P}_{-,n}\tilde{x} = \begin{pmatrix} x_n \\ \Pi_{-,n}x_n \end{pmatrix} \to \begin{pmatrix} x \\ \Pi_- x \end{pmatrix} \text{ (strongly) if } n \to \infty,$$

which means

$$\|x_n - x\| \to 0, \quad \|\Pi_{-,n}x_n - \Pi_- x\| \to 0 \text{ if } n \to \infty. \tag{6.8}$$

On the other hand, with the operator Π_0 introduced above we find for the subsequence (n_j) considered there that $\Pi_{-,n_j}x_{n_j} \to \Pi_0 x$, weakly; here we use the first relation in (6.8) and the weak convergence of the sequence (Π_{-,n_j}). Consequently, $\Pi_- x = \Pi_0 x$ for all $x \in \mathcal{H}$ such that $\begin{pmatrix} x \\ \Pi_- x \end{pmatrix} \in \mathcal{D}(\tilde{A})$. Since $\mathcal{D}(\tilde{A}) \cap \tilde{\mathcal{H}}_-$ is dense in $\tilde{\mathcal{H}}_-$ we obtain finally $\Pi_- = \Pi_0$, and hence Π_- is bounded.

The subspaces $\tilde{\mathcal{H}}_-$ and $\tilde{\mathcal{H}}_+$ admit also representations

$$\tilde{\mathcal{H}}_- = \left\{ \begin{pmatrix} \Lambda_- x \\ x \end{pmatrix} : x \in \operatorname{ran} \Pi_- \right\}, \quad \tilde{\mathcal{H}}_+ = \left\{ \begin{pmatrix} \Lambda_+ x \\ x \end{pmatrix} : x \in \operatorname{ran} \Pi_+ \right\}$$

with a positive self-adjoint operator Λ_- and a negative self-adjoint operator Λ_+. For symmetry reasons, it follows as above that the operator $\Lambda_+ = \Pi_+^{-1}$ is bounded.

In order to prove that the operator $\Pi_- - \Pi_+$ is uniformly positive we observe the relation $\tilde{\mathcal{H}} = \tilde{\mathcal{H}}_- \dotplus \tilde{\mathcal{H}}_+$, which implies for arbitrary $x \in \mathcal{H}$

$$\begin{pmatrix} 0 \\ x \end{pmatrix} = \begin{pmatrix} y \\ \Pi_- y \end{pmatrix} + \begin{pmatrix} -y \\ -\Pi_+ y \end{pmatrix},$$

and hence $\Pi_- y - \Pi_+ y = x$. That is, the range of the positive selfadjoint operator $\Pi_- - \Pi_+$ is the whole space $\mathcal{H}$.

It remains to be shown that $\Pi_- \mathcal{D}(A) \subset \mathcal{D}(A^*)$. To this end we introduce the set

$$\mathcal{D}_0 := \{x \mid x \in \mathcal{D}(A), \Pi_- x \in \mathcal{D}(A^*)\}$$

and observe that

$$\sup \{\Re z : z \in \sigma(A - D\Pi_-)\} = \sup \left\{\Re z : z \in \sigma(\tilde{A}|_{\tilde{\mathcal{H}}_-})\right\} \le -\omega$$

for some $\omega > 0$. Take z with $\Re z > -\omega$. For any $\begin{pmatrix} f \\ \Pi_- f \end{pmatrix} \in \tilde{\mathcal{H}}_-$ there is a unique $x \in \mathcal{D}_0$ such that the following relations hold:

$$\begin{aligned} Ax - zx - D\Pi_- x &= f, \\ -Q - (A^* + z)\Pi_- x &= \Pi_- f. \end{aligned}$$

The first of these equations means that $A - D\Pi_- - z$ maps $\mathcal{D}_0$ bijectively onto $\mathcal{H}$. If $\mathcal{D}_0$ would be a proper subset of $\mathcal{D}(A)$ there would exist an $x_0 \ne 0$ such that

$$(A - D\Pi_- - z)x_0 = 0,$$

which is a contradiction since $z \in \rho(A - D\Pi_-)$ $\qquad \square$

Remark 6.6. *We note that the assumption that A is* mu-*sectorial and the boundedness of $D\Pi_-$ imply* $\sup\{\Re z : z \in \sigma(A - D\Pi_-)\} \leq -\omega'$ *for some $\omega' > 0$ and*

$$\|(A - D\Pi_- - z)^{-1}\| \leq \frac{\gamma}{1 + |z|} \quad \text{if} \quad \Re z > -\omega'.$$

It is easy to see that the projection $\widetilde{P}_-$ onto $\widetilde{\mathcal{H}}_-$ along $\widetilde{\mathcal{H}}_+$ can be expressed by means of the selfadjoint operators Π_- and Π_+ as follows: it is the closure of the operator

$$\begin{pmatrix} I \\ \Pi_- \end{pmatrix} (\Pi_- - \Pi_+)^{-1} \left(-\Pi_+ \quad I \right),$$

which can also be written by means of the bounded operators Π_- and Λ_+:

$$\widetilde{P}_- = \begin{pmatrix} (I - \Lambda_+\Pi_-)^{-1} & (\Lambda_+\Pi_- - I)^{-1}\Lambda_+ \\ \Pi_-(I - \Lambda_+\Pi_-)^{-1} & \Pi_-(\Lambda_+\Pi_- - I)^{-1}\Lambda_+ \end{pmatrix}.$$

7. The Riccati equation

The following theorem, which is an immediate consequence of Theorem 6.5, is the main result of this paper.

Theorem 7.1. *Under the assumptions of Theorem 6.5 the bounded positive operator Π_- satisfies the algebraic Riccati equation*

$$\Pi_- D\Pi_- x - A^*\Pi_- x - \Pi_- Ax - Qx = 0, \quad x \in \mathcal{D}(A). \tag{7.1}$$

Moreover, if Π is any bounded nonnegative operator such that $\Pi\mathcal{D}(A) \subset \mathcal{D}(A^*)$ and it satisfies the equation

$$\Pi D\Pi x - A^*\Pi x - \Pi Ax - Qx = 0, \quad x \in \mathcal{D}(A) \tag{7.2}$$

then $\Pi = \Pi_-$.

Proof. In order to prove (7.1) we observe the relation

$$\begin{pmatrix} A & -D \\ -Q & -A^* \end{pmatrix} \begin{pmatrix} x \\ \Pi_- x \end{pmatrix} = \begin{pmatrix} y \\ \Pi_- y \end{pmatrix}, \quad x \in \mathcal{D}(A).$$

It yields

$$-Qx - A^*\Pi_- x = \Pi_- y = \Pi_-(Ax - D\Pi_- x), \quad x \in \mathcal{D}(A),$$

which is the equation (7.1).

In order to prove the last claim we first show that the spectrum of the operator $A - D\Pi$ lies in the left half plane. To this end we observe that for $x \in \mathcal{D}(A)$ the Riccati equation (7.2) yields

$$\begin{aligned} \Re\left((A - D\Pi)x, \Pi x\right) &= \frac{1}{2}\left((Ax, \Pi x) + (\Pi x, Ax)\right) - (D\Pi x, \Pi x) \\ &= -\frac{1}{2}\left((Qx, x) + (\Pi D\Pi x, x)\right) \leq 0. \end{aligned} \tag{7.3}$$

If $\sigma(A - D\Pi)$ would have a boundary point z_0 in the right half plane, there would exist a sequence of elements x_n, $\|x_n\| = 1$, $n = 1, 2, \ldots$, such that

$$(A - D\Pi - z_0)x_n \to 0, \quad n \to \infty. \tag{7.4}$$

Then

$$\underline{\lim}_{n \to \infty} \|\Pi^{1/2} x_n\| > 0. \tag{7.5}$$

Indeed, otherwise $Ax_n - z_0 x_n \to 0$, $n \to \infty$, which is impossible since $\Re \sigma(A) < 0$. The relation (7.4) implies

$$((A - D\Pi)x_n, \Pi x_n) - z_0(x_n, \Pi x_n) \to 0,$$

which is in contradiction with (7.3) and (7.5).

Now we observe for $x \in \mathcal{H}$, $z \in \rho(\widetilde{A})$ the relation

$$(\widetilde{A} - z)^{-1} \begin{pmatrix} x \\ \Pi x \end{pmatrix} = \begin{pmatrix} (A - D\Pi - z)^{-1} x \\ \Pi(A - D\Pi - z)^{-1} x \end{pmatrix}.$$

The expression on the right hand side is holomorphic on the right half plane, therefore the element $\begin{pmatrix} x \\ \Pi x \end{pmatrix}$ belongs to the spectral invariant subspace $\widetilde{\mathcal{H}}_-$ of $\widetilde{A}$ corresponding to the left half plane, hence $\Pi \subset \Pi_-$, and the equality $\Pi = \Pi_-$ follows. $\square$

Besides the Riccati equation (7.1) also the weak Riccati equation

$$(D\Pi x, \Pi y) - (Ax, \Pi y) - (\Pi x, Ay) - (Qx, y) = 0, \quad x, y \in \mathcal{D}(A) \cap \mathcal{D}(\Pi), \tag{7.6}$$

can be considered. Among all its selfadjoint solutions the solution Π_- has the following extremal property.

Theorem 7.2. *Suppose that the assumptions of Theorem 6.5 are satisfied and that Π is a (possibly unbounded) symmetric solution of the weak Riccati equation (7.6) with the property $\mathcal{D}(A) \subset \mathcal{D}(\Pi)$. If Π_- is as in Theorem 7.1 then the operator $\Pi_- - \Pi$ is nonnegative.*

Proof. It is easy to see that for $x \in \mathcal{D}(A)$ the relation

$$((A - D\Pi_-)x, (\Pi_- - \Pi)x) + ((\Pi_- - \Pi)x, (A - D\Pi_-)x)$$

$$= - (D(\Pi_- - \Pi)x, (\Pi_- - \Pi)x)$$

holds. With $T := A - D\Pi_-$, $\Delta := \Pi_- - \Pi$ this becomes for real t

$$((T - it)x, \Delta x) + (\Delta x, (T - it)x) = -(D\Delta x, \Delta x),$$

or with $y := (T - it)x$:

$$(y, \Delta(T - it)^{-1} y) + (\Delta(T - it)^{-1} y, y) = -(D\Delta(T - it)^{-1} y, \Delta(T - it)^{-1} y), \tag{7.7}$$

and this relation holds for all $y \in \mathcal{H}$. Now we observe that $\sigma(T) = \sigma(\widetilde{A}|\widetilde{\mathcal{H}}_-)$, that this set lies in the open left half plane and that Lemma 6.1 can be applied to T.

Integrating the relation (7.7) along the imaginary axis we get from Lemma 6.1 that

$$2(\Delta y, y) = \frac{1}{\pi} \int_{-\infty}^{+\infty} \left(D\Delta(T - it)^{-1}y, \Delta(T - it)^{-1}y \right) \, dt \geq 0.$$

$\square$

The (possibly unbounded) nonnegative operator $\Lambda_- = \Pi_-^{-1}$ satisfies the Riccati equation

$$Dy - A\Lambda_- y - \Lambda_-(A^* + Q\Lambda_-)y = 0$$

for all $y \in \mathcal{D}(\Lambda_-)$ such that $\Lambda_- y \in \mathcal{D}(A)$. This follows immediately from (7.1) if we observe that $y \in \mathcal{D}(\Lambda_-)$ and $\Lambda_- y \in \mathcal{D}(A)$ imply $y \in \mathcal{D}(A^*)$, $A^*y + Q\Lambda_- y \in \mathcal{D}(\Lambda_-)$.

For symmetry reasons, the bounded nonpositive operator $\Lambda_+ = \Pi_+^{-1}$ satisfies the Riccati equation

$$\Lambda_+ Q\Lambda_+ x + \Lambda_+ A^* x + A\Lambda_+ x - Dx = 0, \quad x \in \mathcal{D}(A^*),$$

and the operator Π_+ the equation

$$Qy + A^*\Pi_+ y + \Pi_+(A - D\Pi_+)y = 0$$

for all $y \in \mathcal{D}(\Pi_+)$ such that $\Pi_+ y \in \mathcal{D}(A^*)$.

Acknowledgements. The authors gratefully acknowledge support by the RTN project HPRN-CT-2000-00116 of the European Community.

References

[AI]   Azizov, T.Ya., and Iokhvidov, I.S., *Linear Operators in Spaces with an Indefinite Metric*, J. Wiley and Sons, Chichester 1989.

[CDW]   Callier, F.M., Dumortier, L., and Winkin, J., On the nonnegative self-adjoint solutions of the operator Riccati equation for infinite dimensional systems, *Integral Equations and Operator Theory* **22** (1995), 162-195.

[CZ]   Curtain, R.F., and Zwart, H., *An Introduction to Infinite-Dimensional Linear Systems Theory*, Springer Verlag, New York 1995.

[DS]   Dijksma, A., and de Snoo, H.S.V., Symmetric and selfadjoint relations in Krein spaces. I, *Operator Theory: Adv. Appl.* **24** (1987), 145-166.

[GGK]   Gohberg, I., Goldberg, S., and Kaashoek, M.A., *Classes of Linear Operators. I*, Birkhäuser Verlag, Basel 1990.

[K]   Kato, T., *Perturbation Theory for Linear Operators*, Springer Verlag, Berlin 1995.

[LR]   Lancaster, P., and Rodman, L., *Algebraic Riccati Equations*, Oxford University Press Inc., New York 1995.

[LT]   Langer, H., and Tretter, C., Diagonalization of certain block operator matrices and applications to Dirac operators, *Operator Theory: Adv. Appl.* **122** (2001), 331-358.

[PS]   Pritchard, A.J., and Salamon, D., The linear quadratic optimal control problem for infinite-dimensional systems with unbounded input and output operators, *SIAM J. Control and Opt.* **25** (1987), 121-144.

[S]   Staffans, O.J., Quadratic optimal control of well-posed linear systems, *SIAM J. Control and Opt.* **37** (1999), 131-169.

[W]   Weiss, M., Riccati equation theory for Pritchard-Salamon systems: a Popov function approach. Distributed parameter systems: analysis, synthesis and applications, Part 1, *IMA J. Math. Control Inform.* **14** (1997), 45-83.

[WW]   Weiss, G., and Weiss, M., Optimal control of weakly regular linear systems, *Math. Control, Signals and Systems* **10** (1997), 287-330.

Technische Universität Wien
Institut für Analysis und Technische Mathematik
Wiedner Hauptstrasse 8-10
A-1040 Wien
Austria
hlanger@mail.zserv.tuwien.ac.at

Vrije Universiteit Amsterdam
Faculteit Exacte Wetenschappen
Divisie Wiskunde en Informatica
De Boelelaan 1081a
1081 HV Amsterdam
The Netherlands
ran@cs.vu.nl

Universiteit Leiden
Mathematisch Instituut
Niels Bohrweg 1
2333 CA Leiden
The Netherlands
barotten@math.leidenuniv.nl

1991 Mathematics Subject Classification. Primary 93B28, 47B50

Received November 24, 2000

Operator Theory:
Advances and Applications, Vol. 130, 255–276

Inertia Bounds for Operator Polynomials

Leonid Lerer and Leiba Rodman

Dedicated to Professor Peter Lancaster on occasion of his 70-th birthday

Abstract. Problems of location of spectra of operator polynomials with respect to the real and imaginary axes are studied. In particular, bounds for inertia of operator polynomials are obtained using generalized Bezoutians and appropriate results on inertia of Hilbert space operators, established in authors' earlier work.

1. Introduction

Let $\mathcal{H}$ be a Hilbert space over the field $\mathbb{C}$ of complex numbers, and let $L(\mathcal{H})$ denote the algebra of linear bounded operators on $\mathcal{H}$. (All operators are assumed to be bounded and linear.) The operator valued function $A : \mathbb{C} \to L(\mathcal{H})$ of the form

$$A(\lambda) = \sum_{j=0}^{\ell} \lambda^j A_j \qquad (A_j \in L(\mathcal{H}),\ j = 0, 1, \ldots, \ell) \qquad (1.1)$$

is referred to as an *operator polynomial*. The set

$$\Sigma(A) = \{\lambda \in \mathbb{C} \,|\, A(\lambda) \text{ is not invertible}\}$$

is called the *spectrum* of the operator polynomial (1.1).

The present paper concerns problems of location of the spectrum $\Sigma(A)$ of (1.1) with respect to the real or imaginary axis. In particular, we focus on bounds for inertia of operator polynomials (roughly speaking, the dimensions of the corresponding spectral subspaces). The paper can be viewed as a continuation of our studies of inertia of matrix and operator polynomials (see [LR1], [LR2], [LR3], [LRT1], [LRT2], [LT2]). The concept of inertia for operator polynomials is crucial in the study of behavior of solutions $x(t)$ of operator differential equations of the form $A(i(d/dt))x(t) = 0$ when $t \to \infty$. In case $A(\lambda) = \lambda I - A$, $A \in L(\mathcal{H})$, the inertia problem for $A(\lambda)$ coincides with the problem of location of the spectrum of the operator A (with respect to the imaginary axis). The latter is characterized by inertia theorems based on equation

$$AX + XA^* = M^*M \ (= W). \qquad (1.2)$$

By means of these theorems, one determines the inertia of A in terms of the inertia of the selfadjoint solution X of (1.2).

Let us recall one inertia theorem. Letting $\mathcal{G}$ be a Hilbert space, an operator $X \in L(\mathcal{G})$ is said to be *dichotomous* (with respect to the imaginary axis) if its spectrum $\sigma(X)$ does not intersect the imaginary axis (including zero). For a dichotomous operator X, the inertia $In(X)$ of X is defined as the ordered triple of nonnegative integers or infinity $(\dim \mathcal{M}_+, \dim \mathcal{M}_-, 0)$, where $\mathcal{M}_+$ (resp., $\mathcal{M}_-$) is the spectral X-invariant subspace corresponding to the part of $\sigma(X)$ in the right (resp., left) halfplane. This concept appears in [C1], [C2]. A pair of Hilbert space operators (A, W), where $A \in L(\mathcal{H})$, $W \in L(\mathcal{G}, \mathcal{H})$ (the Banach space of all linear bounded operators from $\mathcal{G}$ into $\mathcal{H}$) is called *exactly controllable* if the operator

$$[W, \ AW, \ A^2 W, \ \ldots, \ A^{q-1} W] \in L(\mathcal{G}^q, \mathcal{H})$$

is right invertible for some integer q. An operator $W \in L(\mathcal{H})$ is called *positive semidefinite* (notation: $W \geq 0$) if $\langle x, Wx \rangle \geq 0$ for every $x \in \mathcal{H}$.

Theorem 1.1. [Bu] *Let A, $X = X^*$, and $W \geq 0$ be Hilbert space operators such that (1.2) holds true. If (A, W) is exactly controllable, then A and X are dichotomous, and $In(A) = In(X)$.*

In previous works, we have extended the inertia results mentioned above, and in particular Theorem 1.1, to the case of matrix and operator polynomials of type (1.1). In our framework, the place of equation (1.2) is taken over by a polynomial operator equation

$$A(\overline{\lambda})^* A_1(\lambda) + A_1(\overline{\lambda})^* A(\lambda) = M(\overline{\lambda})^* M(\lambda), \tag{1.3}$$

where $A_1(\lambda)$ is an operator polynomial of type (1.1) with coefficients in $L(\mathcal{H})$, and $M(\lambda)$ is an operator polynomial with coefficients in $L(\mathcal{H}, \mathcal{G})$, subject to some additional conditions. In [LR2], we express the inertia of $A(\lambda)$ in terms of the inertia of a certain selfadjoint operator $\frac{1}{i}\mathbb{B}$, where $\mathbb{B}$ is the generalized Bezout operator associated with equality (1.3) (see Section 4 for the definition of $\mathbb{B}$). One of the most essential conditions imposed in [LR2] on equation (1.3) is the right coprimeness of the operator polynomials $A(\lambda)$ and $M(\lambda)$ in the sense of Bezout, i.e., existence of operator polynomials $X_1(\lambda)$ and $X_2(\lambda)$ such that

$$X_1(\lambda) A(\lambda) + X_2(\lambda) M(\lambda) = I, \qquad \lambda \in \mathbb{C}.$$

We note that the above coprimeness condition is a natural counterpart of the exact controllability condition in Theorem 1.1. To explain this, note that Theorem 1.1 can be interpreted as a result on spectrum localization for operator polynomials of first degree. Indeed, rewrite (1.2) in the form

$$B(\overline{\lambda})^* A_1 + A_1^* B(\lambda) = W, \tag{1.4}$$

where $A_1 = iX$, $B(\lambda) = \lambda I + iA^*$. Thus, a characteristic feature of Theorem 1.1 is that the distribution of the spectrum of the operator polynomial $\lambda I + iA^*$ with respect to the real axis is given in terms of the inertia of the selfadjoint operator $\frac{1}{i}A_1$. Moreover, the exact controllability hypothesis in Theorem 1.1 is equivalent to the operator $\begin{bmatrix} B(\lambda) \\ W \end{bmatrix}$ being left invertible for all $\lambda \in \mathbb{C}$ (see [KMR3], [T]).

In turn, the left invertibility of $\begin{bmatrix} B(\lambda) \\ W \end{bmatrix}$ for all $\lambda \in \mathbb{C}$ can be interpreted as coprimeness of the operator polynomials $B(\lambda)$ and W (see Theorem 2.2 of [LR1]).

Recently, the present authors have obtained a generalization of Theorem 1.1, in which the condition of exact controllability (or, which is the same, of coprimeness of $B(\lambda)$ and W) is relaxed. To state this generalization, we need some definitions. Let $A \in L(\mathcal{H})$. A point $\lambda_0 \in \sigma(A)$ is called a normal eigenvalue of A if λ_0 is an isolated point of $\sigma(A)$ and the corresponding spectral subspace

$$\mathcal{R}(A; \lambda_0) = \text{Range } \frac{1}{2\pi i} \int_{|\lambda - \lambda_0| = \varepsilon} (\lambda I - A)^{-1} d\lambda, \qquad (\varepsilon > 0 \text{ small})$$

is finite dimensional. For a non-empty subset $\Sigma \subseteq \mathbb{C}$, define

$$s(A; \Sigma) = \sum_{\lambda_0 \in \sigma(A) \cap \Sigma} \dim \mathcal{R}(A; \lambda_0)$$

if the intersection $\sigma(A) \cap \Sigma$ consists only of a finite number of normal eigenvalues of A; otherwise put $s(A; \Sigma) = \infty$. The ordered triple $\text{In } A = (s(A; \mathbb{C}^+), s(A; \mathbb{C}^-), s(A; i\mathbb{R}))$, where $\mathbb{C}^+$ and $\mathbb{C}^-$ stands for the open right and the open left half-plane, respectively, is called the *inertia* (with respect to the imaginary axis) of the operator A. (This extends the definition of inertia given earlier for dichotomous operators.)

A pair of operators (A, B), where $A \in L(\mathcal{H})$, $B \in L(\mathcal{G}, \mathcal{H})$ is called *almost exactly controllable* if for some positive integer p the linear set

$$\text{Range}[B, AB, \dots, A^{p-1}B] = \sum_{j=0}^{p-1} \text{Range}(A^j B)$$

is closed and has finite codimension in $\mathcal{H}$. Clearly, for an almost exactly controllable pair (A, B) the linear set

$$\sum_{j=0}^{\infty} \text{Range}(A^j B) \tag{1.5}$$

is also closed and has finite codimension. The subspace (1.5) is called the *controllability subspace* of (A, B) and will be denoted by $\mathcal{C}(A, B)$.

In [LR3] we have established the following result:

Theorem 1.2. *Let $A, X \in L(\mathcal{H})$ be such that $X = X^*$, the operator $W := AX + XA^*$ is positive semidefinite, and the pair (A, W) is almost exactly controllable. Then:*

(i) $|s(A; \mathbb{C}^+) - s(X; \mathbb{C}^+)| \leq \text{codim } \mathcal{C}(A, W)$;

(ii) $|s(A; \mathbb{C}^-) - s(X; \mathbb{C}^-)| \leq \text{codim } \mathcal{C}(A, W)$;

(iii) $s(X; i\mathbb{R}) \leq \text{codim } \mathcal{C}(A, W)$; *moreover,* $\text{Ker } X \subseteq \mathcal{C}(A, W)^\perp$;

(iv) $s(A; i\mathbb{R}) < \infty$;

(v) $\sum_{\lambda \in i\mathbb{R}} \dim \text{Ker}(A - \lambda I) \leq \text{codim} \mathcal{C}(A; W)$; *moreover,* $\text{Ker}(A^* - \lambda I) \subseteq \mathcal{C}(A, W)^\perp$ *for every* $\lambda \in i\mathbb{R}$.

Theorem 1.2 extends results obtained in [Ch], [W], [CL], [Bu], under the more restrictive hypothesis of exact controllability. The inertia theorem 1.2 is based on Lyapunov equation $AX + XA^* = W$. Analogous inertia theorems with respect to the unit disk, and based on the Stein equation $X - A^*XA = V$ with positive semidefinite operator V, are also valid (see [LR3]). Inertia theorems based on a more general Stein equation $B^*XB - A^*XA = V$ (however, assuming uniformly positive definite V) have been obtained in [BG].

The main goal of the present paper is to relax the condition of coprimeness of the operator polynomials $A(\lambda)$ and $M(\lambda)$ under which the results of [LR2] and [LRT2] have been obtained regarding operator polynomials that satisfy equation (1.3). To be more specific, instead of coprimeness we assume here that $A(\lambda)$ and $M(\lambda)$ are *almost* (right) *coprime*, which means that there exist two other operator polynomials $X_1(\lambda)$ and $X_2(\lambda)$ such that the spectrum of

$$S(\lambda) := X_1(\lambda)A(\lambda) + X_2(\lambda)M(\lambda)$$

is a finite (or empty) set, and $S(\lambda_0)$ is Fredholm for every $\lambda_0 \in \Sigma(S)$. Of course, in this case one expects to have certain bounds for the inertia characteristics of $A(\lambda)$, in the spirit of Theorem 1.2, rather than precise equalities. We obtain such bounds in the present paper. Note that in the case of equation (1.4), or, which is the same, equation (1.2), the condition of almost coprimeness of $B(\lambda)$ and W means that the pair (A, W) is almost exactly controllable. Thus, Theorem 1.2 can be viewed as a representative of results that we obtain in the present paper for operator polynomials $A(\lambda)$ of arbitrary degree satisfying (1.3).

The rest of the paper is organized as follows. In Section 2 we study the notion of almost coprimeness of operator polynomials. Theorem 1.2 is extended for operator polynomials in terms of linearizations in Section 3. In Section 4 we introduce generalized Bezoutians for operator polynomials and develop their basic properties. Our main results are given in Section 5, providing bounds for inertia of operator polynomials satisfying equation (1.3). We consider separately the case when the leading coefficient of $A(\lambda)$ is Fredholm (see Corollary 5.2). In Section 5 we also apply the main theorem to operator polynomials whose coefficients are block scalars modulo finite rank operators; the results take then a greatly simplified form. Several results of [LR2], [LRT2] are shown to be particular cases of the main results of the present paper. Finally, Section 6 contains applications to operator differential equations.

The following notation is used: $L(\mathcal{X}, \mathcal{Y})$ is the Banach space of operators acting from a Banach space $\mathcal{X}$ into a Banach space $\mathcal{Y}$. The letters $\mathcal{H}$, $\mathcal{G}$, with or without subscripts, stand for Hilbert spaces. We write

$$\text{col } (X_j)_{j=0}^p = \begin{bmatrix} X_0 \\ X_1 \\ \vdots \\ X_p \end{bmatrix}.$$

2. Operator Polynomials of Finite Type and Almost Coprimeness

Let $A(\lambda)$ be an operator polynomial, whose coefficients are operators on a Banach space $\mathcal{X}$. An operator $T \in L(\mathcal{Y})$, where $\mathcal{Y}$ is a Banach space, is called a *linearization* of $A(\lambda)$ if there exist Banach spaces $\mathcal{X}_1$ and $\mathcal{Y}_1$ such that

$$\begin{bmatrix} A(\lambda) & 0 \\ 0 & I_{\mathcal{X}_1} \end{bmatrix} = E(\lambda) \begin{bmatrix} \lambda I - T & 0 \\ 0 & I_{\mathcal{Y}_1} \end{bmatrix} F(\lambda), \quad \lambda \in \mathbb{C} \qquad (2.1)$$

for some operator-valued functions $E(\lambda) \in L(\mathcal{Y} \oplus \mathcal{Y}_1, \mathcal{X} \oplus \mathcal{X}_1)$, $F(\lambda) \in L(\mathcal{X} \oplus \mathcal{X}_1, \mathcal{Y} \oplus \mathcal{Y}_1)$, which are entire (i.e., analytic on $\mathbb{C}$) and have invertible values for every $\lambda \in \mathbb{C}$. (If dim $\mathcal{X} < \infty$, then $E(\lambda)$ and $F(\lambda)$ can be taken to be polynomials in the above definition.) Clearly, if (2.1) holds, then $\Sigma(A) = \sigma(T)$. If (2.1) holds for $\lambda \in \Omega$, $\sigma(T) \subseteq \Omega$, and $E(\lambda)$, $F(\lambda)$ are analytic and invertible operator functions defined on a open set $\Omega \subseteq \mathbb{C}$, then T will be called a *local linearization* of $A(\lambda)$ corresponding to Ω.

Proposition 2.1. [GKL], [KMR1]

(i) $A(\lambda)$ *admits a linearization if and only if the spectrum* $\Sigma(A)$ *is a compact set. Moreover, if* $\mathcal{X}$ *is a Hilbert space and a linearization of* $A(\lambda)$ *exists, then* $\mathcal{Y}$ *(the space on which a linearization acts) can be chosen to be a Hilbert space as well.*

(ii) *Any two linearizations of the same operator polynomial* $A(\lambda)$ *are similar.*

(iii) *Local realization of* $A(\lambda)$ *corresponding to* Ω *exists if and only if* $\Sigma(A) \cap \Omega$ *is compact.*

An operator polynomial $A(\lambda)$ is called of *finite type* if the spectrum $\Sigma(A)$ is a finite set (possibly empty) and $A(\lambda_0)$ is a Fredholm operator for every $\lambda_0 \in \Sigma(A)$. If $A(\lambda)$ is of finite type, then by the stability of the index of Fredholm operators (see [GGK], for example) it follows that $A(\lambda_0)$ has index zero for every $\lambda_0 \in \Sigma(A)$. Moreover (Lemma 3.1 of [GSi1]) the inverse $A(\lambda)^{-1}$ admits the following expansion in a neighborhood U of $\lambda_0 \in \Sigma(A)$:

$$A(\lambda)^{-1} = \Phi + \sum_{j=1}^{p} (\lambda - \lambda_0)^{-j} K_j + (\lambda - \lambda_0) M(\lambda), \quad \lambda \in U \setminus \{\lambda_0\}, \qquad (2.2)$$

where Φ is a Fredholm operator with index zero, $K_1, \ldots, K_p$ are finite rank operators, and the operator-valued function $M(\lambda)$ is analytic in U. Operator polynomials of finite type with respect to an open set of $\mathbb{C}$ are defined analogously.

Proposition 2.2. $A(\lambda)$ *is of finite type if and only if its linearization* T *is a finite-dimensional operator:* $T \in L(\mathbb{C}^m)$.

The number m will be called the *size* of $A(\lambda)$. If T is a local realization of an operator polynomial $A(\lambda)$ which is of finite type with respect to Ω, then the (finite) dimension of the space on which T acts will be denoted $m(A(\lambda), \Omega)$.

Proof. If T is finite dimensional, then clearly $\sigma(T)$, and hence $\Sigma(A)$ is finite. Formula (2.1) guarantees that $A(\lambda_0)$ is Fredholm for every $\lambda_0 \in \Sigma(A)$.

Assume now that $A(\lambda)$ is of finite type. By Theorem 1 of [GSi2], $A(\lambda)$ admits the following factorization:

$$A(\lambda) = E(\lambda) \left[I - \sum_{k=1}^{m} P_k + \sum_{k=1}^{m} d_k(\lambda) P_k \right] F(\lambda), \qquad (2.3)$$

where the operator valued functions $E(\lambda)$ and $F(\lambda)$ are analytic and invertible in $\mathbb{C}$, the operators $P_1, \ldots, P_m$ are rank one mutually orthogonal projections, and $d_k(\lambda)$ are (scalar) polynomials. (A more general result is proved in [L]; see Theorem 5.6 there.) Clearly, the polynomials $A(\lambda)$ and

$$D(\lambda) := \sum_{k=1}^{m} d_k(\lambda) P_k \in L \left(\sum_{k=1}^{m} \text{Range } P_k \right),$$

have the same linearizations (this follows from (2.3). But $D(\lambda)$ is an operator polynomial acting on a finite dimensional space, with the property that $\det D(\lambda) \not\equiv 0$, hence its linearization is finite dimensional (this fact can be easily deduced from the Smith form for matrix polynomials; see, e.g., [GLR2]). $\qquad\square$

Operator polynomials $A_1(\lambda)$, $A_2(\lambda)$, with coefficients in $L(\mathcal{X},\mathcal{Y})$ and $L(\mathcal{X},\mathcal{Z})$, respectively (here $\mathcal{X}$, $\mathcal{Y}$, $\mathcal{Z}$ are Banach spaces) are called *m-almost right coprime* if there exist operator polynomials (not necessarily with compact spectrum) $X_1(\lambda)$, $X_2(\lambda)$ such that $X_1(\lambda)A_1(\lambda) + X_2(\lambda)A_2(\lambda)$ is an operator polynomial with coefficients in $L(\mathcal{X},\mathcal{X})$ of finite type having size $\leq m$. If, in addition, $(X_1(\lambda)A_1(\lambda) + X_2(\lambda)A_2(\lambda))^{-1}$ has *at most polynomial growth at infinity*, i.e.,

$$(X_1(\lambda)A_1(\lambda) + X_2(\lambda)A_2(\lambda))^{-1} \leq |\lambda|^r$$

for all sufficiently large $|\lambda|$, where the positive constant r is independent of λ, then we say that $A_1(\lambda)$ and $A_2(\lambda)$ are *strongly m-almost right coprime*. Operator polynomials $A_1(\lambda)$, $A_2(\lambda)$ will be called *almost right coprime* (resp., *strongly almost right coprime*) if they are m-almost right coprime (resp., strongly m-almost right coprime) for some positive integer m.

Let $A(\lambda)$ be an operator polynomial, and let $\lambda_0 \in \Sigma(A)$. We say that λ_0 is a *point of finite multiplicity m* of $A(\lambda)$ if λ_0 is an isolated point of $\Sigma(A)$ and the local realization T_{λ_0} of $A(\lambda)$ corresponding to a small disc centered at λ_0 acts on a finite-dimensional space of dimension m: $T_{\lambda_0} \in L(\mathbb{C}^m)$. Otherwise, we say that λ_0 is a *point of infinite multiplicity* of $A(\lambda)$. We denote by $m(A(\lambda); \lambda_0) \leq \infty$ the multiplicity of $\lambda_0 \in \Sigma(A)$.

In case $m(A(\lambda); \lambda_0) < \infty$ we have the following formula:

$$m(A(\lambda); \lambda_0) = \operatorname{rank} \begin{bmatrix} A_{-q} & A_{-q+1} & \cdots & A_{-1} \\ 0 & A_{-q} & \cdots & A_{-2} \\ \vdots & \vdots & \ddots & \vdots \\ 0 & 0 & \cdots & A_{-q} \end{bmatrix}, \tag{2.4}$$

where $A(\lambda)^{-1} = \sum_{j=-q}^{\infty} (\lambda - \lambda_0)^j A_j$ in a deleted neighborhood of λ_0. Formula (2.4) is known in the finite dimensional case ($\dim \mathcal{X} < \infty$), see, e.g., Theorem 3.2 in [GR2]; if $\dim \mathcal{X} = \infty$, the proof is easily reduced to the finite dimensional case, using a local version of formula (2.3).

To formulate the main result of this section, we need the notion of a right spectral pair. Let $B(\lambda) = \sum_{j=0}^{p} \lambda^j B_j$, $B_j \in L(\mathcal{X})$, be an operator polynomial with a linearization $T \in L(\mathcal{Y})$. Then there exists an operator $X \in L(\mathcal{Y}, \mathcal{X})$ such that

$$\sum_{j=0} B_j X T^j = 0 \tag{2.5}$$

and the operator $\operatorname{col} (XT^j)_{j=0}^{p=1}$ is left invertible. The pair (X, T) is called a *right spectral pair* of $B(\lambda)$. It always exists provided $B(\lambda)$ has compact spectrum, and is unique up to similarity: if (X_1, T_1) is another right spectral pair of $B(\lambda)$, then $X_1 = XS$, $T_1 = S^{-1}TS$ for some invertible operator S. See [KMR1], [KMR2], Chapter 6 in [R] for more details.

We now state the main result of this section.

Theorem 2.3. *Let $A_1(\lambda)$ be an operator polynomial with compact spectrum. Let $A_2(\lambda) = \sum_{j=0}^{q} \lambda^j A_{2j}$ with $A_{2j} \in L(\mathcal{X}, \mathcal{Y})$. Then (i) $\Rightarrow$ (ii) $\Rightarrow$ (iii), where the statements (i), (ii), and (iii) are given below:*

(i) *$A_1(\lambda)$ and $A_2(\lambda)$ are m-almost right coprime;*

(ii) *the operator polynomial*

$$\tilde{A}(\lambda) = \begin{bmatrix} A_1(\lambda) \\ A_2(\lambda) \end{bmatrix} \in L(\mathcal{X}; \mathcal{X} \oplus \mathcal{Y})$$

has the properties that for every $\lambda \in \mathbb{C}$, the range of $\tilde{A}(\lambda)$ is closed and

$$\sum_{\lambda \in \mathbb{C}} \dim \operatorname{Ker} \tilde{A}(\lambda) \le m.$$

(iii) *Let $Z = \sum_{j=0}^{q} A_{2j} X T^j$, where (X, T) is a right spectral pair for $A_1(\lambda)$ with respect to $\mathbb{C}$; then for some positive integer r the operator $\tilde{Z} = \operatorname{col} (ZT^j)_{j=0}^{r-1}$ has the properties that Range $\tilde{Z}$ is closed and $\dim \operatorname{Ker} \tilde{Z} \le m_0 r$, where*

$$m_0 = \max_{\lambda_0} \dim \operatorname{Ker} \tilde{A}(\lambda_0).$$

Moreover, under the additional hypothesis that

$$\|A_1(\lambda)^{-1}\| \le |\lambda|^{\alpha}, \tag{2.6}$$

for $|\lambda|$ *sufficiently large and for some constant* $\alpha > 0$ *independent of* λ, *we have* (iii) $\Rightarrow$ (iv), *where*

(iv) $A_1(\lambda)$ *and* $A_2(\lambda)$ *are strongly 2t-almost right coprime, where*

$$t = \max\{(m_0 + 1)r, \text{degree of } A_1\};$$

the integers m_0 *and* r *are taken from* (iii).

Proof. (i) $\Rightarrow$ (ii). Let $X_1(\lambda)$, $X_2(\lambda)$ be such that

$$B(\lambda) := X_1(\lambda)A_1(\lambda) + X_2(\lambda)A_2(\lambda)$$

is of finite type having size $\leq m$. Then clearly Range $\tilde{A}(\lambda)$ is closed, and Ker $\tilde{A}(\lambda) = \{0\}$, for every $\lambda \notin \Sigma(B)$. For $\lambda_0 \in \Sigma(B)$ we have that

$$B(\lambda_0) = [X_1(\lambda_0), X_2(\lambda_0)]\tilde{A}(\lambda_0)$$

is Fredholm with zero index. Hence the restriction of $\tilde{A}(\lambda_0)$ to a subspace of finite codimension is left invertible; therefore, this restriction has closed range, and consequently $\tilde{A}(\lambda_0)$ itself has closed range. Finally, let $T \in L(\mathbb{C}^m)$ be a linearization of $B(\lambda)$. Then

$$E(\lambda)\begin{bmatrix} X_1(\lambda) & X_2(\lambda) & 0 \\ 0 & 0 & I_{H_1} \end{bmatrix}\begin{bmatrix} \tilde{A}(\lambda) & 0 \\ 0 & I_{H_1} \end{bmatrix}F(\lambda) = \begin{bmatrix} \lambda I - T & 0 \\ 0 & I_{G_1} \end{bmatrix}$$

for some entire invertible operator functions $E(\lambda)$ and $F(\lambda)$. Since

$$\sum_{\lambda \in \mathbb{C}} \dim \text{Ker}(\lambda I - T) \leq m,$$

we also have

$$\sum_{\lambda \in \mathbb{C}} \dim \text{Ker } \tilde{A}(\lambda) \leq m.$$

(ii) $\Rightarrow$ (iii). Let

$$m_0 = \max_{\lambda_0} \dim \text{Ker } \tilde{A}(\lambda_0).$$

Let $A_3(\lambda) = \Sigma\lambda^j A_{3j} \in L(\mathcal{X}, \mathbb{C}^{m_0})$ be an operator polynomial such that $A_3(\lambda_0)x \neq 0$ for every $x \in \text{Ker } \tilde{A}(\lambda_0) \setminus \{0\}$ and for every $\lambda_0 \in \Sigma(\tilde{A})$. (The existence of such $A_3(\lambda)$ is obvious; one can take $A_3(\lambda)$ to be a constant polynomial.) Then

$$\text{Ker}\begin{bmatrix} A_1(\lambda) \\ A_2(\lambda) \\ A_3(\lambda) \end{bmatrix} = \{0\}$$

for every $\lambda \in \mathbb{C}$. Let $Z_3 = \sum_j A_{3j}XT^j$. By Theorem 8.3.2 in [R], see also the remark after the proof of that theorem, we have that for some r, the operator

$$\begin{bmatrix} \tilde{Z} \\ \text{col } (Z_3T^j)_{j=0}^{r-1} \end{bmatrix}$$

is left invertible. This implies (iii).

(iii) $\Rightarrow$ (iv). Denote $\tilde{Z}_s = \text{col}\,(ZT^j)_{j=0}^{s-1}$. Let s_0 be the smallest integer $\geq r$ such that

$$\text{Ker}\,\tilde{Z}_{s_0} = \text{Ker}\,\tilde{Z}_{s_0+1}.$$

In view of (iii), such s_0 exists. Moreover, $s_0 \leq (m_0 + 1)r$. Then $\text{Ker}\,\tilde{Z}_{s_0}$ is T-invariant. Also, $\dim \text{Ker}\,\tilde{Z}_{s_0} \leq m_0 r$ (again by (iii)). Select $Z_3 : \text{Ker}\,\tilde{Z}_{s_0} \to \mathbb{C}$ such that

$$\text{col}\,(Z_3 T^j)_{j=0}^{t_0-1} : \text{Ker}\,\tilde{Z}_{s_0} \to \mathbb{C}^{t_0}$$

is invertible, where $t_0 = \dim \text{Ker}\,\tilde{Z}_{s_0}$. Fix any integer

$$t_1 \geq \max\{s_0, t_0, \text{degree of } A_1(\lambda)\}.$$

Denoting $X = \begin{bmatrix} Z \\ Z_3 \end{bmatrix}$, we have that the operator $\text{col}\,(XT^j)_{j=0}^{t_1-1}$ is left invertible. Let $Z_3' : \mathcal{Y} \to \mathbb{C}$, where $\mathcal{Y}$ is the domain of T, be any operator with the only property that $Z_3' x = Z_3 x$ for every $x \in \text{Ker}\,\tilde{Z}_{s_0}$. Then there exist operators A_{3j} such that

$$Z_3' = \sum_{j=0}^{t_1-1} A_{3j} X T^j.$$

We now let

$$A_2'(\lambda) = \sum_{j=0}^{t_1-1} \lambda^j \begin{bmatrix} A_{2j} \\ A_{3j} \end{bmatrix} = \sum_{j=0}^{t_1-1} \lambda^j A_{2j}',$$

and

$$Z' = \sum_{j=0}^{t_1-1} A_{2j}' X T^j.$$

Then

$$\text{Ker}\,\text{col}\,(Z'T^j)_{j=0}^{t_1-1} = \{0\},$$

and $\text{Im}\,\text{col}\,(Z'T^j)_{j=0}^{t_1-1}$ is closed. Denoting $A_3(\lambda) = \sum_{j=0}^{t_1-1} \lambda^j A_{3j}$, by Theorem 8.3.2 in [R] we have

$$X_1(\lambda)A_1(\lambda) + X_2(\lambda)A_2(\lambda) + X_3(\lambda)A_3(\lambda) \equiv I$$

for some operator polynomials $X_1(\lambda)$, $X_2(\lambda)$, and $X_3(\lambda)$ (here the hypothesis (2.6) is used). So

$$X_1(\lambda)A_1(\lambda) + X_2(\lambda)A_2(\lambda) = I - X_3(\lambda)A_3(\lambda).$$

Now the proof of Theorem 8.3.2 in [R] shows that the degree of $X_3(\lambda)$ is at most t_1. The degree of $A_3(\lambda)$ is at most $t_1 - 1$, so the degree of $I - X_3(\lambda)A_3(\lambda)$ is at most $2t_1 - 1$. On the other hand, notice that $1 - A_3(\lambda)X_3(\lambda)$ is a scalar polynomial of degree at most $2t_1 - 1$, and therefore it is of finite type having size at most $2t_1 - 1$. Using the well-known formulas (see, e.g., Section III.2 in [GGK]; for brevity we omit the variable λ)

$$\begin{bmatrix} 1 - A_3 X_3 & 0 \\ 0 & I \end{bmatrix} \begin{bmatrix} -A_3 & 1 \\ I - X_3 A_3 & X_3 \end{bmatrix} = \begin{bmatrix} -A_3 & 1 - A_3 X_3 \\ I & X_3 \end{bmatrix} \begin{bmatrix} I - X_3 A_3 & 0 \\ 0 & 1 \end{bmatrix},$$

together with

$$\begin{bmatrix} -A_3 & 1 \\ I - X_3 A_3 & X_3 \end{bmatrix}^{-1} = \begin{bmatrix} -X_3 & I \\ 1 - A_3 X_3 & A_3 \end{bmatrix},$$

$$\begin{bmatrix} -A_3 & 1 - A_3 X_3 \\ I & X_3 \end{bmatrix}^{-1} = \begin{bmatrix} -X_3 & I - X_3 A_3 \\ 1 & A_3 \end{bmatrix},$$

we see that the operator polynomial $I - X_3(\lambda)A_3(\lambda)$ is also of finite type having size at most $2t_1 - 1$. The formula

$$(I - X_3(\lambda)A_3(\lambda))^{-1} = I + X_3(\lambda)(I - A_3(\lambda)X_3(\lambda))^{-1}A_3(\lambda)$$

shows that $(I - X_3(\lambda)A_3(\lambda))^{-1}$ has at most polynomial growth at infinity. □

3. Spectral Quantities of Operator Polynomials

Let $A(\lambda)$ be an operator polynomial whose coefficients are operators on a Hilbert space $\mathcal{H}$. We define the spectral quantity $s(A(\lambda); \Omega)$ of $A(\lambda)$ with respect to a set $\Omega \subseteq \mathbb{C}$. Assume that the intersection $\Omega \cap \Sigma(A)$ is a finite set, and for every $\lambda_0 \in \Omega \cap \Sigma(A)$ the operator $A(\lambda_0)$ is Fredholm. Then we let

$$s(A(\lambda); \Omega) = \sum_{\lambda_0 \in \Omega \cap \Sigma(A)} m(A(\lambda); U_{\lambda_0}),$$

where U_{λ_0} is a sufficiently small neighborhood of λ_0. Otherwise, we let

$$s(A(\lambda); \Omega) = \infty.$$

Of particular interest for us will be the spectral quantities with respect to the imaginary axis $i\mathbb{R}$ and the half-planes $\mathrm{Re}\, z > 0$, $\mathrm{Re}\, z < 0$. We obviously have

$$s(A(\lambda); \Omega) = s(\lambda I - T; \Omega),$$

where T is a linearization of $A(\lambda)$.

The following result represents an extension of Theorem 1.2 to operator polynomials of arbitrary degree with compact spectrum. Using linearizations, it is obtained immediately from Theorem 1.2 and from the definition of $s(A(\lambda); \Omega)$.

Theorem 3.1. *Let $A(\lambda)$ be an operator polynomial with compact spectrum and linearization $T \in L(\mathcal{G})$. Assume that there exists a selfadjoint operator $X \in L(\mathcal{G})$ such that the operator $W := TX + XT^*$ is positive semidefinite and the pair (T, W) is almost exactly controllable. Then*

(i) $|s(A(\lambda); \mathrm{Re}\, z > 0) - s(\lambda I - X; \mathrm{Re}\, z > 0)| \leq \mathrm{codim}\, \mathcal{C}(T, W);$

(ii) $|s(A(\lambda); \mathrm{Re}\, z < 0) - s(\lambda I - X; \mathrm{Re}\, z < 0)| \leq \mathrm{codim}\, \mathcal{C}(T, W);$

(iii) $s(\lambda I - X; i\mathbb{R}) \leq \mathrm{codim}\, \mathcal{C}(T, W);$ *moreover,* $\mathrm{Ker}\, X \subseteq \mathcal{C}(T, W)^{\perp};$

(iv) $s(A(\lambda); i\mathbb{R}) < \infty;$

(v) $\sum_{\lambda \in i\mathbb{R}} \dim \mathrm{Ker}(A(\lambda)) \leq \mathrm{codim}\, \mathcal{C}(T; W).$

4. Generalized Bezoutians

Consider two collections of operator polynomials

$$K_j(\lambda) = \sum_{i=0}^{k_j} \lambda^i K_{ij}$$

and

$$A_j(\lambda) = \sum_{i=0}^{\ell_j} \lambda^i A_{ij} \quad (j = 1, \ldots, s)$$

with coefficients $K_{ij} \in L(\mathcal{G}_j, \mathcal{H})$, and $A_{ij} \in L(\mathcal{H}, \mathcal{G}_j)$. Assume that the equality

$$\sum_{j=1}^{s} K_j(\lambda) A_j(\lambda) = 0, \quad \lambda \in \mathbb{C}, \qquad (4.1)$$

holds. Then the operator valued function

$$\Gamma(x, y) := (x - y)^{-1} \sum_{j=1}^{s} K_j(x) A_j(y)$$

is an operator polynomial in the scalar variables x and y. Choose integers $k \geq \max(k_1, \ldots, k_s)$ and $\ell \geq \max(\ell_1, \ldots, \ell_s)$, and write

$$\Gamma(x, y) = \sum_{i,j=0}^{k-1, \ell-1} x^i y^j \Gamma_{ij}.$$

The $k \times \ell$ block operator matrix

$$\begin{aligned} \mathbb{B} &= \mathbb{B}(K_1, \ldots, K_s; A_1, \ldots, A_s; k, \ell) \\ &= [\Gamma_{ij}]_{i,j=0}^{k-1, \ell-1} \in L(\mathcal{H}^\ell, \mathcal{H}^k) \end{aligned} \qquad (4.2)$$

will be referred to as the *Bezoutian associated with* (4.1). Essentially, this is the concept introduced and studied in [LT1] for matrix polynomials (see also [LRT1]). For two pairs of matrix polynomials the Bezoutian was introduced in [AJ] and studied in [BKAK], [LT1], [LT2] in connection with common divisors and also in [LT1] in connection with root separation problems; the Bezoutian for two pairs of operator polynomials and its various applications are studied in [LRT2], [CK], [LM].

This concept of Bezoutian was introduced and studied in [LT1] for the case of matrix polynomials. More general notions of Bezoutians that subsume the definition (4.2) were developed in [LT4] and [GS], in connection with fast inversion of structured matrices. Note that in the scalar case (i.e., when all Hilbert spaces are one-dimensional) the Bezoutian defined in (4.2) for $s = 2$ coincides with the classical Bezoutian for two scalar polynomials $K_1(\lambda) = A_2(\lambda)$ and $K_2(\lambda) = -A_1(\lambda)$. Some other matrix generalizations of the classical Bezoutian have appeared also in the literature [BL], [H].

We now introduce some useful notational devices. Let $F(\lambda) = \sum_{i=0}^{t} \lambda^i F_i$ be an operator polynomial with coefficients $F_j \in L(\mathcal{G}, \mathcal{H})$. For a pair (X, T) of operators $X \in L(\mathcal{G}_1, \mathcal{G})$, $T \in L(\mathcal{G}_1)$, we let

$$F(X, T)_r = \sum_{i=0}^{\ell} F_i X T^i \in L(\mathcal{G}_1, \mathcal{H}).$$

For a pair (V, Z) of operators $Z \in L(\mathcal{H}, \mathcal{H}_1)$, $V \in L(\mathcal{H}_1)$, we let

$$F(V, Z)_\ell = \sum_{i=0}^{\ell} V^i Z F_i \in L(\mathcal{G}, \mathcal{H}_1).$$

(The subscripts "r" and "ℓ" stand for "right" and "left", respectively.) We need the following results which are proved in [LT1] (see also [LRT1], [LT2]) for the matrix polynomial case and can be extended to operator polynomials without any difficulties.

Theorem 4.1. *Let $\mathbb{B}$ be the Bezoutian associated with (4.1), and let k and ℓ be positive integers such that*

$$k \geq \max\{k_j \mid j = 1, \ldots, s\};$$
$$\ell \geq \max\{\ell_j \mid j = \ell, \ldots, s\}.$$

Then:

(i) *for every pair of operators $X \in L(\mathcal{H}_0, \mathcal{H})$, $T \in L(\mathcal{H}_0)$, we have*

$$\mathbb{B} \operatorname{col}[XT^{j-1}]_{j=1}^{i} = \sum_{j=1}^{s} \begin{bmatrix} K_{1j} & K_{2j} & \cdots & K_{kj} \\ K_{2j} & & \cdot^{\cdot^{\cdot}} & 0 \\ \vdots & \cdot^{\cdot^{\cdot}} & 0 & \\ K_{kj} & & & \end{bmatrix} \operatorname{col}[A_j(X, T)_r T^{i-1}]_{i=1}^{k};$$

(ii) *for every pair of operators $Z \in L(\mathcal{H}, \mathcal{H}_0)$, $V \in L(\mathcal{H}_0)$ we have*

$$-\operatorname{row}[V^{j-1} Z]_{j=1}^{k} \mathbb{B} = \sum_{j=1}^{s} \operatorname{row}[V^{i-1} M_j(V, Z)_\ell]_{i=1}^{\ell} \begin{bmatrix} A_{1j} & A_{2j} & \cdots & A_{\ell j} \\ A_{2j} & & \cdot^{\cdot^{\cdot}} & 0 \\ \vdots & \cdot^{\cdot^{\cdot}} & 0 & \\ A_{\ell j} & & & \end{bmatrix};$$

(iii) *for every integer $p \geq \max\{\ell, k\}$, the following representations of the Bezoutian are valid:*

$$
\mathbb{B} = \sum_{j=1}^{s}
\begin{bmatrix}
K_{pj} & K_{p-1,j} & \cdots & K_{1j} \\
 & \ddots & & \vdots \\
 & & \ddots & K_{p-1,j} \\
0 & & & \\
 & & & K_{pj}
\end{bmatrix}
\begin{bmatrix}
 & & & A_{0j} \\
0 & & \cdots & A_{1j} \\
 & & \cdots & \vdots \\
 & \cdots & & \vdots \\
A_{0j} & A_{1j} & \cdots & A_{p-1,j}
\end{bmatrix},
$$

$$
-\mathbb{B} = \sum_{j=1}^{s}
\begin{bmatrix}
 & & & K_{0j} \\
0 & \cdots & K_{ij} \\
 & \cdots & & \vdots \\
 & \cdots & & \vdots \\
K_{0j} & K_{ij} & \cdots & K_{p-1,j}
\end{bmatrix}
\begin{bmatrix}
A_{pj} & & & \\
A_{p-1,j} & \ddots & 0 \\
\vdots & & \ddots \\
\vdots & & & \ddots \\
A_{1j} & \cdots & A_{p-1,j} & A_{pj}
\end{bmatrix}.
$$

Here and elsewhere we put $K_{ij} = 0$ if $i > k_j$, and $A_{ij} = 0$ if $i > \ell_j$.
Assume that the equality (4.1) holds, and let

$$
\tilde{\mathbb{B}} := \tilde{\mathbb{B}}(V, Z; X, T) = \text{row}[V^{j-1}Z]_{j=1}^{k}\, \mathbb{B}\, \text{col}[XT^{j-1}]_{j=1}^{t}, \tag{4.3}
$$

where B denotes the Bezoutian associated with (4.1) and $X, T, V,$ and Z are operators. For convenience the matrix $\tilde{\mathbb{B}}$ in (4.3) will be called a *modified Bezoutian* associated with B (this terminology was introduced in [LT2]).

Theorem 4.2. *The modified Bezoutian defined in (4.3) satisfies the equation*

$$
V\tilde{\mathbb{B}} - \tilde{\mathbb{B}}T = \sum_{j=1}^{s}(K_j(V, Z)_\ell)(A_j(X, T)_r). \tag{4.4}
$$

The proof is based on the formulas of Theorem 4.1 (see [LT2]).
Consider now the Bezoutian associated with an equation of the form

$$
A^*(\lambda)Y(\lambda) + Y^*(\lambda)A(\lambda) - M^*(\lambda)M(\lambda) = 0, \tag{4.5}
$$

where $X^*(\lambda) = (X(\bar{\lambda}))^*$. It turns out that the Bezoutian associated with (4.5) enjoys a certain symmetry.

Proposition 4.3. *Let $A(\lambda) = \sum_{i=0}^{\ell} \lambda^i A_i$, $Y(\lambda) = \sum_{i=0}^{\ell} \lambda^i Y_i$ and $M(\lambda) = \sum_{i=0}^{\ell} \lambda^i M_i$ be operator polynomials that satisfy (4.5). Then the Bezoutian $\mathbb{B}$ associated with (4.5) is skew-adjoint.*

Proposition 4.3 is proved in [LRT1] for matrix polynomials. The same proof works for operator polynomials as well.

268 L. Lerer and L. Rodman

5. Main Results

We now state one of the main results of this paper. It represents a generalization of Theorem 1.3 in [LR2].

Theorem 5.1. *Let $A(\lambda)$ be an operator polynomial with coefficients in $L(\mathcal{H})$ of degree $\leq \ell$ with compact spectrum. Assume that there exists an operator polynomial $A_1(\lambda)$ of degree $\leq \ell$ such that*

$$(A(\bar{\lambda}))^* A_1(\lambda) + (A_1(\bar{\lambda}))^* A(\lambda) = (M(\bar{\lambda}))^* M(\lambda), \tag{5.1}$$

for some operator polynomial $M(\lambda)$ of degree $\leq \ell$ with coefficients in $L(\mathcal{H}, \mathcal{G})$ such that $M(\lambda)$ and $A(\lambda)$ are almost right coprime. Let $\mathbb{B}$ be the Bezoutian associated with the equality

$$(A(\bar{\lambda}))^* A_1(\lambda) + (A_1(\bar{\lambda}))^* A(\lambda) - (M(\bar{\lambda}))^* M(\lambda) = 0. \tag{5.2}$$

Then:

(i) $\mathbb{B}$ *is skew-adjoint:* $\mathbb{B} = -\mathbb{B}^*$;

(ii) *all spectral points of $A(\lambda)$ on the real line are points of finite multiplicity,*

(iii) *let $S_1 = Z^* \left(\frac{1}{i}\mathbb{B}\right) Z$, where $Z = \mathrm{col}[X_A T_A^{j-1}]_{j=1}^{\ell}$, and (X_A, T_A) is the right spectral pair for $A(\lambda)$; then*

$$|s(A(\lambda); \mathrm{Im}\,\lambda > 0) - s(\lambda I - S_1; \mathrm{Re}\,\lambda > 0)|$$

$$\leq \dim \left(\bigcap_{j=0}^{\infty} \mathrm{Ker} \left(M(X_A, T_A)_r T_A^j \right) \right) < \infty; \tag{5.3}$$

$$|s(A(\lambda); \mathrm{Im}\,\lambda < 0) - s(\lambda I - S_1; \mathrm{Re}\,\lambda < 0)|$$

$$\leq \dim \left(\bigcap_{j=0}^{\infty} \mathrm{Ker} \left(M(X_A, T_A)_r T_A^j \right) \right) < \infty. \tag{5.4}$$

Proof. We follow the proof of Theorem 1.3 in [LR2]. Assume the hypotheses of the theorem, and use the notation introduced in (iii).

By Theorem 4.2, we have

$$\begin{aligned} T_A^*(iS_1) - (iS_1)T_A &= (A(X_A, T_A)_r)^* A_1(X_A, T_A)_r \\ &\quad + (A_1(X_A, T_A)_r)^* A(X_A, T_A)_r \\ &\quad - (M(X_A, T_A)_r)^* M(X_A, T_A)_r. \end{aligned} \tag{5.5}$$

The first two terms on the right are zeros because $A(X_A, T_A)_r = 0$ (as it follows from (2.5). Thus (5.5) can be rewritten in the form

$$T_1 S_1 + S_1 T_1^* = -(M(X_A, T_A)_r)^* M(X_A, T_A)_r, \tag{5.6}$$

where $T_1 = iT_A^*$.

By (i) $\Rightarrow$ (iii) of Theorem 2.3, the pair $(M(X_A, T_A)_r, T_A)$ is almost exactly observable, i.e., the pair $(T_A^*, (M(X_A, T_A)_r)^*)$ is almost exactly controllable. It is easy to see that

$$(T_A^*, (M(X_A, T_A)_r)^* M(X_A, T_A)_r)$$

is almost controllable as well, and moreover

$$\mathcal{C}(T_A^*, (M(X_A, T_A)_r)^*) = \mathcal{C}(T_A^*, (M(X_A, T_A)_r)^* M(X_A, T_A)_r).$$

Now apply Theorem 3.1. The skew-adjointness of $\mathbb{B}$ follows from Proposition 4.3. $\square$

For some special classes of operator polynomials the statement of Theorem 5.1 may be simplified, or put in a different form (using $\mathbb{B}$ rather than S_1). We consider three such classes in the following corollaries.

Corollary 5.2. *Assume the hypotheses and notation of Theorem 5.1. If the leading coefficient of $A(\lambda)$ is Fredholm, then the range of Z is closed and has finite codimension q, the operator $\mathbb{B}$ is Fredholm, and the inequalities*

$$\left| s(A(\lambda); \operatorname{Im} \lambda > 0) - s(\lambda I - \frac{1}{i} \mathbb{B}; \operatorname{Re} \lambda > 0) \right|$$

$$\leq \quad q + \dim \left(\bigcap_{j=0}^{\infty} \operatorname{Ker} \left(M(X_A, T_A)_r T_A^j \right) \right) \tag{5.7}$$

and

$$\left| s(A(\lambda); \operatorname{Im} \lambda < 0) - s(\lambda I - \frac{1}{i} \mathbb{B}; \operatorname{Re} \lambda < 0) \right|$$

$$\leq \quad q + \dim \left(\bigcap_{j=0}^{\infty} \operatorname{Ker} \left(M(X_A, T_A)_r T_A^j \right) \right) \tag{5.8}$$

hold true. In particular, if the leading coefficient of $A(\lambda)$ is invertible, then $q = 0$.

Proof. The statement about the range of Z follows using Theorem 2.3 of [LR2]; the theorem states that if (X, T) is a right spectral pair of an operator polynomial of degree ℓ with compact spectrum and Fredholm leading coefficient, then the range of the operator $\operatorname{col}[XT^{j-1}]_{j=1}^{\ell}$ is closed and has finite codimension. The operator S_1 is Fredholm (as it follows from Theorem 3.1), and therefore so is $\mathbb{B}$. Now (5.7) and (5.8) can be easily deduced from (5.3) and (5.4), respectively, and from the equation $S_1 = Z^* \left(\frac{1}{i} \mathbb{B} \right) Z$; compare the proof of Corollary 4.2 in [LR2]. Finally, if the leading coefficient of $A(\lambda)$ is invertible, then Z is known to be invertible as well (see [GLR1] or Section 2.1 in [R]). $\square$

To formulate the second corollary, we need additional notation. Fix an infinite dimensional Hilbert space $\mathcal{H}_0$, and write

$$\mathcal{H}_0^m = \mathcal{H}_0 \oplus \mathcal{H}_0 \oplus \cdots \oplus \mathcal{H}_0, \tag{5.9}$$

where $\mathcal{H}_0$ appears m times in the right-hand side. Denote by $F_{m,n}$ the class of operators $S : \mathcal{H}_0^m \to \mathcal{H}_0^n$ which have the operator matrix representation with respect to decomposition (5.9) of the form

$$S = [\alpha_{i,j} I_{\mathcal{H}_0} + K_{i,j}]_{i=1;j=1}^{n;m}, \qquad (5.10)$$

where $\alpha_{i,j} \in \mathbb{C}$, and where $K_{i,j} \in L(\mathcal{H}_0)$ are finite rank operators. For $S \in F_{m,n}$ given by (5.10), we let

$$\widehat{S} = [\alpha_{i,j}]_{i=1;j=1}^{n;m} \in L(\mathbb{C}^m, \mathbb{C}^n).$$

For an operator polynomial $A(\lambda) = \sum_{j=0}^{\ell} \lambda^j A_j$ with coefficients in $F_{m,n}$, we let

$$\widehat{A}(\lambda) = \sum_{j=0}^{\ell} \lambda^j \widehat{A}_j, \quad \widetilde{A}(\lambda) = \sum_{j=0}^{\ell} \lambda^j \left(\widehat{A}_j \otimes I_{\mathcal{H}_0} \right).$$

Note that the difference $A(\lambda) - \widetilde{A}(\lambda)$ has finite rank; more precisely, there exists a finite dimensional subspace $\mathcal{M} \subset \mathcal{H}_0^n$ such that $\text{Range}\,(A(\lambda) - \widetilde{A}(\lambda)) \subseteq \mathcal{M}$ for every $\lambda \in \mathbb{C}$. Note also that the hypothesis that $\dim \mathcal{H}_0 = \infty$ is needed to distinguish the spectral properties of $A(\lambda)$ from those of $\widehat{A}(\lambda)$ (cf. statement (c) in the next proposition).

Proposition 5.3. *Let be given operator polynomials $M(\lambda) \in F_{n,m}$ and $A(\lambda) \in F_{n,n}$, and assume that $\text{Ker}\, A(\lambda_0) = \{0\}$ for some $\lambda_0 \in \mathbb{C}$. Then the following statements are equivalent:*

(a) *$M(\lambda)$ and $A(\lambda)$ are almost right coprime;*
(b) *$M(\lambda)$ and $A(\lambda)$ are strongly almost right coprime;*
(c) $\qquad \text{Ker}\,\widehat{A}(\lambda) \cap \text{Ker}\,\widehat{M}(\lambda) = \{0\}, \quad$ *for every $\lambda \in \mathbb{C}$.* $\qquad (5.11)$

Proof. The implication (a) $\Rightarrow$ (c) follows from Theorem 2.3, (i) $\Rightarrow$ (ii). Indeed, if

$$\text{Ker}\,\widehat{A}(\lambda') \cap \text{Ker}\,\widehat{M}(\lambda') \neq \{0\}$$

for some $\lambda' \in \mathbb{C}$, then

$$\dim \left(\text{Ker}\,\widetilde{A}(\lambda') \cap \text{Ker}\,\widetilde{M}(\lambda') \right) = \infty,$$

and because the differences $A(\lambda') - \widetilde{A}(\lambda')$, $M(\lambda') - \widetilde{M}(\lambda')$ have finite ranks, we have also

$$\dim\,(\text{Ker}\,A(\lambda') \cap \text{Ker}\,M(\lambda')) = \infty,$$

a contradiction with (ii) of Theorem 2.3.

Conversely, assume (5.11) holds. By Proposition 6.1 of [LR2], the spectrum of $A(\lambda)$ is a finite set, and $A(\lambda)$ has at most polynomial growth at infinity (here the condition $\text{Ker}\,A(\lambda_0) = \{0\}$ was used). By Theorem 2.3, we need only to show that the condition (ii) of Theorem 2.3 is satisfied, i.e., that

$$\text{Range} \begin{bmatrix} A(\lambda) \\ M(\lambda) \end{bmatrix} \qquad (5.12)$$

is closed for every $\lambda \in \mathbb{C}$, and

$$\sum_{\lambda \in \mathbb{C}} \dim \operatorname{Ker} \begin{bmatrix} A(\lambda) \\ M(\lambda) \end{bmatrix} < \infty. \tag{5.13}$$

The closedness of (5.12) is evident because $\operatorname{Im} \begin{bmatrix} \widetilde{A}(\lambda) \\ \widetilde{M}(\lambda) \end{bmatrix}$ is closed and the differ-

ence $\begin{bmatrix} A(\lambda) \\ M(\lambda) \end{bmatrix} - \begin{bmatrix} \widetilde{A}(\lambda) \\ \widetilde{M}(\lambda) \end{bmatrix}$ has finite rank. As for (5.13), note that since $\Sigma(A)$ is finite, we only have to check that

$$\dim \operatorname{Ker} \begin{bmatrix} A(\lambda') \\ M(\lambda') \end{bmatrix} < \infty \tag{5.14}$$

for every $\lambda' \in \mathbb{C}$. But (5.11) implies that $\begin{bmatrix} \widetilde{A}(\lambda') \\ \widetilde{M}(\lambda') \end{bmatrix}$ is left invertible, and again

using the finite rank property of $\begin{bmatrix} A(\lambda') \\ M(\lambda') \end{bmatrix} - \begin{bmatrix} \widetilde{A}(\lambda') \\ \widetilde{M}(\lambda') \end{bmatrix}$, (5.14) follows. $\square$

Thus, for operator polynomials with coefficients in $F_{n,m}$, the statement of Theorem 5.1 can be greatly simplified:

Corollary 5.4. *Let $A(\lambda) \in F_{n,n}$ be an operator polynomial of degree $\leq \ell$ such that $\operatorname{Ker} A(\lambda_0) = \{0\}$ for some $\lambda_0 \in \mathbb{C}$. Assume that (5.1) holds for some operator polynomials $A_1(\lambda)$ and $M(\lambda)$ of degrees $\leq \ell$, where $M(\lambda) \in F_{n,m}$ is such that $\operatorname{Ker} \widetilde{A}(\lambda) \cap \operatorname{Ker} \widetilde{M}(\lambda) = \{0\}$ for every $\lambda \in \mathbb{C}$. Then all conclusions of Theorem 5.1 hold true.*

Theorem 6.3 of [LR2] appears as a particular case of Corollary 5.4.

Consider now the case when $\mathcal{H}$ and $\mathcal{G}$ are finite dimensional: $\mathcal{H} = \mathbb{C}^n$, $\mathcal{G} = \mathbb{C}^p$. Thus, the operator polynomials that appear in (5.1) are in fact matrix polynomials. First note the following well-known fact:

Proposition 5.5. *Let $A(\lambda)$ (with coefficients of size $q \times n$) and $B(\lambda)$ (with coefficients of size $r \times n$) be matrix polynomials. Assume that the columns of $\begin{bmatrix} A(\lambda_0) \\ B(\lambda_0) \end{bmatrix}$ are linearly independent for some $\lambda_0 \in \mathbb{C}$. Then $A(\lambda)$ and $B(\lambda)$ are m-almost right coprime if and only if*

$$m \geq \text{degree} \left(\det D(A(\lambda), B(\lambda)) \right),$$

where we denote by $D(A(\lambda), B(\lambda))$ the greatest right common divisor of $A(\lambda)$ and $B(\lambda)$.

Note that under the hypotheses of Proposition 5.5 the greatest right common divisor $D(A(\lambda), B(\lambda))$, which is a matrix polynomial of size $n \times n$, exists, it is unique up to a multiplication on the left by a matrix polynomial having a constant non-zero determinant, and $\det D(A(\lambda), B(\lambda)) \not\equiv 0$ (see [GLR2] and [GKLR] for these and additional properties of greatest common divisors of matrix polynomials).

A proof of Proposition 5.5 can be easily derived from the following two facts: (1) The equation

$$X(\lambda)A(\lambda) + Y(\lambda)B(\lambda) = S(\lambda)$$

is solvable for matrix polynomials $X(\lambda)$ and $Y(\lambda)$ if and only if $D(A(\lambda), B(\lambda))$ is a right divisor of the matrix polynomial $S(\lambda)$ (see [LW], where additional relevant references are found as well); and (2) The linearization of $S(\lambda)$ is a matrix of size $s \times s$, where $s = \text{degree}\,(\det S(\lambda))$ (this can be easily deduced from the Smith form of matrix polynomials; see, for example, [GLR2]).

Next, it is convenient to formulate a proposition.

Proposition 5.6. *If $A(\lambda)$ is an $n \times n$ matrix polynomial with determinant not identically zero, and $M(\lambda)$ is a $p \times n$ matrix polynomial, then*

$$\dim \left(\bigcap_{j=0}^{\infty} \text{Ker}\, \left(M(X_A, T_A)_r T_A^j \right) \right) = \text{degree}\,(\det D(M(\lambda), A(\lambda))). \tag{5.15}$$

Proof. Denote

$$\mathcal{N} = \bigcap_{j=0}^{\infty} \text{Ker}\, \left(M(X_A, T_A)_r T_A^j \right).$$

The subspace $\mathcal{N}$ is the largest T_A-invariant subspace contained in $\text{Ker}\, M(X_A, T_A)_r$. Let $E(\lambda)$ be an $n \times n$ matrix polynomial whose right spectral pair is similar to $(X_A|_\mathcal{N}, T_A|_\mathcal{N})$; a construction of such polynomial $E(\lambda)$ is given in [GR1]. Since $A(X_A, T_A)_r = 0$, also $A(X_A|_\mathcal{N}, T_A|_\mathcal{N})_r = 0$, and therefore by the division theorem for matrix polynomials (Theorem 2.1 in [GKLR] or Theorem 7.10 in [GLR2]) $E(\lambda)$ is a right divisor of $A(\lambda)$. Since $\mathcal{N} \subseteq \text{Ker}\, M(X_A, T_A)_r$, we have $M(X_A|_\mathcal{N}, T_A|_\mathcal{N})_r = 0$, and therefore by the same division theorem $E(\lambda)$ is a right divisor of $M(\lambda)$. (Note that although the division theorem is stated in [GKLR], [GLR2] for the case when the dividend matrix polynomial is of square size with not identically zero determinant, the same proof works also in the more general case when the dividend may be of rectangular size.) Thus, $E(\lambda)$ is common right divisor of $A(\lambda)$ and of $M(\lambda)$. In fact, $E(\lambda)$ is a greatest right common divisor of $A(\lambda)$ and $M(\lambda)$, as one can easily see using the division theorem again. Now it remains to observe that the dimension of the subspace $\mathcal{N}$, the space on which the linearization of $E(\lambda)$ acts as a linear transformation, coincides with the degree of $\det E(\lambda)$ (cf. the fact (2) above). $\qquad\square$

Note that in case when $M(\lambda)$ is of square size, the result of Proposition 5.6 can be found in [LW] (see Remark 2.2 there).

Using (5.15), Theorem 5.1 leads to the following result:

Corollary 5.7. *Let $A(\lambda)$ be an $n \times n$ matrix polynomial of degree $\leq \ell$ with $\det A(\lambda) \not\equiv 0$. Assume that there exists an $n \times n$ matrix polynomial $A_1(\lambda)$ of degree $\leq \ell$ such that (5.1) holds for some $p \times n$ matrix polynomial $M(\lambda)$ of degree $\leq \ell$. Let $\mathbb{B}$ be the Bezoutian associated with the equality (5.2), and define S_1 as in Theorem 5.1. Then S_1 is a Hermitian matrix, and the equalities*

$$|s(A(\lambda); \operatorname{Im} \lambda > 0) - s(\lambda I - S_1; \operatorname{Re} \lambda > 0)| \leq \text{degree} \det D(A(\lambda), M(\lambda)) \quad (5.16)$$

and

$$|s(A(\lambda); \operatorname{Im} \lambda < 0) - s(\lambda I - S_1; \operatorname{Re} \lambda < 0)| \leq \text{degree} \det D(A(\lambda), M(\lambda)) \quad (5.17)$$

hold true. If the leading coefficient of $A(\lambda)$ is invertible, then S_1 can be replaced by $\frac{1}{i}\mathbb{B}$ in (5.16) and (5.17).

Regarding Corollary 5.7, note that any two matrix polynomials of appropriate sizes are m-almost right coprime, where m is the degree of the determinant of their greatest right common divisor (provided one of the polynomials has square size and its determinant is not identically zero).

Returning to the general operator case of Theorem 5.1, note that if the operator polynomials $A(\lambda)$ and $M(\lambda)$ are right coprime in the usual Bézout sense, i.e., they are 0-almost right coprime in the terminology of the present paper, and if $A(\lambda)^{-1}$ has at most polynomial growth at infinity, then from Theorem 2.2 of [LR2] we know that

$$\bigcap_{j=0}^{\infty} \operatorname{Ker}\left(M(X_A, T_A)_r T_A^j\right) = \{0\}.$$

Using this observation, Theorem 1.3 of [LR2] may be deduced from Theorem 5.1.

6. Almost Stability of Operator Differential Equations

Consider the differential equation

$$A\left(i\frac{d}{dt}\right)x(t) = 0, \qquad t \geq 0, \quad (6.1)$$

where $A(\lambda) = \sum_{j=0}^{\ell} \lambda^j A_j$ is an operator polynomial with coefficients in $L(\mathcal{H})$. We will assume throughout this section that $A(\lambda)$ has compact spectrum, and that $A(\lambda)^{-1}$ has at most polynomial growth at infinity. The latter condition means that

$$\|A(\lambda)\| \leq C|\lambda|^{\alpha} \quad (6.2)$$

for all $\lambda \in \mathbb{C}$ having sufficiently large absolute value, where the constants $C > 0$ and α (assumed to be integer) are independent of λ. The $\mathcal{H}$-valued function $x(t)$, $t \geq 0$, is said to be a solution of (6.1) if $x(t)$ has all derivatives up to the order $\max\{\ell, 2\ell + \alpha\}$, where α is taken from (6.2), and satisfies (6.1). The equation (6.1) will be called *almost exponentially stable* if for every linear set $\mathcal{V}$ of solutions of (6.1) having a sufficiently large finite dimension, there is a non-trivial exponentially decaying solution $x(t) \in \mathcal{V}$.

Theorem 6.1. *Assume the hypotheses and notation of Theorem 5.1. If the spectral subspace corresponding to the positive part of the spectrum of the selfadjoint operator S_1 is finite dimensional, then the equation* (6.1) *is almost exponentially stable.*

If, in addition, the leading coefficient of $A(\lambda)$ is Fredholm, then one can replace S_1 by $\frac{1}{i}\mathbb{B}$ in the preceding paragraph.

Proof. Recall that by Theorem 1.7 of [LR2] the general solution of the operator differential equation $A(\frac{d}{dt})x(t) = 0$ is given by the formula

$$x(t) = Xe^{tT}x_0, \qquad (6.3)$$

where (X,T) is a right spectral pair of $A(\lambda)$, and $x_0 \in \mathcal{H}$ is fixed. Moreover, the formula (6.3) provides a one-to-one correspondence between the solutions $x(t)$ and vectors x_0. Now apply Theorem 5.1, and, after some simple algebra, the result follows. For the statement in the second paragraph of Theorem 6.1, use Corollary 5.2 instead. □

Analogously, Corollary 5.4 may be used to obtain results on almost exponential stability of differential equations (6.1). We leave the formulation and proof of such results to interested readers.

Acknowledgements. The research of both authors is partially supported by United States–Israel Binational Science Foundation Grant 9400271. The research of the second author is partially supported by the NSF Grant DMS-9800704.

References

[AJ]   Anderson, B. D. O., Jury, E. I., Generalized Bezoutian and Sylvester matrices in multivariable linear control, *IEEE Trans. Automatic Control* **AC-21** (1976), 551–556.

[BL]   Barnett, S., Lancaster, P., Some properties of the Bezoutian for polynomial matrices, *Linear and Multilinear Algebra* **9** (1980), 99–111.

[BG]   Ben-Artzi, A., Gohberg, I., Inertia theorems for operator pencils and applications, *Integral Eq. and Operator Theory* **21** (1995), 270–318.

[BKAK] Bitmead, R. R., Kung, S. Y., Anderson, B. D. O., Kailath, T., Greatest common divisors via generalized Sylvester and Bezout matrices, *IEEE Trans. Automatic Control* **AC-23** (1978), 1043–1047.

[Bu]   Bunce, J. W., Inertia and controllability of infinite dimensions, *J. Math. Analysis and Appl.* **129** (1988), 569–580.

[C1]   Cain, B., An inertia theory for operators on a Hilbert space, *J. Math. Analysis and Appl.* **4** (1973), 97–114.

[C2]   ———, Inertia theory, *Linear Algebra and Appl.* **30** (1980), 211–240.

[CL]   Carlson, D., Schneider, H., Inertia theorems for matrices: the semidefinite case, *J. Math. Anal. Appl.* **6** (1963), 430–446.

[Ch]   Chen, C. T., A generalization of the inertia theorem, *SIAM J. Appl. Math.* **25** (1973), 158–161.

[CK]   Clancey, K., Kon, B. A., The Bezoutian and the algebraic Riccati equation, *Linear and Multilinear Algebra* **15** (1984), 265–278.

[GGK]   Gohberg, I., Goldberg, S., Kaashoek, M. A., *Classes of Linear Operators. Vol. I,* Operator Theory: Advances and Applications, Vol. *49*, Birkhäuser Verlag, Basel 1990.

[GKL]   Gohberg, I. C., Kaashoek, M. A., Lay, D.C., Equivalence, linearization, and decomposition of holomorphic operator functions, *J. of Functional Analysis* **28** (1978), 102–144.

[GKLR]   Gohberg, I., Kaashoek, M. A., Lerer, L., Rodman, L., Common multiples and common divisors of matrix polynomials, I. Spectral method, *Indiana Univ. Math. J.* **30** (1981), 321–356.

[GLR1]   Gohberg, I., Lancaster, P., Rodman, L., Representation and divisibility of operator polynomials, *Canadian J. of Math.* **30** (1978), 1045–1069.

[GLR2]   _____ , *Matrix Polynomials*, Academic Press 1982.

[GR1]   Gohberg, I., Rodman, L., On spectral analysis of non-monic matrix and operator polynomials, II. Dependence on the finite spectral data, *Israel Journal of Mathematics* **30** (1978), 321–334.

[GR2]   _____ , Interpolation and local data for meromorphic matrix and operator functions, *Integral Eq. and Operator Theory* **9** (1986), 60–94.

[GS]   Gohberg, I., Shalom, T., On Bezoutian of non-square matrix polynomials, *Linear Algebra and Appl.* **137/138** (1990), 249–323.

[GSi1]   Gohberg, I. C., Sigal, E. I., An operator generalization of the logarithmic residue theorem and Rouché's theorem, *Math. Sbornik, N. S.* **84(126)** (1971), 607–629. (Russian)

[GSi2]   _____ , Global factorization of meromorphic operator functions and some of its applications, *Matematicheskiye Issledovaniya* **VI** (1971), 63–82. (Russian)

[H]   Heinig, G., Bezoutian, resultant and spectral distribution problems for operator polynomials, *Mathematische Nachrichten* **91** (1979), 23–43.

[KMR1]   Kaashoek, M. A., van der Mee, C. V. M., Rodman, L., Analytic operator functions with compact spectrum, I. Spectral nodes, linearization and equivalence, *Integral Eq. and Operator Theory* **4** (1981), 504–547.

[KMR2]   _____ , Analytic operator functions with compact spectrum, II. Spectral pairs and factorizations, *Integral Eq. and Operator Theory* **5** (1982), 791–827.

[KMR3]   _____ , Analytic operator functions with compact spectrum, III. Hilbert space case and applications, *J. Operator Theory* **10** (1983), 219–250.

[LM]   Lancaster, P., Maroulas, J., The kernel of the Bezoutian for operator polynomials, *Linear and Multilinear Algebra* **17** (1985), 181–201.

[L]   Leiterer, J., Local and global equivalence of meromorphic operator functions, Part II, *Mathematische Nachrichten* **84** (1978), 145–170.

[LR1]   Lerer, L., Rodman, L., Spectrum separation and inertia for operator polynomials, *J. Math. Analysis and Appl.* **169** (1992), 260–282.

[LR2]   ——— , Inertia of operator polynomials and stability of differential equations, *J. Math. Analysis and Appl.* **192** (1995), 579–606.

[LR3]   ——— , Inertia theorems for Hilbert space operators based on Lyapunov and Stein equations, *Mathematische Nachrichten* **198** (1999), 131–148.

[LRT1] Lerer, L., Rodman, L., Tismenetsky, M., Inertia theorems for matrix polynomials, *Linear and Multilinear Algebra* **30** (1991), 157–182.

[LRT2] ——— , Bezoutian and the Schur-Cohn problem for operator polynomials, *J. Math. Analysis and Appl.* **103** (1984), 83–102.

[LT1]   Lerer, L., Tismenetsky, M., Generalized Bezoutian and matrix equations, *Linear Algebra and Appl.* **99** (1988), 123–160.

[LT2]   ——— , The eigenvalue separation problem for matrix polynomials, *Integral Eq. and Operator Theory* **5** (1982), 386–455.

[LT3]   ——— , The Bezoutian and eigenvalue separation problem for matrix polynomials, *Integral Eq. and Operator Theory* **5** (1982), 386–445.

[LT4]   ——— , Generalized Bezoutian and the inversion problem for block matrices. I. General scheme, *Integral Eq. Operator Theory* **9** (1986), 790–819.

[LW]   Lerer, L., Woerdeman, H. J., Resultant operators and Bezout equations for analytic matrix functions, I, *J. Math. Analysis and Appl.* **125** (1987), 531–552.

[R]   Rodman, L., *An Introduction to Operator Polynomials, Operator Theory: Advances and Applications*, V o.l. 38 Birkhäuser, Boston 1989

[T]   Takahashi, K., On relative primeness of operator polynomials, *Linear Algebra and Appl.* **50** (1983), 521–526.

[W]   Wimmer, H., Inertia theorems for matrices, controllability, and linear vibrations, *Linear Algebra and Appl.* **8** (1974), 337–343.

Department of Mathematics
Technion-Israel Institute of Technology
32000 Haifa, Israel

Department of Mathematics
The College of William and Mary
Williamsburg
VA 23187–8795 USA

1991 Mathematics Subject Classification. 47A56

Received May 31, 2000

Operator Theory:
Advances and Applications, Vol. 130, 277–281
© 2001 Birkhäuser Verlag Basel/Switzerland

A Note on the Level Sets of a Matrix Polynomial and Its Numerical Range

Panayiotis J. Psarrakos

Dedicated with admiration to my teacher and friend Peter Lancaster on the occasion of his 70th birthday.

Abstract. Let $P(\lambda)$ be an $n \times n$ matrix polynomial with bounded numerical range $W(P)$ and let $n > 2$. If Ω is a connected subset of $W(P)$, then the set

$$\bigcup_{\omega \in \Omega} \{x \in \mathbb{C}^n : x^* P(\omega)x = 0, \; x^*x = 1\}$$

is also connected. As a consequence, if $P(\lambda)$ is selfadjoint, then every $\omega \in (\overline{W(P)\backslash\mathbb{R}}) \cap \mathbb{R}$ is a multiple root of the equation $x_\omega^* P(\lambda)x_\omega = 0$ for some unit $x_\omega \in \mathbb{C}^n$.

1. Introduction

Consider the $n \times n$ *matrix polynomial*

$$P(\lambda) = A_m \lambda^m + \cdots + A_1 \lambda + A_0, \tag{1.1}$$

where A_j $(j = 0, 1, \ldots, m)$ are $n \times n$ matrices, with $A_m \neq 0$, and λ is a complex variable. If $A_m = I$, then $P(\lambda)$ is called *monic* and if the coefficients A_j $(j = 0, 1, \ldots, m)$ are Hermitian matrices, then $P(\lambda)$ is called *selfadjoint*.

The *numerical range* of $P(\lambda)$ is defined by

$$W(P) = \{\lambda \in \mathbb{C} : x^* P(\lambda)x = 0 \text{ for some nonzero } x \in \mathbb{C}^n\}. \tag{1.2}$$

Evidently, $W(P)$ is always closed and it contains the *spectrum* $\sigma(P) = \{\lambda \in \mathbb{C} : \det P(\lambda) = 0\}$ of $P(\lambda)$. For $P(\lambda) = I\lambda - A$, $W(P)$ coincides with the *classical numerical range (field of values)* of the matrix A,

$$F(A) = \{x^* Ax \in \mathbb{C} : x \in \mathcal{S}\},$$

where $\mathcal{S} = \{x \in \mathbb{C}^n : x^*x = 1\}$ is the unit sphere in $\mathbb{C}^n$. It is known that $W(P)$ is bounded if and only if $0 \notin F(A_m)$, [2]. Moreover, if $W(P)$ is bounded, then it has no more than m connected components. If $W(P)$ is unbounded, then it may have as many as $2m$ connected components. The closure and the interior of $W(P)$ are denoted by $\overline{W(P)}$ and $\text{Int}W(P)$, respectively.

If $n > 2$, then for every $\omega \in W(P)$, the corresponding *level set*
$$\mathcal{L}(\omega) = \{x \in \mathcal{S} : x^* P(\omega)x = 0\}$$
is path-connected (see Main Theorem in [4]). In this paper we continue the study of this subject working on connected subsets of $W(P)$. An interesting result on selfadjoint matrix polynomials is also obtained.

2. Connectivity of level sets

Let $P(\lambda)$ be an $n \times n$ matrix polynomial, as in (1.1), and assume that the numerical range $W(P)$ in (1.2) is bounded and $n > 2$. If we consider a connected subset of $W(P)$, then it follows that the union of the corresponding level sets is also connected.

Theorem 2.1. *Let $P(\lambda)$ be an $n \times n$ matrix polynomial with bounded numerical range and $n > 2$. If Ω is a connected subset of $W(P)$, then the set $\cup_{\omega \in \Omega} \mathcal{L}(\omega)$ is a connected subset of $\mathcal{S}$.*

Proof. Assume that $\cup_{\omega \in \Omega} \mathcal{L}(\omega)$ is not connected. Then there exist two disjoint open sets $\mathcal{A}, \mathcal{B} \subset \mathcal{S}$ such that
$$\mathcal{A} \cap [\cup_{\omega \in \Omega} \mathcal{L}(\omega)] \neq \emptyset, \quad \mathcal{B} \cap [\cup_{\omega \in \Omega} \mathcal{L}(\omega)] \neq \emptyset$$
and
$$\cup_{\omega \in \Omega} \mathcal{L}(\omega) \subseteq \mathcal{A} \cup \mathcal{B}.$$
For every $\omega \in \Omega$, the level set $\mathcal{L}(\omega)$ is path-connected and thus,
$$\mathcal{L}(\omega) \subset \mathcal{A} \quad \text{or} \quad \mathcal{L}(\omega) \subset \mathcal{B}.$$
Moreover, consider the sets
$$\Omega_{\mathcal{A}} = \{\omega \in \Omega : \mathcal{L}(\omega) \subseteq \mathcal{A}\} = \Omega \cap \{\omega \in W(P) : \mathcal{L}(\omega) \subseteq \mathcal{A}\}$$
and
$$\Omega_{\mathcal{B}} = \{\omega \in \Omega : \mathcal{L}(\omega) \subseteq \mathcal{B}\} = \Omega \cap \{\omega \in W(P) : \mathcal{L}(\omega) \subseteq \mathcal{B}\},$$
which are open in the relative topology of Ω. Then it follows that
$$\Omega \cap \Omega_{\mathcal{A}} \neq \emptyset, \quad \Omega \cap \Omega_{\mathcal{B}} \neq \emptyset$$
and
$$\Omega = \Omega_{\mathcal{A}} \cup \Omega_{\mathcal{B}}.$$
Since $\mathcal{A} \cap \mathcal{B} = \emptyset$, we have that $\Omega_{\mathcal{A}} \cap \Omega_{\mathcal{B}} = \emptyset$ and consequently, Ω is not connected, that is a contradiction. $\square$

Assume that Ω is an open subset of $W(P)$. For any $x_0 \in \cup_{\omega \in \Omega} \mathcal{L}(\omega)$, there exists an $\omega_0 \in \Omega$ satisfying the equation $x_0^* P(\omega_0)x_0 = 0$. Since Ω is open, there is a real $\varepsilon > 0$ such that the disk $S(\omega_0, \varepsilon)$, with centre ω_0 and radius ε, belongs to Ω. By the continuous dependence of the roots of polynomials on their coefficients, [6], there exists an $r > 0$ such that for every $x \in S(x_0, r)$, the equation $x^* P(\lambda)x = 0$ has a root in $S(\omega_0, \varepsilon) \subset \Omega$. Thus, $S(x_0, r) \subset \cup_{\omega \in \Omega} \mathcal{L}(\omega)$ and $\cup_{\omega \in \Omega} \mathcal{L}(\omega)$ is an open subset of $\mathcal{S}$. It is also easy to verify that if Ω is a closed subset of $W(P)$, then $\cup_{\omega \in \Omega} \mathcal{L}(\omega)$ is a closed subset of $\mathcal{S}$.

Corollary 2.2. *Suppose that $P(\lambda)$ is an $n \times n$ matrix polynomial with bounded numerical range and $n > 2$. If Ω is an open path-connected subset of $W(P)$, then the set $\cup_{\omega \in \Omega} \mathcal{L}(\omega)$ is an open path-connected subset of $\mathcal{S}$.*

Proof. Since the sets $\cup_{\omega \in \Omega} \mathcal{L}(\omega)$ and Ω are both open (see the discussion above), the notions of connectivity and path-connectivity are equivalent (see Corollary 26.7 in [7]). □

Proposition 2.3. *Let $u(t) : [0,1] \to \mathbb{C}$ be a continuous rectifiable curve in the interior of $W(P)$ and let δ be any positive real number. Then there exist a continuous vector-curve $y_\delta(t) : [0,1] \to \mathcal{S}$ and a continuous curve $u_\delta(t) : [0,1] \to \mathrm{Int} W(P)$ such that $u(0) = u_\delta(0)$, $u(1) = u_\delta(1)$ and for every $s \in [0,1]$,*

$$y_\delta(s)^* P(u_\delta(s)) y_\delta(s) = 0 \quad and \quad \min\{|u_\delta(s) - u(t)| : t \in [0,1]\} < \delta.$$

Proof. Let $\Gamma \subset \mathrm{Int} W(P)$ be the image of u(t). For any $\varepsilon > 0$, there exists a finite number of points $\omega_1, \omega_2, \ldots, \omega_k \in \Gamma$ such that

$$\Omega_\varepsilon = \bigcup_{j=1}^{k} \mathrm{Int} S(\omega_j, \varepsilon)$$

is an open covering of Γ. The set Ω_ε is path-connected, and for ε sufficiently small, Ω_ε lies in the interior of $W(P)$. Moreover, by Corollary 2.2, the set $\cup_{\omega \in \Omega_\varepsilon} \mathcal{L}(\omega)$ is also path-connected. Hence, there is a vector-curve $y_\varepsilon(t) : [0,1] \to \cup_{\omega \in \Omega_\varepsilon} \mathcal{L}(\omega)$ such that $y_\varepsilon(0)^* P(u_\varepsilon(0)) y_\varepsilon(0) = y_\varepsilon(1)^* P(u_\varepsilon(1)) y_\varepsilon(1) = 0$. Thus, by the continuous dependence of the roots of the equation $y_\varepsilon(t)^* P(\lambda) y_\varepsilon(t) = 0$ on $t \in [0,1]$, the proof is complete. □

3. Selfadjoint matrix polynomials

Let $P(\lambda) = A_m \lambda^m + \cdots + A_1 \lambda + A_0$ be an $n \times n$ selfadjoint matrix polynomial. It is easy to see that the numerical range $W(P)$ is symmetric with respect to the real axis. As a consequence, the points of $(\overline{W(P) \backslash \mathbb{R}}) \cap \mathbb{R}$ are of particular interest (see [1] and [5]). The results of the previous section yield a generalization of Proposition 4 in [1] and Theorem 3.1 in [5].

Theorem 3.1. *Let $P(\lambda) = A_m \lambda^m + \cdots + A_1 \lambda + A_0$ be an $n \times n$ selfadjoint matrix polynomial with bounded numerical range $W(P)$ and assume that $n > 2$. For every $\omega \in (\overline{W(P) \backslash \mathbb{R}}) \cap \mathbb{R}$, there exists a vector $x_\omega \in \mathcal{S}$ such that ω is a multiple root of the equation $x_\omega^* P(\lambda) x_\omega = 0$.*

Proof. Suppose that $\omega \in (\overline{W(P) \backslash \mathbb{R}}) \cap \mathbb{R}$.

If ω is an isolated point of $(\overline{W(P) \backslash \mathbb{R}}) \cap \mathbb{R}$, then by Theorem 3.1 (and its proof) in [5], there exists a vector $x_\omega \in \mathcal{S}$ such that ω is a multiple root of the equation $x_\omega^* P(\lambda) x_\omega = 0$.

If ω is not an isolated point of $(\overline{W(P)\backslash\mathbb{R}}) \cap \mathbb{R}$, then consider the set

$$T_0 = \{\lambda \in W(P) : x^*P(\lambda)x = x^*P'(\lambda)x = 0 \text{ for some } x \in \mathcal{S}\},$$

where $P'(\lambda)$ is the derivative of $P(\lambda)$. Since the sets $(\overline{W(P)\backslash\mathbb{R}})\cap\mathbb{R}$ and T_0 are both closed, it is enough to show that $T_0 \cap (\overline{W(P)\backslash\mathbb{R}}) \cap \mathbb{R}$ is dense in $(\overline{W(P)\backslash\mathbb{R}}) \cap \mathbb{R}$. By the symmetry of $W(P)$ with respect to the real axis, it follows that for every real $r > 0$, $S(\omega, r) \cap \text{Int}W(P) \cap \mathbb{R} \neq \emptyset$. Hence, without lost of generality assume that $\omega \in (\overline{W(P)\backslash\mathbb{R}}) \cap \mathbb{R} \cap \text{Int}W(P)$. Then there is a continuous rectifiable curve

$$u(t) : [0,1] \rightarrow \{\lambda \in W(P) : \text{Im}\lambda \geq 0\} \cap \text{Int}W(P)$$

such that $u(1) = \omega$ and $\text{Im}u(t) > 0$ for every $t \in [0,1)$. For any $\delta > 0$, by Proposition 2.3, there exist a continuous vector-curve $y_\delta(t) : [0,1] \rightarrow \mathcal{S}$ and a continuous curve $u_\delta(t) : [0,1] \rightarrow \text{Int}W(P)$ such that $u(0) = u_\delta(0)$, $u(1) = u_\delta(1)$ and for every $s \in [0,1]$,

$$y_\delta(s)^*P(u_\delta(s))y_\delta(s) = 0$$

and

$$\min\{|u_\delta(s) - u(t)| : t \in [0,1]\} < \delta.$$

Moreover, there exists a $s_0 \in [0,1]$ such that $u_\delta(s_0) \in \mathbb{R}$ and for every $s \in [0, s_0)$, $u_\delta(s) \notin \mathbb{R}$. Thus, for every $s \in [0, s_0)$, the polynomial $y_\delta(s)^*P(\lambda)y_\delta(s)$ is written in the form $y_\delta(s)^*P(\lambda)y_\delta(s) = (\lambda - u_\delta(s))(\lambda - \overline{u_\delta(s)})g_s(\lambda)$, where $g_s(\lambda)$ is a polynomial of $(m-2)$-th degree and its coefficients are continuous on s. By the continuity of the root $u_\delta(s)$ on s, we have

$$\lim_{s \to s_0} u_\delta(s) = \lim_{s \to s_0} \overline{u_\delta(s)} = \omega_\delta,$$

where $|\omega - \omega_\delta| < \delta$. Hence, the set $T_0 \cap (\overline{W(P)\backslash\mathbb{R}}) \cap \mathbb{R}$ is dense in $(\overline{W(P)\backslash\mathbb{R}}) \cap \mathbb{R}$ and the proof is complete. $\qquad\square$

Corollary 3.2. *Let $P(\lambda)$ be an $n \times n$ $(n > 2)$ selfadjoint matrix polynomial with bounded numerical range $W(P)$ and suppose that for every unit vector $x \in \mathbb{C}^n$, the equation $x^*P(\lambda)x = 0$ has m distinct roots. Then every connected component of $W(P)$ either has no real points or it is a closed real interval.*

(The above corollary follows also from [3, Theorem 1] and the symmetry of $W(P)$.)

Corollary 3.3. *Suppose that $P(\lambda)$ is an $n \times n$ $(n > 2)$ selfadjoint matrix polynomial with bounded numerical range $W(P)$, and for any $n \times n$ matrix B consider the matrix polynomial $Q_B(\lambda) = P(\lambda) + B$. Then*

$$(\overline{W(Q_B)\backslash\mathbb{R}}) \cap \mathbb{R} \subset W(P') \cap \mathbb{R}.$$

Note that if $P(\lambda) = A_2\lambda^2 + A_1\lambda + A_0$ is a quadratic selfadjoint matrix polynomial, such that the numerical range of the derivative $P'(\lambda)$ does not coincide with $\mathbb{C}$, then $W(P') = \{-(x^*A_1x)/(2x^*A_2x) \in \mathbb{C} : x \in \mathbb{C}^n, \text{ with } x^*A_2x \neq 0\} \subseteq \mathbb{R}$. In this case,

$$(\overline{W(P)\backslash\mathbb{R}}) \cap \mathbb{R} \subseteq \overline{\text{Re}(W(P)\backslash\mathbb{R})} \subset W(P').$$

References

[1] Lancaster, P., Psarrakos, P., *The numerical range of selfadjoint quadratic matrix polynomials*, preprint (2000).

[2] Li, C.-K., Rodman, L., Numerical range of matrix polynomials, *SIAM J. Matrix Anal. Applic.* **15** (1994), 1256–1265.

[3] Lyubich, Y., Separation of roots of matrix and operator polynomials, *Integral Equations and Operator Theory* **29** (1998), 52–62.

[4] Lyubich, Y., Markus, A.S., Connectivity of level sets of quadratic forms and Hausdorff-Toeplitz type theorems, *Positivity* **1** (1997), 239–254.

[5] Maroulas, J., Psarrakos, P., A connection between numerical ranges of selfadjoint matrix polynomials, *Linear and Multilinear Algebra* **44** (1998), 327–340.

[6] Ostrowski, A.M., *Solutions of equations in Euclidean and Banach spaces*, Academic Press, New York 1973.

[7] Willard, S., *General Topology*, Addison-Wesley Publ. Company 1970.

Department of Mathematics and Statistics
University of Regina
Regina, Saskatchewan, S4S 0A2
CANADA
Email: panos@math.uregina.ca

1991 Mathematics Subject Classification. Primary 15A60; Secondary 47A12

Received May 10, 2000